Foodborne Pathogens
and Food Safety

Foodborne Pathogens and Food Safety

Editors

Md. Latiful Bari
Center for Advanced Research in Sciences
University of Dhaka
Dhaka, Bangladesh

and

Dike O. Ukuku
Food Safety Intervention & Technology
Eastern Regional Research Center
Wyndmoor, PA, USA

CRC Press
Taylor & Francis Group
Boca Raton London New York

CRC Press is an imprint of the
Taylor & Francis Group, an **informa** business

A SCIENCE PUBLISHERS BOOK

CRC Press
Taylor & Francis Group
6000 Broken Sound Parkway NW, Suite 300
Boca Raton, FL 33487-2742

First issued in paperback 2020

© 2016 by Taylor & Francis Group, LLC
CRC Press is an imprint of Taylor & Francis Group, an Informa business

No claim to original U.S. Government works

ISBN-13: 978-1-4987-2408-1 (hbk)
ISBN-13: 978-0-367-73752-8 (pbk)

Visit the Taylor & Francis Web site at
http://www.taylorandfrancis.com

and the CRC Press Web site at
http://www.crcpress.com

Preface to the Series

Food is the essential source of nutrients (such as carbohydrates, proteins, fats, vitamins, and minerals) for all living organisms to sustain life. A large part of daily human efforts is concentrated on food production, processing, packaging and marketing, product development, preservation, storage, and ensuring food safety and quality. It is obvious therefore, our food supply chain can contain microorganisms that interact with the food, thereby interfering in the ecology of food substrates. The microbe-food interaction can be mostly beneficial (as in the case of many fermented foods such as cheese, butter, sausage, etc.) or in some cases, it is detrimental (spoilage of food, mycotoxin, etc.). The *Food Biology* series aims at bringing all these aspects of microbe-food interactions in form of topical volumes, covering food microbiology, food mycology, biochemistry, microbial ecology, food biotechnology and bio-processing, new food product developments with microbial interventions, food nutrification with nutraceuticals, food authenticity, food origin traceability, and food science and technology. Special emphasis is laid on new molecular techniques relevant to food biology research or to monitoring and assessing food safety and quality, multiple hurdle food preservation techniques, as well as new interventions in biotechnological applications in food processing and development.

The series is broadly broken up into food fermentation, food safety and hygiene, food authenticity and traceability, microbial interventions in food bio-processing and food additive development, sensory science, molecular diagnostic methods in detecting food borne pathogens and food policy, etc. Leading international authorities with background in academia, research, industry and government have been drawn into the series either as authors or as editors. The series will be a useful reference resource base in food microbiology, biochemistry, biotechnology, food science and technology for researchers, teachers, students and food science and technology practitioners.

Ramesh C. Ray
Series Editor

Preface

Foodborne disease and food safety is a large and growing public health problem worldwide. Millions of words of advice and millions of dollars spent but the problem is getting worse in many developed and developing countries. A range of reasons have been presented in this book explaining why food poisoning has increased in developed countries. The ancient food history, food culture, scientific developments in food microbiology, food laws, several multidisciplinary and one health approach have been discussed to reduce the risk of foodborne illness, future risk associated with food and growing population has also been explained in this book.

This book serves as a comprehensive resource on the foodborne microbial agents and food safety that includes topics such as: foodborne illness and microbial event, emerging foodborne pathogens; health problems/disease caused by foodborne pathogens; emergence of drug resistance in foodborne pathogens; methods and technology for rapid and accurate detection of foodborne pathogens; survival, injury and inactivation of pathogens in foods; risk and hazard analysis of food; strategies to destroy or control foodborne pathogens in food production and processing environments; Innovative measures for ensuring food safety in the food value chain. A major part of the book is devoted to foodborne bacterial pathogens. These chapters include information about each organism's reservoirs, infectious dose, associated outbreaks, and resistance to antimicrobial agents; survival, injury and inactivation of non-thermal food processing discussed the emphasis in most of these chapters is on pathogenesis and of the factors that determine virulence.

The basic science of foodborne microorganisms and their relationship to the food micro-environment underlies current food safety measures. One chapter discusses the hazard analysis of food while another covers innovative measures for ensuring food safety in the food value chain. The novel strategies for the prevention and control of plant and animal diseases that impact food safety are also covered. Dr. Richard Bonne discussed Food Quality Management System—a tool for controlling mycotoxin contamination and discussion on HACCP approach for ensuring food safety in developing countries made this book more attractive to the reader. Therefore, this book contributes in different ways to the overall objective in improving the safety of food. In addition, last chapter discuss on the food bio-security issues and its implications of new regulatory guidelines. This chapter points out the need to reduce conflicts of interest in agencies whose mandates are both to inspect food products and to support markets for them.

The intent of this book is to define and categorize the real and perceived safety issues surrounding food, to provide scientifically non-biased perspectives on these issues, and to provide assistance to the reader in understanding these issues. The editors would like to acknowledge people who provided valuable input and assistance and to express their sincere appreciation for their efforts. This appreciation is especially extended to R.C. Ray, Didier Montet, Francisco B. Elegado, Sabina Yeasmin, Nadine Zakhia-Rozis, Malik A. Hussain, Dinesh Babu, Dam Sao Mai and Yasuhiro Inatsu, for their enthusiasm and diligence in serving preparation this book and to all of the numerous authors of the chapters.

Md. Latiful Bari
Dike O. Ukuku

Contents

1

History and Safety of Food: Past, Present and Future

Takashi Uemura[1] *and Md. Latiful Bari*[2,]*

Introduction

Food history is an interdisciplinary field that examines the history of food and its cultural, economic, environmental, and sociological impact. History of food is considered distinct from the more traditional field of culinary history, which focuses on the origin and re-creation of specific recipes. Food historians look at food as one of the most important elements of cultures, reflecting the social and economic structure of society. The history of food safety binds all civilizations. Living things seek food for energy required for survival and breeding. But human beings differ from other living things, and expect physical, mental and social functions as well (Table 1).

Table 1. What people expect from food.

Stage	Object	Acceptor	Components
1st (Basic)	Calories, Nutrition, Safety	Body	Carbohydrate, Protein, Fat, Vitamins, Mineral, Fiber, Fresh, Clean
2nd (Additional)	Suitability, Additional quality	Brain	Color, Appearance, Shape, Texture, Crunchy, Taste, Scent, Sound
3rd (Expanding)	Food culture, Quality of life	Heart	Atmosphere, Relaxation, Superiority, Pride, Reliability

[1] Shijonawate Gakuen University, Kadoma-shi, Osaka 571-0075, Japan.
[2] Center for Advanced Research in Sciences, University of Dhaka, Dhaka-1000, Bangladesh.
* Corresponding author: latiful@du.ac.bd

As time for people for acquisition and cooking of food became shortened, they could now use the leisure time for improvement in the quality of their life, as well as in culture and spiritual activities.

Ancient Food Laws

Food laws in one form or another, such as religious tenets or prohibition, were inherent in all ancient civilizations and have come down to us from early times. But it was not until the late 19th century, with the urbanization of societies and the de-population of rural areas, that food laws, as understood today, were made. Food laws can be traced back to times of the earliest societies. Ancient food regulations are referred to in Egyptian, Chinese, Hindu, Greek and Roman literatures. In the Middle Ages, the trade guilds exerted a powerful influence on the regulation of food trade and the prevention of adulteration of food products (Miller and Taylor, 1989). All these early food laws were designed to protect consumers from fraud—this was the predominant legal concern. Although there was no stated intention to protect public health, it was fortunate, that health protection happened, in many instances, to be almost synonymous with protection against fraud. Many other references to food safety are found in the Bible—the dietary laws of Moses; the Book of Leviticus prohibiting the use of pork or the meat of any scavenger or deceased animal, and perhaps the best-known Biblical reference to food safety, the phrase "death in the pot", that became a rallying cry against food adulteration and other food "proponents" of Fredrick Accum (Hutt and Hutt, 1984). Islam religion (570 AC) first clarifies the food safety by providing the word *Haram* and *Halal* intended to protect human health. The Holy Quran warns: '*O mankind! Eat of that which is lawful and wholesome in the earth, and follow not the footsteps of the devil. Lo! he is an open enemy for you*' (Al-Baqara 2/168).

Many Asian countries have traditional beliefs that contributed to food safety. Some of these beliefs have originated from mythological stories and were transmitted orally from one generation to another. Foods were recognized for their health-giving properties and ancient texts often mention about the practice of traditional medicine. This has given rise to attitudes toward food and food safety that continues to the present day. Most of the early history of food safety science was in the domain of human experience and judgment. Later better analytical methods were used in food control to mainly detect adulteration. The Arab scientist Al Chazini, constructed a high sensitivity balance for measuring purposes, and this innovation prompted in reducing fraudulent practices in food trade in the ancient world.

Food Safety—A Brief History

The history of food safety (Table 2) is probably nearly as old as human history itself and may have started with the recognition and subsequent avoidance of foods that were naturally toxic. Early humans, probably by trial and error, also started to develop basic forms of food preservation in 4000 BC, which possibly also made food safer, e.g., drying, salting, fermentation, etc. The Chinese, in prehistoric times reportedly preserved vegetables by fermentation and Plinius has mentioned of preservation of white cabbage in earthenware pots in Italy in the 1st century (Montville and Matthews, 2005). As human eating patterns, habits and foods changed, food safety laws became more formalized.

Table 2. A timeline of food safety (adopted from Griffith, 2006).

50000 BC	Early humans	Self-preservation
4000 BC	Various	Early food fermentation
2000 BC	Leviticus	Religious beliefs
AD 1676	Antonie van Leeuwenhoek	Origins of microbiology
AD 1810	Nicholas Appert	Basis of commercial heat processing
AD 1857	Pasteur	Early food microbiology
AD 1880	Gartner	First isolation of Salmonella from food poisoning outbreak
1880 to date	Salmon, Russell, Frazier and many others	Start of Golden Age of food microbiology

In ancient Israel laws included advice on foods to be avoided, methods of preparation and the importance of food hygiene. In 2000 BC the book of Leviticus tells us that Moses introduced laws, such as, the washing of clothes and bathing after the sacrificial slaughter of animals to protect his people from food-related diseases. The Egyptians, Greeks and Romans also expressed similar concerns (Mossel et al., 1995). Subsequently, throughout history, the food laws or legislation was based upon the need to prevent adulteration of food (primarily chemical) and to ensure safety of food products. Some historical developments in the UK and EU food safety legislation are presented in Table 3.

Responsibility for the safety of food supply has always been a major concern for governments in all parts of the world often based on religious or quasi-religious concepts. Chinese medicine has a long history of prevention in which food and diet played a central role (Miller, 1993). For example, in 2nd century AD, Chang Chung-Ching prepared a manual for Safety Regulations for Food (Chin Kuei Yao Lueh) (Needham, 1962) that incorporated many of the earlier Confucian prohibitions. Western European concern for food safety lagged much behind that of the East. In 1266, Parliament in UK enacted the Assize of Bread, which prohibited the sale of any staple food product that is "not wholesome for man's body". No modern statute has found better or a more inclusive language to convey the legislative directive to prohibit unsafe food (Hutt and Hutt, 1984). Nevertheless, it was not until the 19th century, that comprehensive food sanitation and safety legislations were adopted.

History of US Food Law

In 1906, after many previous failures, US Congress enacted the first national legislation regulating the economic integrity, safety and labeling of food. Although the 1906 act addressed some of the more flagrant abuses and problems, science and technology soon made it obsolete. By 1930, a variety of pesticides and other agricultural chemicals were in common use and additives were becoming common in processed foods. An

Table 3. Some historical developments in UK and EU food safety legislation (modified from Griffith, 2006).

Date (AD)	Food legislation
1266	Assize of Bread and Ale
1760–1830	Processes for the preservation of food invented
1820	A treatise on adulterations of food and culinary poisons
1848	First Public Health Act
1860	Adulteration of Food Act
1872	Adulteration of Food, Drink and Drugs Act
1875	Sale of Food and Drugs Act
1928	Food and Drugs (Adulteration) Act
1938	Food and Drugs Act
1955	Food and Drugs Act
1976	Food and Drugs (Amendment) Act
1984	Food Act
1990	Food Safety Act
1991	Organic Food Products Regulation (EEC) No 2092/91
1993	Council Directive EC 93/43
1995	Food Safety (General Food Hygiene) Regulations
2003	GMOs traceability, labeling & derived food & feed 1829 & 1830/2003
2004	Hygiene of Foodstuffs EC 852/2004
2005	Microbiological criteria for foodstuffs regulations 2073/2005
2009	Food labeling regulation 41/2009
2011	Food contact materials regulation 10/2011

increasingly urbanized, industrialized society developed a greater dependence on and greater concern for a more sophisticated food industry to ensure an abundant and economical food supply. Figure 1 shows this history by providing a timeline of selected important dates for food safety legislation in the United States.

American federal laws give food manufacturers, distributors, and retailers the basic responsibility for assuring that foods were wholesome, safe, and handled under sanitary conditions. A number of federal agencies, cooperating with state, local, and international entities, play a major role in regulating food quality and safety under these laws. The combined efforts of the food industry and the regulatory agencies are credited in making the U.S. food supply very safe. Nonetheless, each year an estimated 48 million people become sick, and there are 128,000 cases of hospitalization and 3,000 deaths due to foodborne illnesses caused by contamination from a number of microbial pathogens. In addition to these, experts have cited numerous other health hazards, including the use of unapproved veterinary drugs, pesticides, and other dangerous substances in food commodities.

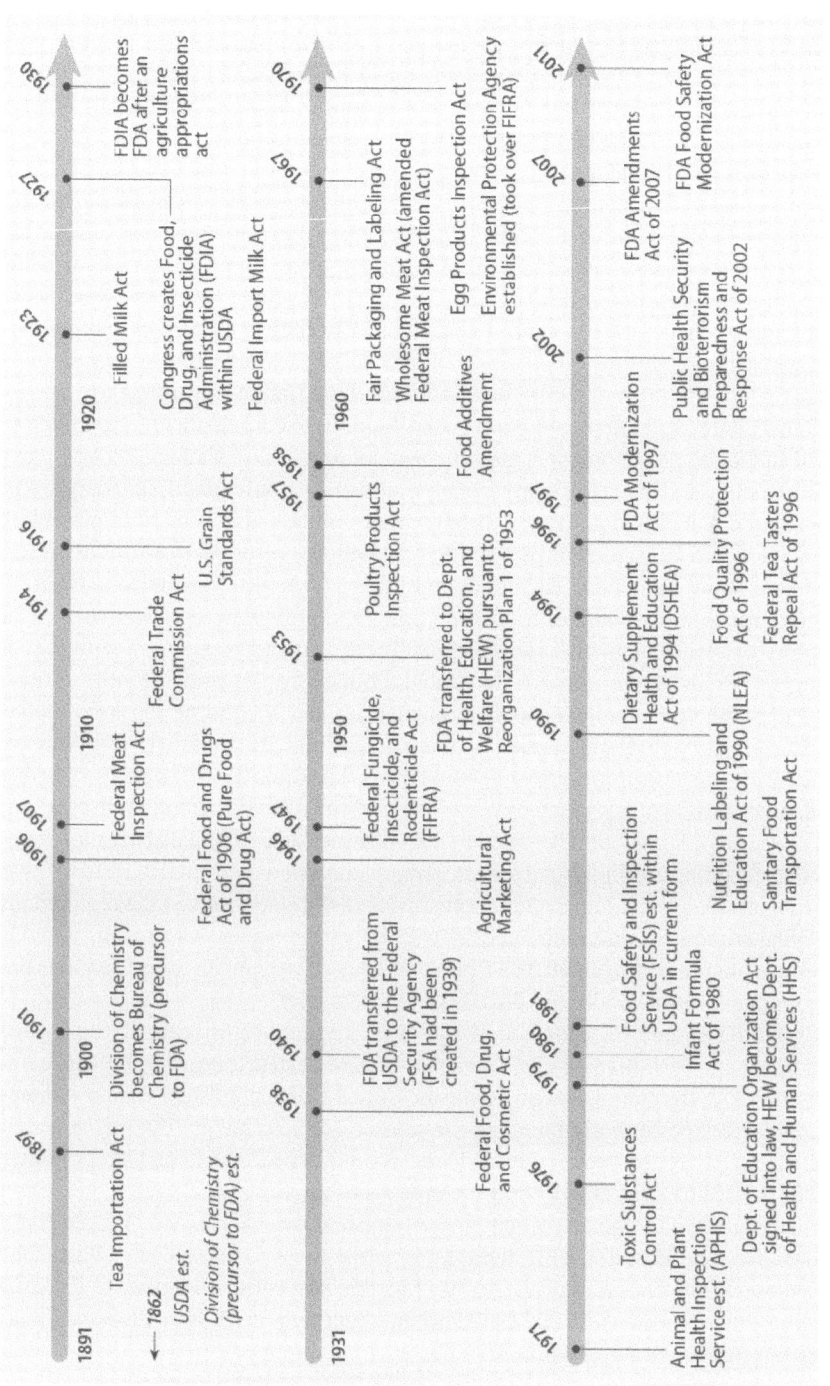

Figure 1. Selected Important Dates for Food Safety in the United States, 1862–2011 (adopted from Renée Johnson, 2014).

History of Foodborne Bacteria

Scientists did not begin to understand bacteria, and their relationship to disease, until the late 19th century. People did recognize that food spoils, but the reasons for that and the potential for becoming ill from food were not known. The history of food safety is the history of the numerous discoveries, inventions, and regulations that all led to the present knowledge.

Food preservation methods such as drying, smoking, freezing, marinating, salting, and pickling began thousands of years ago. Whether these methods were employed solely to keep food for later use, to improve flavor, or for some other reasons is not known, but whatever might be the reason they were developed and had the effect of keeping food safer. Even cooking can be viewed as an ancient method of making food safer. The Chinese Confucian Analects of 500 B.C.E. warned against consumption of sour rice, spoiled fish or flesh, food kept for a long time, or insufficiently cooked food. The Chinese disliked eating uncooked food, believing that anything boiled or cooked cannot be poisonous. It is possible that the practice of drinking tea originated because tea required using hot water, which would make it safer than using non-heated contaminated water (Zaman et al., 2014). Other cultures also had some prescribed methods to avoid the causes or prevention of foodborne disease. Most of the present knowledge about pathogens and foodborne illnesses is built on the foundation of scientific discoveries spanning over three centuries. Following the discovery of the ubiquitous existence of microorganisms (mainly bacteria and yeasts) by Leeuwenhoek around the 1670s, some individuals started associating the possible role of these organisms with food spoilage, food fermentation, and foodborne diseases. Italian scientists Francisco Redi and Lazzaro Spallanzani performed experiments that dispelled the theory of spontaneous generation of organisms. The discovery of bacteria in the late 19th century provided increased understanding of the role of bacteria in disease, and the realization of connection between human diseases and animal diseases led to the ideas that cleanliness is important and unsanitary conditions can contribute to disease. The Hungarian physician Ignaz Semmelweiss in 1847, first observed the requirement of hospital doctors to wash their hands before delivering babies. As a result, maternal death rates plummeted from 10 to 1.5 percent. Lack of personal hygiene remains one of the main causes of foodborne illnesses 150 years later.

Louis Pasteur first elucidated the link between spoilage, disease, and microorganisms with his work on fermentation and pasteurization in the 1860s and 1870s. In 1872 German scientist Ferdinand Julius Cohn published a three-volume book on bacteria, essentially founding the science of bacteriology. But this new field of bacteriology needed bacteria on which to conduct experiments and study. Robert Koch in the 1880s, was able to find the process of growing pure strains of bacteria in the laboratory and later on Koch also established the criteria and showed that a specific microbe causes a specific disease. These are now known as Koch's Postulates. Using these criteria scientists can identify bacteria that cause a number of diseases, including foodborne diseases. In 1947 Joshua Lederberg and Edward Lawrie Tatum discovered that bacteria reproduce sexually, opening up a whole new field of bacterial genetics (Asimov, 1972).

Even though Antonie van Leeuwenhoek, the Dutch biologist and microscopist, had improved the microscope to the degree that small microscopic organisms could be seen as far back as 1673, the discovery of foodborne diseases caused by microorganisms developed slowly. Although James Paget and Richard Owen described the parasite *Trichinella spiralis* for the first time in 1835, and German pathologists Friedrich Albert von Zenker and Rudolph Virchow noted the clinical symptoms of trichinosis in 1860, the association between trichinosis and the parasite *Trichinella spiralis* was not realized until much later. The English scientist William Taylor showed in 1857 that milk can transmit typhoid fever. In 1885, United States Department of Agriculture (USDA) veterinarian Daniel Salmon described a microorganism that caused gastroenteritis with fever when ingested in contaminated food. The bacterium was eventually named Salmonella (Asimov, 1972). Same year, Theodor Escherich isolated *Bacterium coli* (later named *Escherichia coli*) from faeces and suggested that some strains were associated with infant diarrhea. In 1888, August Gärtner, a German scientist, was the first to isolate *Bacillus enteritidis* (later known as *Salmonella enteritidis*) from a patient with food poisoning. In 1896, Marie von Ermengem proved that *Salmonella enteritidis* caused a fatal disease in humans who consumed contaminated sausage (Table 4). In 1895, Emilie Pierre-Mare van Ermengem, a Belgian bacteriologist, was the first to isolate the bacteria that causes botulism, *Clostridium botulinum*, from uncooked, salted ham served at a restaurant in Belgium—twenty-three people became ill, and three died with this bacterium. In 1914, *Staphylococcus aureus* was detected as food poisoning bacteria by M.A. Barber who visited a particular farm in the Philippines and drank milk cream, and became ill after each visit. He isolated a bacterium from the milk, placed it in a germ-free container of milk, waited for a while, and then convinced two volunteers to drink the milk with him, they all became ill with the same symptoms (Asimov, 1972). In 1945 *Clostridium perfringens* was first recognized as a cause of foodborne illnesses. It was not until the years 1975 to 1985 that scientists first recognized some of today's major foodborne pathogens—*C. jejuni*, *Y. enterocolitica*, *E. coli* O157:H7, and *Vibrio cholerae*. Some of the major developments in foodborne pathogens in the 19th and 20th century are briefly listed in Table 4.

The major developments of ideas on the possible role of microorganisms in foods and their scientific proof were initiated by Louis Pasteur in the 1870s, followed by many other scientists before the end of the 19th century. This paved the way for the establishment of early food microbiology in the 20th century. History reveals that advances in science and technology stimulate greater agricultural productivity and the growth of a food-processing industry, both essential for the growth of urban multiplexes. These in turn increase dependence of the consumer on a complex, highly structured, and hopefully, safe food supply. This inturn puts increasing pressure on science and technology to assure that safety (Griffith, 2006). The result has been the development of a complex, often archaic set of regulations based upon varying legislation, precedents and a rapidly changing science, elaborated with the goal of making certain that the consumer receives exactly what is promised and does so in such a way that health is not impaired.

Table 4. Some of the major developments in foodborne pathogens and timeline.

Timeline	Development events
1820	Justin Kerner described food poisoning from eating blood sausage (due to botulism). Fatal disease from eating blood sausage was recognized as early as A.D. 900.
1835	James Paget and Richard Owen described the parasite *Trichinella spiralis* for the first time.
1849	John Snow suggested the spread of cholera through drinking water contaminated with sewage. In 1854, Filippo Facini named the cholera bacilli as *Vibrio cholerae*, which was isolated in pure form by Robert Koch in 1884.
1856	William Budd suggested that water contamination with faeces from infected person spread typhoid fever and advocated the use of chlorine in water supply to overcome the problem. In 1800, G. de Morveau and W. Cruikshank advocated the use of chlorine to sanitize potable water.
1870–1885	Various food processing.
1885	Theodor Escherich isolated *Bacterium coli* (later named *Escherichia coli*) from the faeces and suggested that some strains were associated with infant diarrhea.
1888	A.A. Gartner isolated *Bacterium* (later *Salmonella*) *enteritidis* from the organs of a diseased man as well as from the meat the man ate. In 1896, Marie von Ermengem proved that *Salmonella enteritidis* caused a fatal disease in humans who consumed contaminated sausage.
1890s	Various Dairy Bacteriology.
1892	Koch, Salmon and others Veterinary Microbiology.
1894	J. Denys associated pyogenic *Staphylococcus* with death of a person who ate meat prepared from a diseased cow.
1895	Marie von Ermengem isolated *Bacillus botulinus* (*Clostridium botulinum*) from fermented ham and proved that it caused botulism.
1895	Harry Russell showed that gaseous swelling with bad odors in canned peas was due to the growth of heat-resistant bacteria (spores).
1900–1910	Russell's First textbook of Dairy Bacteriology published.
1914	*Staphylococcus aureus* was detected as food poisoning bacteria.
1930s/1940s	Various food microbiology.
1944	Frazier one of the first food microbiology textbooks.
1945	*Clostridium perfringens* was first recognized as a cause of foodborne illness.
1947	Joshua Lederberg and Edward Lawrie Tatum discovered that bacteria reproduce sexually, opening up a new field of bacterial genetics.
1957	Scott and others food processing/Microbiology.
1975	Scientists first recognized some major foodborne pathogens—*C. jejuni*, *Y. enterocolitica*, and *Vibrio cholerae*.
1984	*Escherichia coli* O157 H7.

Food Safety: The Present Scenario

Scientists and engineers in the food science use and share the same concept of some technical terms as follows:

Food Security (World Summit on Food Security, 2009): Food security exists when all people, at all times, have physical, social and economic access to sufficient, safe and nutritious food to meet their dietary needs and food preferences for an active and healthy life. The four pillars of food security are availability, access, utilization and stability.

Food Hygiene (FAO/WHO/Codex, 2005): Food hygiene is all conditions and measures necessary to ensure the safety and suitability of food at all stages of the food chain.

Food Hygiene (WHO, 1955): Food hygiene means all necessary measure for the safety, wholesomeness, and soundness of food at all stages from its growth, production or manufacture until its final consumption.

Food Safety (FAO/WHO/Codex, 2003): Food safety is the assurance that food will not cause harm to the consumer when it is prepared and/or eaten according to its intended use.

With several thousand years of experience of food safety combined with over 150 years of food microbiology and the latest molecular biology techniques it might be assumed that problems of food safety would have been resolved. In fact, the opposite is true with increased incidents of foodborne disease (Griffith, 2000; Redmond and Griffith, 2003).

Foodborne diseases caused by microbiological hazards are a growing public health problem till today. Most countries that developed systems for reporting foodborne disease have documented significant increases (FAO/WHO, 2012). Scientists are concerned that foodborne illnesses associated with microbial pathogens, biotoxins and chemical contaminants in food present a serious threat to the health of millions of people in the world (FAO, 2014). Millions of words of advice and millions of pounds have been spent but the problem is getting worse. Furthermore, problems of food safety have increased in developed and developing countries. In some developing countries having epidemiological data collection systems, have reported increased incidence of foodborne diseases. Data collection is very important to understand the nature and extent of the problem. For example, Egypt typically reports only about three cases per 100,000 populations, whereas Sweden reports 5770. The difference is not due to better hygiene, in reality, the reverse is true, but due to better reporting and data collection. A range of reasons have been presented explaining why food poisoning has increased in developed countries:

1. Changing patterns of food consumption.
2. Proper/incorrect use of new cooking equipment.
3. More varied cuisine, including ethnic foods.
4. Change in cooking/shopping practices: weekly or monthly shops rather than daily.
5. Decreased use of preservatives and less processing.
6. Greater eating out.
7. Reduced consumer immunity: greater number of higher risk individuals in the population.
8. Lengthened gap between production and consumption.
9. Greater consumer awareness of foodborne disease.
10. Changes in farming practices.

11. Evolution of existing and new food pathogens.
12. Failure to accept responsibility.
13. Government failures.
14. Food industry failures.
15. Consumer negligence/ignorance.
16. Failures in management practices.
17. Lack of multidisciplinary research approach.

Historically, based on end product testing, strategic changes towards a more preventative approach to food safety management started as early as the 1920s (Mossel et al., 1995), although these strategies were largely unsuccessful. There was a renewed emphasis on preventative food safety in the 1930s, however, in 1970s an improved approach was adopted leading to the use of HACCP (Bauman, 1994). In HACCP approach, food business operators are responsible for ensuring the safety of their foods at all stages of production that are under their control. They are required to put in place a procedure based on HACCP, a proactive control system created to ensure food safety by controlling all stages of production. It has seven principles: (1) identification of—any hazards that must be prevented, eliminated or reduced to acceptable levels; (2) identification of the critical control points at the step(s) at which control is necessary—to prevent or eliminate a hazard or to reduce it to acceptable levels; (3) establishment of—critical limits at critical control points which separate acceptability from unacceptability for the prevention, elimination or reduction of identified hazards; (4) establishment and implementation of—effective monitoring procedures at critical control points; (5) establishment of—corrective actions when monitoring indicates that a critical control point is not under control; (6) establishment of—procedures, which shall be carried out regularly, to verify that [measures (1) to (5)] are working effectively; and (7) establishment of—documents and records commensurate with the nature and size of the food business to demonstrate the effective application of the measures outlined in [provisions (1) to (6)].

There is evidence that businesses adopting a food safety management approach based on HACCP and pre-requisite programmes (PRPs) produce better microbiological quality food (Little et al., 2002; Little et al., 2003). However, considering management systems is only part of the multidisciplinary approach that needs adoption. Documented food systems explain how things "should be done" but what people "actually do" is a manifestation of the food safety organisational culture (Griffith, 2000). This is a complete integration of the individual food handlers' knowledge, attitudes and practices with the culture or standards set by the manager/owner of the business. Food safety organizational culture is influenced by many things, including the facilities available (e.g., for hand-washing), as well as the time available to implement food safety practices. Within the food service context, this can be a greater problem, where staff may need to supply food to order rather than from stock.

There had been many great risk of food security, food hygiene and/or food safety and people so often faced the food crisis, and most of the crises was handled carefully by people's wisdom and cooperative efforts. The past success, however, do not guarantee the solution of future food problems. At present, we are holding many problems in foods which need to be resolved, e.g., radioactive contamination in foods, emerging and re-emerging infectious diseases, GMO's foods, etc.

Food safety risks in the future

In the future, more serious and high-level risks might occur. Recent years have witnessed a host of large-scale disasters of various kinds throughout the world: hugely damaging windstorms and floods in Europe and ice storms in Canada; new diseases infecting both humans (AIDS, the Ebola virus) and animals (BSE, Nipah), tsunami and radiation leakage episodes of Japan, war in Arab countries. These are just some of the extremely costly disasters that are beyond society's capacity to manage. The forces shaping these changes are many and varied. For example, weather conditions appear to be becoming increasingly extreme. The population density in urban centres and concentrations of economic activity in certain regions are rising, rendering these areas more vulnerable. Globalization in all its dimensions—economic, technological, cultural, and environmental—is growing apace and increasing interdependence, making it easier for dangerous pathogens, pollutants to spread. Equally important, the frontiers of scientific discovery and technological innovation are expanding at breathtaking speed, confronting society with unknown (indeed, unknowable) impacts, and therefore immensely affect the systems on which society depends. However, the risk in future may be on a much larger scale and their effects may spread much further. Important changes to major risks are expected to take place in the coming decades. Five categories of such risks are: natural disasters, industrial accidents, infectious diseases, terrorism, and food safety. To learn from the past experience and the battle against present food problem, and hope for the best in future will make the food risk lower. Fighting for being released from fear brought by defective and/or harmful foods does not have an end. We, scientists, engineers and consumers have been fighting in order to realize the state of risk-off, though it seems very difficult.

People have offered wisdom for improvement in the qualities of security, hygiene, and safety of food resulting in continuous increase of world population. In this century, world' population will surpass 10 billion (Table 5) and it is necessary to increase the food production by 1.5 times in near future.

Table 5. Projected World Population by 2100.

Year	1	1000	1650	1804	1927	1960	1975	1990	1999	2011	2024	2040	2062	2100
Population	2 m	2.75 m	500 m	1 b	2 b	3 b	4 b	5.3 b	6 b	7 b	8 b	9b	10 b	10.8b

m = million b = billion

It is expected that many risks will arise with increase in food production, climate change, or increased human activities in nature and consequently more unexpected events may occur. The world's reality today is:

1. Number of people with obesity and/or diabetes are higher than that of starving.
2. Obesity-risk for poor people is higher than those of rich people in developed countries.

Many people in the world suffer from food shortage, while some others are eating much and harming their health. People's gourmet taste does not decline. Malnutrition as well as the imbalance nutrition ingestion may cause people to become sick. This type of food risk will increase in the future and needs to be addressed properly.

The Tsunami and Radiation Leakage Episodes of Japan in 2011 Pose Future Food Risks

Japan experienced a devastating 9.0 magnitude earthquake on March 11, 2011. The earthquake has struck north east coast of Honshu area. The earthquake and tsunami have destroyed coastal areas of Tohoku. The tsunami destroyed the excellent cultivated lands in these areas. Such destructive earthquake and tsunami might happen once in thousand years but it left behind a trail of destructions and demolished the ecological balance and crop lands.

Unexpected disaster related to food, however, have often destroyed the smooth going life of people. Unseasonable weather of makes yields of crops unstable. Typhoon hit some districts of Japan every year, where rice fields are severely damaged. Rinderpest had afflicted European people for thousands of years. Although it is not pathogenic to people, the mortality rate of infected animals is high, and caused severe starvation in Europe. In the 18th century, when the human population of the world was only ca. 700 million total ca. 200 million heads of cattle were infected and died.

Radioactive material leakage from the Fukushima Daiichi Nuclear Power Plant in Japan have diffused in natural environment and polluted the land, mountains and ocean. Unless this present situation is improved from viewpoints of food security as well as food safety, human beings' prosperity cannot be sustained. Various risks in food supply have afflicted people and will continue to do so. For example:

(a) Natural disasters:
  • Earthquake, typhoon (cyclone, hurricane), flood, drought, cold or heat wave, etc.
(b) Infectious diseases in animals and plants:
  • rinderpest, mouth and foot disease, etc.
  New infectious diseases:
  • Avian flu, nipah virus infection, etc.
(c) Side effects of modern agriculture and industry:
  • BSE (bovine spongeform encephalitis), *S. enteritidis*, organic mercury poisoning, radiation injury, etc.

In the last 20 years, researchers, have recognized that food safety is a multidisciplinary problem requiring fresh approaches (Mortlock et al., 2000). Whilst food microbiology is an important component of food safety, the development of rapid microbiological techniques and the identification of new pathogens, identifying the genome of a pathogen may only be of limited value in prevention.

A multidisciplinary range of food safety systems or approaches has been suggested by Griffin (2006) which includes human behavior, auditing to HACCP principle as presented in Table 6. This may be combined with the development of international standards for HACCP (ISO 22000) or as part of global food quality systems such as

Table 6. A multidisciplinary range of food safety techniques (adapted from Griffin, 2006).

A	Auditing
FP	Food processing
M	Microbiology
RT	Rapid testing
PRPs	Pre-requisite programmes
HACCP	Hazard analysis critical control points
RA	Risk assessment
RR	Risk Re-enactment
NA	Notational analysis
SCM	Social cognition models
HEA	Human error analysis
SM	Social marketing
OC	Organizational culture/climate
CBA	Cost benefit analysis
TNA	Training needs analysis
WTP	Willingness to pay

SQF or BRC. By virtue of their external auditing component, it may provide greater retailer and consumer confidence in international trade.

The recognition of human behavior linked to organizational culture has triggered new approaches to improving food safety. These include the application of psychological models, e.g., social cognition models and human error analysis, to understand, possibly predict and then to change/improve behavior. Training may involve a greater social marketing element where the information communicated considers the recipient of the information more fully (Redmond et al., 2000). The same is likely to be true of consumer food safety campaigns, many of which have been relatively unstructured and uncoordinated in the past. Initiatives based on general food safety awareness, coupled with social marketing approaches for specific target groups, are likely to become more common. Convincing managers to improve food safety, and that it is not a cheap option to ignore, will involve greater use of cost-benefit analysis and studies on consumer willingness to pay. This may be linked to consumer choice and publication of inspection reports as well as outbreak data (Griffith, 2006). Recognizing the link between individual and corporate behavior means greater use and understanding of models of food safety organizational culture and the role of managers in setting standards.

One Health Concept for Food Safety

The One Health concept recognizes that the health of humans is connected to the health of animals and the environment (As shown in Fig. 2). One Health is not a new concept, but it has become more important in recent years because many factors have changed the interactions among humans, animals, and the environment. These changes have caused the emergence and re-emergence of many diseases. Human populations are growing and expanding into new geographic areas and as a result, more people live in close contact with wild and domestic animals. This close contact provides more opportunities for diseases to transmit between animals and people.

The One Health Triad

Figure 2. One Health approach traid includes Healthy environment, healthy animals and healthy people.

The earth has experienced changes in climate and land use such as deforestation and intensive farming practices. Disruptions in environmental conditions and habitats provide new opportunities for diseases to pass onto animals. International travel and trade have increased; as a result, diseases can spread quickly across the globe. Successful public health interventions require the cooperation of human health, veterinary health, and environmental health communities. By promoting this collaboration, optimal health outcomes for both people and animals can be achieved.

Summary/Conclusions

Food is essential to life and all foods harbor one or more types of microorganisms, some of them have desirable roles in food, such as in the production of naturally fermented food, whereas others cause food spoilage and foodborne diseases. The morbidity associated with the millions of cases of foodborne illnesses worldwide has significant social and economic consequences. The recent improvement in science and technology, detection methods of pathogen foodborne illness is, in many countries, at or near an all time high. Whilst it is difficult to predict events, strategies and research in the future are likely to recognize not only the importance of food safety management systems, but the role of individuals, working with their peers and superiors, within a business food safety culture. These new fields of study still lack essential data, and the use of additional research techniques to understand and achieve behavioral change is required. Almost implicit is that this actively involves the businesses and managers themselves and they must consider the active and subliminal food safety messages to communicate to their employees. Failure to do so will result in stubbornly high levels of foodborne diseases.

References

Bauman, H.E. 1994. The origin of the HACCP systems and subsequent evaluation. Food Science and Technology Today, 8: 66–72.
Bian Yongmin. 2004. The Challenges for Food Safety in China. China Perspectives, 53: 1–15.

Cartwright, R.Y. 2003. Food and waterborne infections associated with package holidays. Journal of Applied Microbiology, 94: 1–12.

Clayton, D. and C.J. Griffith. 2004. Observation of food safety practices in catering using notational analysis. British Food Journal, 106: 211–27.

Clayton, D., C.J. Griffith, P. Price and A.C. Peters. 2002. Food handlers' beliefs and self-reported practices. International Journal of Environmental Health, 12: 25–39.

Griffith, C.J. 2006. Food safety: where from and where to? British Food Journal, 108(1): 6–15.

Hartman, P. 2001. The evolution of food microbiology. Food Microbiology: Fundamentals and Frontiers, 2nd edn., ASM Press, Washington DC, pp. 3–12.

Hutt, P.B. and P.B. Hutt, II. 1984. A history of government regulation and misbranding of food. Food, Drug Cosmet. Law J., 39: 3.

Lu Gwei-Djen and J. Needham. 1951. A contribution to the history of Chinese dietetics. Isis, 42: 13–20.

Marth, E.H. 2001a. The emergence of food microbiology: its origin in dairy microbiology. Dairy, Food and Environmental Sanitation, 21: 818–24.

Marth, E.H. 2001b. The emergence of food microbiology: from dairy microbiology to food microbiology. Dairy, Food and Environmental Sanitation, 21: 886–96.

Merrill, R. 1985. FDA's implementation of the Delaney clause: repudiation of congressional choice or reasoned adaptation to scientific progress? Yale J. Reg., 5: 1.

Miller, S.A. 1993. Science, Law and Society: The Pursuit of Food Safety. Symposium: Historical Overview of the Safety of the Food Supply. American Institute of Nutrition. J. Nutr., 123: 279–284.

Miller, S.A. and M.R. Taylor. 1989. Historical development of food regulation in international food regulation handbook. *In*: R.G. Middlekauf and P. Shubik (eds.). Marcel Dekker, Inc., New York, NY.

Montville, T.J. and K.R. Matthews. 2005. Food Microbiology: An Introduction, ASM Press, Washington DC.

Mortlock, M.P., A.C. Peters, C.J. Griffith and D. Lloyd. 2000. Evaluating impacts of food safety control on retail butchers. *In*: N.H. Hooker and E.A. Murano (eds.). Interdisciplinary Food Safety Research, CRC Press, Boca Raton, FL.

Mossel, D.A.A., J.E.L. Corry, C.B. Struijk and R.M. Baird. 1995. Essentials of the Microbiology of Foods: A Textbook for Advanced studies, John Wiley & Sons, Chichester.

Needham, J. 1962. Hygiene and preventive medicine in ancient China. J. Hist. Med. Allied Sci., 17: 429–478.

Renée Johnson. 2014. The Federal Food Safety System: A Primer. Congressional Research Service, CRS 7-5700 RS22600 and available at www.crs.gov.

Redmond, E.C., C.J. Griffith and A.C. Peters. 2000. Use of social marketing techniques in the prevention of specific cross contamination actions in the domestic environment. 2nd NSF Conference on Food Safety, Savannah, GA, October.

Redmond, E., C.J. Griffith, J. Slader and T. Humphrey. 2004. Microbiological and observational analysis of cross contamination risks during domestic food preparation, British Food Journal, 106: 581–97.

Schmidt, K. 1995. WHO Surveillance Programme for Control of Foodborne Infections and Intoxications in Europe, Federal Institute for Health Protection of Consumers and Veterinary Medicine, Berlin, 6th Report 1990–1992.

2

Foodborne Illness and Microbial Agents: General Overview

Md. Latiful Bari[1],* *and Dike O. Ukuku*[2]

Introduction

'Foodborne illnesses result from the consumption of food containing microbial agents such as bacteria, viruses, parasites or food contaminated by poisonous chemicals or biotoxins' (World Health Organization [WHO], 2011a). Due to the nutrient composition of foods, they are capable of supporting the growth of microorganisms including human bacterial pathogens that lead to foodborne diseases—contaminated foods and food products (Guyader and Atmar, 2008). More than 250 different types of viruses, bacteria, parasites, toxins, metals, and prions are associated with foodborne diseases in humans (Linscott, 2011). Although viruses are responsible for more than 50% of all foodborne illnesses, generally hospitalizations and deaths are due to bacterial agents. The diseases range from mild gastroenteritis to life-threatening neurologic, hepatitic, and renal syndromes caused by either toxins from the "disease-causing" microbes, or by the human body's reaction to the microbes itself (Teplitski et al., 2009). It is difficult to determine the exact mortality rate associated with foodborne illnesses (Helms et al., 2003). However, worldwide an estimated 2.0 million deaths have occurred till now due to gastrointestinal illnesses (Fleury et al., 2008). Although majority of the foodborne illness cases are mild and self-limiting, severe cases can occur in high-risk groups resulting in high mortality and morbidity in this group. The high-risk groups for foodborne diseases include infants, young children, the elderly and the immunocompromised persons (Fleury et al., 2008).

---

[1] Food Analysis and Research Laboratory, Center for Advanced Research in Sciences, University of Dhaka, Dhaka-1000, Bangladesh.
[2] Food Safety Intervention & Technology, Eastern regional Research Center, 600 East Mermaid Lane, Wyndmoor, PA 19038, USA.
* Corresponding author: latiful@du.ac.bd

The war against foodborne diseases is facing new challenges everyday due to the globalization of the food market, climate change and changing patterns of human consumption as fresh and minimally processed foods are currently preferred by consumers (Schelin et al., 2011). Foodborne illnesses have a negative impact on the public health as well as on the economy of a country (Helms et al., 2003). They also have a negative impact on the trade and industries of the affected countries. Identification of a contaminated food product can result in recalling of that specific food product leading to economic loss to the industry. Foodborne outbreaks may lead to closure of the food outlets or food industry resulting in job losses for workers, affecting the individuals as well as the communities. Localized foodborne illness outbreaks may become a global threat. The health of people in many countries can be affected by consuming contaminated food products, and may negatively impact a country's tourism industry. The foodborne illness outbreaks are reported frequently at national as well as international levels underscoring the importance of food safety (WHO, 2011b).

Several factors are necessary for foodborne illnesses to occur, chief among them being: (1) a pathogen; (2) a food vehicle; (3) conditions that allow the pathogen to survive, reproduce, or produce a toxin; and (4) a susceptible person who ingests enough of the pathogen or its toxin to cause illness. The symptoms often are similar to those associated with flu—nausea, vomiting, diarrhea, abdominal pain, fever, headache. Most people have experienced a foodborne illness sometime in their life, even though they might not recognize it as such, instead blaming it on the stomach flu or a twenty-four-hour bug. Usually symptoms disappear within a few days, but in some cases there can be more long-lasting effects such as joint inflammation or kidney failure. In the most severe cases a person can die from a foodborne illness.

Classification of Foodborne Causative Agents

In general, a foodborne illness can be caused by contamination of food with some biological agents or pathogens (e.g., viruses, bacteria, parasites, prions), chemical agents (e.g., toxins, metals, drug residues, pesticides), or physical agents (e.g., glass fragments, bone chips, stone). Pathogens are the most significant cause in more than 200 known diseases being transmitted through food. With the exception of certain parasites, nearly all foodborne pathogens are microscopic in nature. In increasing order of size, these pathogens include viruses, bacteria, protozoa and other parasites.

Viruses are particulate in nature and multiply only in other living cells. Thus, they are incapable of surviving for long periods outside the host. While more than a 100 types of enteric viruses have been associated with foodborne illness, the most common foodborne viruses are the norwalk virus, norovirus (formerly known as Norwalk-like viruses [NLVs]), rotavirus, astrovirus, and hepatitis A.

Bacteria are one-celled microorganisms with a cell wall but no nucleus. They exist in a variety of shapes, types and properties. Some pathogenic bacteria are capable of spore formation and are thus, highly heat-resistant (e.g., *Clostridium botulinum, C. perfringens, Bacillus subtilus, B. cereus*). Others are capable of producing heat-resistant toxins (e.g., *Staphylococcus aureus*). Most pathogens are mesophilic with optimal growth temperature ranging from 20–45°C (68–113°F). However, certain foodborne pathogens (termed psychrotrophs) are capable of growth under refrigerated conditions or temperatures less than 10°C (50°F). The most well-documented psychrotrophic foodborne pathogens are *Listeria monocytogenes* and *Yersinia enterocolitica*. *Listeria monocytogenes*, for example, will grow (albeit slowly) at temperatures just above freezing point (approx. 33–34°F). Certain strains or serotypes of *Bacillus cereus, Clostridium botulinum, Salmonella* spp., *E. coli* O157:H7, and *Staphylococcus aureus* may also grow slowly under refrigeration conditions.

Parasitic Protozoa are one-celled microorganisms without a rigid cell wall, but with an organized nucleus. They are larger than bacteria. Like viruses, they do not multiply in foods but only in their hosts. The transmissible form of these organisms is termed *cyst*. Protozoa that have been associated with food and water-borne infections include *Entamoeba histolytica, Toxoplasma gondii, Giardia lamblia, Cryptosporidium parvum* and *Cyclospora cayatenensis*.

Multicellular Parasites are animals that live at the expense of the host. They may occur in foods in the form of eggs, larvae, or other immature forms. Trichinosis has been an important reportable pathogen associated with undercooked pork. Other parasites of concern include flatworms or nematodes (associated with fish), cestodes or tapeworms (usually associated with beef, pork, or fish) and trematodes or flukes (more a concern outside the US) (Schmidt et al., 2009).

Classification of Foodborne Illnesses Caused by Pathogens

Foodborne Illness occurs when a food contaminated with a bacteria pathogen is ingested and the pathogen establishes itself and multiplies in the human host, or when a toxigenic pathogens in a food product produces a toxic product that is ingested by the human host. Thus, foodborne illness is generally classified into two main categories, (1) *foodborne infection* and (2) *foodborne toxicity*:

(1) **Foodborne Infections:** "Infection" is caused by the oral ingestion of viable microorganisms in adequate amounts to build up infection and the commencement of symptoms is normally delayed, reflecting the time required for an infection to develop. Examples of food-poisoning that cause infection are enteric viruses, *Salmonella, Campylobacter* and *Vibrio* species. The two basic categories of foodborne infections are:

  (a) *Invasive infections,* which are caused by pathogens that invade body tissues and organs. Included in this group are viruses, parasitic protozoa, other parasites, and invasive bacteria (e.g., *Salmonella, Aeromonas, Campylobacter, Shigella, Vibrio parahaemolyticus, Yersinia* and enteric-type *Escherichia coli*).

  (b) *Toxic infections,* which are caused by infective bacteria that are not considered invasive in nature, but are capable of multiplication or colonization in the human intestinal tract and produce toxins. Included in this group are: *Vibrio cholerae, Bacillus cereus* (diarrheal-type), *C. botulinum* (in infants), *C. perfringens* and verotoxigenic *E. coli* (*E. coli* O157:H7 and others).

(2) **Foodborne toxicity:** On the other hand, is caused by the ingestion of toxins that have been pre-formed in the food. There is no need for live organisms to be present and the onset of the symptoms is rapid. Examples are *Bacillus cereus* and *Staphylococcus aureus*. Other non-bacterial toxins that cause illness are:

- Paralytic shellfish toxin (caused by the consumption of mussels, clams and scallops which have ingested toxic dinoflagellates).
- Ciguatera toxins (associated with certain tropical fishes).
- Scombroid toxins (results from the production of histamine due to bacteria spoilage of fish).
- Fungal toxins or mycotoxins that can be of long-term carcinogenic concern with consumption of mold-contaminated foods (e.g., aflatoxins in contaminated corn, peanuts, or other foods and patulin from contaminated apples or other fruit products).

Mode of Action of Foodborne Pathogens

Pathogens enter the body through the digestive tract. These organisms differ in how they establish themselves (and may or may not cause illness), and the number of microorganisms (infective dose) or quantity of toxin (toxic dose) required to cause illness. A major factor in the ability of these organisms to cause illness, is the characteristics of the host. Humans can be generally characterized as low-risk and high-risk groups in terms of susceptibility. Low-risk or healthy individuals may be resistant to many (but not all) of the foodborne illnesses. However, high-risk individuals (e.g., those who are immunocompromised, immunosuppressed, infants, elderly, small children, etc.) have much lowered (and highly varied) resistance.

A person infected with a foodborne pathogen can be a carrier as well as a source of infection to others. Even infected individuals who do not show signs of illness can be chronic carriers. While this is especially true for persons infected with viruses (e.g., hepatitis A), chronic carrier states occur for many infective foodborne bacteria as well which can be excreted during the illness period (e.g., *Salmonella* spp., *Shigella* spp.). While the carrier state usually ceases after several weeks, it can last for longer time periods in certain individuals and for certain pathogens (e.g., *Salmonella typhi*).

While in most cases, the symptoms of foodborne illness cease after a relatively short time, complications (e.g., chronic sequalae) can occur in certain infected individuals. While these complications occur in less than 5% of cases, the occurrence is relatively rare but can be severe. Examples of chronic sequalae include: reactive arthritis, Reiter's syndrome, Guillain-Barre syndrome, ankylosing spondylitis, meningitis, rheumatoid arthritis, septic arthritis, septicemia, and cardiac manifestations (Schmidt et al., 2009).

Remarkable Foodborne Disease Outbreaks

Foodborne outbreak is defined as an incidence, observed under given circumstances, of two or more human cases of the same disease and/or infection, or a situation in which the observed number of cases exceeds the expected number and where the cases are linked, or are probably linked, to the same food source. Foodborne diseases are prevalent in all parts of the world, and the toll in terms of human life and suffering is enormous. Contaminated food contributes to 1.5 billion cases of diarrhea in children each year, resulting in more than three million premature deaths, according to the World Health Organization [WHO, 2011b]. Those deaths and illnesses are shared by both developed and developing nations. For example, CDC estimates that each year roughly 1 in 6 Americans (or 48 million people) gets sick, 128,000 are hospitalized, and 3,000 die of foodborne diseases (CDC, 2011). In France, it is estimated that these pathogens cause 10,200–17,800 hospitalizations yearly (Vaillant et al., 2005). Several devastating foodborne outbreaks have been reported on the African continent; in 2004, Kenya experienced an acute aflatoxicosis outbreak which was attributed to maize (WHO, 2005). In South East Asia, approximately one million children under five years of age die every year from diarrheal diseases after consuming contaminated food and water (CSPI, 2005). Usually, most foodborne outbreaks often go unrecognized, unreported or uninvestigated and may only be visible if connected to major public health or economic impact. Many countries have not yet established adequate surveillance or reporting mechanisms to identify and track foodborne illnesses (Nyenje and Ndip, 2013). Therefore, data on foodborne diseases are extremely scarce and improvements are needed to better identify the causes of them. A list of remarkable foodborne bacterial outbreaks worldwide from 1963–2014 are presented in Table 1.

Characteristics of Viruses, Bacteria and Parasites

Viruses

Viruses are very small microorganisms, ranging in size from 0.02 to 0.4 micrometres in diameter and are smaller than bacteria and parasites. Viruses can only reproduce within living cells in the body of the host and cannot multiply in foods; therefore the number of viral particles in food does not increase and sensory features of the contaminated and non-contaminated food will be identical (Koopmans and Duizer, 2004). However, some viruses remain infectious in the environment and are thus transported through food. Viruses that are associated with foodborne diseases are characterized by their growth in the intestinal cells and subsequent excretion in the faeces. More than 100 types of enteric viruses exist, although only a few have been proved to cause foodborne disease. These include rotavirus, norovirus, enteric adenovirus, hepatitis A virus (HAV), enterovirus, human astrovirus, aichivirus, torovirus, coronavirus and picobirnavirus (Chitimbar et al., 2012). However, overwhelming majority of cases is due to norovirus and HAV (Koopmans and Duizer, 2004). Although other viruses such as adenovirus can cause gastrointestinal illness, the mode of transmission is believed to be primarily person-to-person. Foodborne viruses cause infection and not toxicity. Documentation

Table 1. Foodborne bacterial outbreaks (1963–2014) worldwide.

Year	Event	Agent	Vehicle	Company	Infected	Deaths	References
1963	botulism outbreak from canned tuna	*Clostridium botulinum*	canned tuna	Washington Packing	3	2	Johnston et al., 1963
1971	botulism outbreak from Bon Vivant soup	*Clostridium botulinum*	vichyssoise soup	Bon Vivant	2	1	NYT, 1971
1985	salmonellosis outbreak from milk	*Salmonella*	milk	Hillfarm Dairy	5,295	9	Caroline et al., 1987
1985	California listeriosis outbreak from cheese	*Listeria*	queso fresco	Jalisco Cheese	> 86	47 or 52	Neuman, 2011; Segal et al., 1988
1993	*E. coli* outbreak Jack in the Box restaurant	*E. coli* O157:H7	undercooked hamburgers	Jack-in-the-Box	> 700	4	Davis (1993)
1996	*E. coli* outbreak from Odwalla juice	*E. coli* O157:H7	unpasteurized apple juice	Odwalla	66	1	Drew et al., 1988
1996	*E. coli* O157:H7 outbreaks from radish sprouts from Japan	*E. coli* O157:H7	radish sprouts	A Japanese company	> 6000	13	Michino et al., 1999
1998	United States: listeriosis outbreak from infected cold cuts	*Listeria*	cold cuts and hot dogs	Bil Mar Foods	> 100	18 or 21	Mead et al., 2006
2002	listeriosis outbreak from poultry	*Listeria*	poultry	Pilgrim's Pride	> 50	8	CDC, 2002
2003	hepatitis A outbreak	Hepatitis A	green onions	A restaurant	555	3	CDC, 2003
2005	South Wales: *E. coli* O157 outbreak	*E. coli* O157	meat	Local butcher	157	1	Pennington (2009)
2006	North American: *E. coli* O157:H7 outbreak from infected spinach	*E. coli* O157:H7	spinach	Dole Foods	> 205	3	Grant et al., 2008

Table 1. contd....

Table 1. contd....

Year	Event	Agent	Vehicle	Company	Infected	Deaths	References
2008	Canada: listeriosis outbreak from infected cold cuts	*Listeria*	cold cuts	Maple Leaf Foods	> 50	22	Amir Attaran et al., 2008
2008	United States: salmonellosis outbreak from peanuts	*Salmonella*	peanuts	Peanut Corporation of America	> 200	9	CDC, 2009
2011	United States: listeriosis outbreak from infected cantaloupes	*Listeria*	cantaloupe	Jensen Farms	146	30	CDC, 2011
2011	Germany: *E. coli* O104:H4 outbreak	*E. coli* O104:H4	fenugreek sprouts	A restaurant in Lübeck, Schleswig–Holstein	> 3,950	53	Frank et al., 2011
2014	2013–2014 Denmark: listeriosis outbreak	*Listeria*	Spiced lamb roll, pork, sausages, bacon, liver pâté, etc.	Jorn A. Rullepolser	34	13	HPSWR, 2014

of viral foodborne disease is scant. This is because of diverse symptoms (often mild illness), difficulty of detection of viruses in food, and difficulty of routine, conclusive diagnosis through stool specimens. Food usually becomes contaminated when it is handled by a person infected with a virus who has poor personal hygiene or when the food comes in contact with virus-laden sewage. It does not take a large quantity of virus for infection. For example, a person with rotavirus diarrhea may excrete approximately a trillion infectious particles per milliliter of stool, but as few as 10 particles can cause illness. Additionally, excretion of viruses in faeces may occur even if a person has no symptoms of GI illness. Viruses are increasingly being recognized as significant causes of foodborne illnesses in the United States (Table 2). For example in the USA, viruses account for 67% of food-related illnesses, compared to 9.7% and 14.2% for *Salmonella* and *Campylobacter*, respectively (Vasickova et al., 2005).

Table 2. Common Foodborne Diseases Caused by Viruses.

Disease (Causative agent)	Onset (Duration)	Principal symptoms	Typical Foods	Mode of contamination	Prevention of disease
Hepatitis A (Hepatitis A virus)	15–50 days (weeks to months)	Fever, weakness, nausea, discomfort, often jaundice	Raw or undercooked shellfish; sandwiches, salads, etc.	Human faecal contamination, via water or directly	Cook shellfish thoroughly; carefulness about general sanitation
Viral gastroenteritis (Norwalk-like viruses)	1–2 days (1–2 days)	Nausea, vomiting, diarrhea, pain, headache, mild fever	Raw or undercooked shellfish, sandwiches, salads, etc.	Human faecal contamination, via water or directly	Cook shellfish thoroughly; general carefulness about general sanitation
Viral gastroenteritis (rotaviruses)	1–3 days (4–6 days)	Diarrhea, especially in infants and children	Raw or mishandled food	Probably human faecal contamination	General sanitation

Bacterial agents

Bacteria are one-celled living microorganisms which possess cell walls. Bacterial cells vary in shape and range in sizes from about 1 micrometer (μm), which equals one millionth of a meter, to 5 or 10 micrometers in length. In contrast to viruses, bacteria can be seen with a conventional microscope. Bacterial cells multiply when each cell divides into two, which grow to full size and divides into two again (two-fold division). Unlike viruses or parasites, bacteria are able to multiply in or on food. Under optimum conditions, large numbers can easily be achieved. Some pathogenic bacteria, including *Bacillus cereus*, *Clostridium botulinum*, and *Clostridium perfringens*, form spores that can survive extreme environmental conditions. The spores germinate to form viable cells that multiply to large numbers. Spore-forming pathogens are significant because when the spores occur in foods, they are more difficult to kill. For example, although *Bacillus cereus* bacteria survive up to 122°F, much higher temperatures are required to kill the spores of *B. cereus*.

Pathogenic bacteria can cause foodborne infections or toxicity. *Salmonella* is the leading documented cause of foodborne infections in the 20th century followed by entero-pathogenic *Escherichia coli, Campylobacter, Listeria* (Nyenje et al., 2012b). The bacteria that produce foodborne toxicity most often in the United States include *Bacillus cereus, Clostridium botulinum* and *Staphylococcus aureus.* Foodborne bacterial agents are the leading cause of severe and fatal foodborne illnesses. For instance in France, in the last decade of the 20th century, *Salmonella* was the most frequent cause of bacterial foodborne illness (5,700–10,200 cases), followed by *Campylobacter* (2,600–3,500 cases) and *Listeria* (304 cases) (Vaillant et al., 2005). In South Africa, species of *Listeria, Enterobacter* and *Aeromonas* were the most prevalent bacteria in ready-to-eat foods (Nyenje et al., 2012b) (Table 3).

Parasites

Parasites are single or multi-celled organisms that need a host to survive. They are larger than viruses and bacteria, with dimensions usually greater than 10 micrometers (μm). One-celled parasites are commonly termed "parasitic protozoa", however for the purpose of simplicity; "parasites" will be used throughout this chapter. With regard to foodborne illness, parasites only cause infection, not toxicity, and similar to viruses, parasites do not multiply in foods, but can survive in the environment and thus be transported through food. Often, parasites undergo structural changes during their life-cycle. The structural form transmissible through food often is a cyst that is inert and resistant to the environment outside the host, similar to a bacterial spore, but less resistant to heat. Once the cyst, enters the body of a new host, through food, it can multiply.

One-celled parasites occurring in foodborne outbreaks in the United States include *Entamoeba histolytica, Toxoplasma gondii* and *Giardia lamblia. Cryptosporidium parvum* is becoming more common and is also a problem in immunocompromised people, e.g., patients with acquired immune deficiency syndrome (AIDS) (Khan et al., 2007; Dorny et al., 2009). *Cyclospora cayetanensis* is a newly identified parasite that was first reported in the medical literature in 1979. Cases have been identified and reported with increased frequency since the mid-1980s. During the summers of 1996 and 1997, nationwide outbreaks of cyclosporiasis occurred from the consumption of imported contaminated berries.

The multi-celled parasites found in food may occur as eggs, larvae, or other forms. They can be ingested into the body where they may hatch, leading to the development of new parasites. *Trichinella spiralis* is reported to cause a few cases of foodborne illness (trichinosis) in the United States each year. Formerly, this was an important pathogen associated with undercooked pork. Tapeworm species occurring in the United States include the beef tapeworm (*Taenia saginata*), the pork tapeworm (*Taenia solium*), and the fish tapeworm (*Diphyllobothrium* species). However, infections from these parasites are rare. The most common foodborne diseases caused by protozoa and parasites and their onset in typical food system is shown in Table 4.

Table 3. Common Foodborne Diseases Caused by Bacteria.

Disease (causative agent)	Latency period (durration)	Principal symptoms	Typical foods	Mode of contamination	Prevention of disease
food poisoning, diarrhea (*Bacillus cereus*)	8–16 hr (12–24 hr)	Diarrhea, cramps, occasional vomiting	Meat products, soups, sauces, vegetables	From soil or dust	Thorough heating and rapid cooling of foods
food poisoning, emetic (*Bacillus cereus*)	1–5 hr (6–24 hr)	Nausea, vomiting, sometimes diarrhea and cramps	Cooked rice and pasta	From soil or dust	Thorough heating and rapid cooling of foods
Botulism; food poisoning (heat labile toxin of *Clostridium botulinum*)	12–36 hr (months)	Fatigue, weakness, double vision, slurred speech, respiratory failure, sometimes death	Types A & B: vegetables, fruits, meat, fish, and poultry products, condiments; Type E: fish and fish products	Types A & B: from soil or dust; Type E: water and sediments	Thorough heating and rapid cooling of foods
Food poisoning infant infection (*C. botulinum* spores)	12–36 hr 2–10 months	Constipation, weakness, respiratory failure, sometimes death	Honey; soil	Ingested spores from soil, dust or honey colonize the intestine	Do not feed honey to infants
Campylobacteriosis (*Campylobacter jejuni*)	3–5 days (2–10 days)	Diarrhea, abdominal pain, fever, nausea, vomiting	Infected food-source—animals	Chicken, raw milk	Cook chicken thoroughly; avoid cross-contamination; irradiate chicken; pasteurized milk
Cholera (*Vibrio cholera*)	2–3 days hours to days	Profuse, watery stools; sometimes vomiting, dehydration; often fatal if untreated	Raw or undercooked seafood	Human faeces in marine environment	Cook seafood thoroughly; general sanitation
food poisoning (*Clostridium perfringens*)	8–22 hr (12–24 hr)	Diarrhea, cramps, rarely nausea and vomiting	Cooked meat and poultry	Soil, raw foods	Thorough heating and rapid cooling of foods
foodborne infections enterohemorrhagic (*Escherichia coli*)	12–60 hr (2–9 days)	Watery, bloody diarrhea	Raw or undercooked beef, raw milk	Infected cattle	Cook beef thoroughly; pasteurize milk
enteroinvasive (*Escherichia coli*)	at least 18 hr (uncertain)	Cramps, diarrhea, fever, dysentery	Raw foods	Human faecal contamination, direct or via water	Cook foods thoroughly; general sanitation

Table 3. contd....

Table 3. contd....

Disease (causative agent)	Latency period (duration)	Principal symptoms	Typical foods	Mode of contamination	Prevention of disease
foodborne infection: enterotoxigenic (*Escherichia coli*)	10–72 hr (3–5 days)	Profuse watery diarrhea; sometimes cramps, vomiting	Raw foods	Human faecal contamination, direct or via water	Cook food thoroughly; be mindful of general sanitation
Listeriosis (*Listeria monocytogenes*)	3–70 days	Meningoencephalitis; stillbirths; septicemia or meningitis in newborns	Raw milk, cheese and vegetables	Soil or infected animals, directly or via manure	Pasteurization of milk; cooking
Salmonellosis (*Salmonella* species)	5–72 hr (1–4 days)	Diarrhea, abdominal pain, chills, fever, vomiting, dehydration	Raw and undercooked eggs; raw milk, meat and poultry	Infected food-source animals; human faeces	Cook eggs, meat, and poultry thoroughly; pasteurize milk; irradiate chickens
Shigellosis (*Shigella* species)	12–96 hr (4–7 days)	Diarrhea, fever, nausea; sometimes vomiting, and cramps	Raw foods	Human faecal contamination, direct or via water	General sanitation; cook food thoroughly
Staphylococcal food poisoning (heat-stable enterotoxin of *Staphylococcus aureus*)	1–6 hours (6–24 hours)	Nausea, vomiting, diarrhea, cramps	Ham, meat and poultry products, cream-filled pastries, whipped butter, cheese	Handlers with colds, sore throat or infected cuts, food slicers	Thorough heating and rapid cooling of foods
Streptococcal foodborne infection (*Streptococcus pyogenes*)	1–3 days (varies)	Various symptoms including sore throat, erysipelas, scarlet fever	Raw milk, deviled eggs	Handlers with sore throats, other "strep" infections	General sanitation, pasteurize milk
Vibrio parahaemolyticus foodborne infection	12–24 hr (4–7 days)	Diarrhea, cramps; sometimes nausea, vomiting, fever headache	Fish and seafoods	Marine coastal environment	Cook fish and seafoods thoroughly
Vibrio vulnificus foodborne infection	In persons with high serum iron: 1 day	Chills, fever, prostration, often death	Raw oysters and clams	Marine coastal environment	Cook shellfish thoroughly
Yersiniosis (*Yersinia enterocolitica*)	3–7 days (2–3 weeks)	Diarrhea, pains mimicking appearance of appendicitis, fever, vomiting, etc.	Raw or undercooked pork and beef; tofu packed in spring water	Infected animals especially swine; contaminated water	Cook meats thoroughly, use chlorinated water

Table 4. Common Foodborne Diseases Caused by Protozoa and Parasites.

Disease (Causative agent)	Onset (duration)	Principal symptoms	Typical foods	Mode of contamination	Prevention of disease
(PROTOZOA) Amebic dysentery (*Entamoeba histolytica*)	2–4 weeks (varies)	Dysentery, fever, chills; sometimes liver abscess	Raw or mishandled foods	Cysts in human faeces	General sanitation; thorough cooking
Cryptosporidiosis (*Cryptosporidium parvum*)	1–12 days (1–30 days)	Diarrhea; sometimes fever, nausea, and vomiting	Mishandled foods	Oocysts in human faeces	General sanitation; thorough cooking
Giardiasis (*Giardia lamblia*)	5–25 days (varies)	Diarrhea with greasy stools, cramps, bloating	Mishandled foods	Cysts in human and animal faeces, directly or via water	General sanitation; thorough cooking
Toxoplasmosis (*Toxoplasma gondii*)	10–23 days (varies)	Resembles mononucleosis; fetal abnormality or death	Raw or undercooked meats; raw milk; mishandled foods	Cysts in pork or mutton, rarely beef; oocysts in cat faeces	Cook meat thoroughly; pasteurize milk; general sanitation
(ROUNDWORMS, Nematodes) Anisakiasis (*Anisakis simplex*, *Pseudoterranova decipiens*)	Hours to weeks (varies)	Abdominal cramps, nausea, vomiting	Raw or undercooked marine fish, squid or octopus	Larvae occur naturally in edible parts of seafoods	Cook fish thoroughly or freeze at –4°F for 30 days
Ascariasis (*Ascaris lumbricoides*)	10 days–8 weeks (1–2 years)	Sometimes pneumonitis, bowel obstructions	Raw fruits or vegetables that grow in or near soil	Eggs in soil, from human faeces	Sanitary disposal of faeces; cooking food
Trichinosis (*Trichinella spiralis*)	8–15 days (weeks, months)	Muscle pain, swollen eyelids, fever; sometimes death	Raw or undercooked pork or meat or carnivorous animals (e.g., bears)	Larvae encysted in animal's muscles	Thorough cooking of meat; freezing pork at 5°F for 30 days; irradiation
(TAPEWORMS, Cestodes) Beef tapeworm (*Taenia saginata*)	10–14 weeks (20–30 years)	Worm segments in stool; sometimes digestive disturbances	Raw or undercooked beef	"Cysticerci" in beef muscles	Cook beef thoroughly or freeze below 23°F
Fish tapeworm (*Diphyllobothrium latum*)	3–6 weeks (years)	Limited; sometimes vitamin B_{12} deficiency	Raw or undercooked fresh-water fish	"Plerocercoids" in fish muscles	Heat fish 5 minutes at 133°F or freeze 24 hours at 0°F
Pork tapeworm (*Taenia solium*)	8 weeks–10 years (20–30 years)	Worm segments in stool; sometimes "cysticercosis" of muscles, organs, heart, or brain	Raw or undercooked pork; any food mishandled by a *T. solium* carrier	"Cysticerci" in pork muscle; any foodhuman faeces with *T. solium* eggs	Cook pork thoroughly or freeze below 23°F; general sanitation

Fungi

Filamentous fungi and moulds are able to produce an enormous number of secondary metabolites, including antibiotics and mycotoxins. The term "mycotoxin" refers to those secondary metabolites which, even at low concentrations, are toxic to humans and animals (Sánchez-Hervás et al., 2008). The existence of mycotoxin-producing fungi in plants is not always favorable for contamination with mycotoxins. In order for fungi to produce these secondary metabolites, they have to be stressed by some factor, such as nutritional imbalance, drought or water excess (Dutton, 2009). Mycotoxins have been implicated as causative agents of human foodborne toxicity, as well as human hepatitic and extra-hepatitic carcinogenesis (Wild and Gong, 2010); clinical symptoms include diarrhea, liver and kidney damage, pulmonary oedema, vomiting, haemorrhaging and tumours (Bryden, 2012). The most frequent toxigenic fungi are *Aspergillus*, *Penicillium* and *Fusarium* species (Sánchez-Hervás et al., 2008). The foodborne mycotoxins of greatest significance in Africa and other tropical countries are the fumonisins (FB), aflatoxins (AFs) and trichothecenes (Wagacha and Muthomi, 2008). These toxins contaminate various food stuffs, including maize, cereals, groundnuts and tree nuts feed during production, harvest, storage or processing (Sánchez-Hervás et al., 2008); mycotoxins can also occur in milk, meat and their products as a result of animals consuming mycotoxin contaminated feeds (Wild and Gong, 2010). The toxins frequently occur in maize, a staple food in most parts of Africa, Asia and Latin America; hence, their contamination translates to high-level chronic exposure in these countries (Wild and Gong, 2010).

The most common foodborne disease, symptoms and mode of contamination by fungi other than mushroom is shown in Table 5.

Table 5. Common Foodborne Diseases Caused by Fungi other than Mushrooms.

Disease (causative agent)	Latency Period (duration)	Principal symptoms	Typical foods	Mode of contamination	Prevention of disease
Aflatoxicosis ("aflatoxins" of *Aspergillus flavus* and related molds)	Varies with dose	Vomiting, abdominal pain, liver damage; liver cancer (mostly Africa and Asia)	Grains, peanuts, milk	Molds grow on grains and peanuts in field or storage; cows fed with moldy grain	Prevent mold growth; don't eat or feed moldy grain or peanuts; treat grain to destroy toxins
Alimentary toxic aleukia ("trichothecene" toxin of *Fusarium* molds)	1–3 days (weeks to months)	Diarrhea, nausea, vomiting; destruction of skin and bone marrow; sometimes death	Grains	Mild growth on grain, especially if left in the field through winter	Harvest grain in the fall; don't use moldy grain
Ergotism (toxins of *Claviceps purpurea*)	Varies with dose	Gangrene (limbs die and drop off); or convulsions and dementia; abortion (now not seen in the U.S.)	Rye; or wheat, barley, and oats	Fungus grows on grain in the field; grain kernel is replaced by a "sclerotium"	Remove sclerotia from harvested grain

Mushroom

Toxins

Mushroom poisoning is caused by consumption of raw or cooked fruiting bodies (mushrooms, toadstools) of a number of species of higher fungi. The term "toadstool" is commonly used for poisonous mushrooms. For individuals who are not experts in identifying mushrooms are, generally, no easily recognizable differences between poisonous and nonpoisonous species. Folklore notwithstanding, there is no reliable rule of thumb for distinguishing edible mushrooms from poisonous ones. The toxins involved in mushroom poisoning are produced naturally by the fungi themselves. Most mushrooms which cause human poisoning cannot be made nontoxic by cooking, canning, freezing, or any other means of processing. Thus, the only way to avoid poisoning is to avoid consumption of toxic species. Mushroom poisoning is generally acute, although onset of symptoms may be greatly delayed in some cases, and are manifested by a variety of symptoms and prognosis, depending on the amount and species consumed. Each poisonous species contains one or more toxic compounds that are unique to few other species. Therefore, cases of mushroom poisoning generally do not resemble each other, unless they are caused by the same or very closely related mushroom species.

The chemical constitution of many mushroom toxins (especially the less deadlier ones) is still unknown, and identification of mushrooms is well high difficult or impossible; however, mushroom poisoning is generally categorized by their physiological effects. Cultivated commercial mushrooms of various species have not been implicated in poisoning outbreaks, although they may result in other problems such as bacterial food poisoning associated with improper canning. Mushroom poisoning is almost always caused by ingestion of wild mushrooms collected by nonspecialists. Illnesses have occurred after ingestion of fresh, raw mushrooms, stir-fried mushrooms, home-canned mushrooms, mushrooms cooked in tomato sauce (which can render the sauce itself toxic, even when no mushrooms are consumed), and mushrooms that were blanched and frozen at home. Accurate figures on the relative frequency of mushroom poisoning are difficult to obtain, and the fact that some cases are not reported must be taken into account. During the 10-year period, from 2001 to 2011, 83,140 mushroom ingestions were reported to the US Poison Control Centers; of these, 64,534 (77.6%) were pediatric ingestions and 48,437 (58.3%) occurred in children younger than six years. The toxin group was identified in 12,147 (14.6%) of ingestions (Hatten et al., 2012). Between 1959 and 2002, there were more than 28,000 reported mushroom poisonings around the world, resulting in 133 deaths (Diaz, 2005a). In April 2008, an outbreak of mushroom poisoning in Upper Assam, India claimed more than 30 lives (Dutta et al., 2013). Known cases are sporadic, and large outbreaks are rare. Mushroom poisonings tend to be grouped in the spring and fall, when most mushroom species are at the height of their fruiting stage. Unfortunately, a number of factors (not discussed here) often make identification of the causative mushroom impossible. In such cases, diagnosis must be based on symptoms alone. To rule out other types of food poisoning and to conclude that the mushrooms were the cause of poisoning, it must be established that everyone who ate the suspected mushrooms became ill and that no one who did not eat the mushrooms became ill. Wild mushrooms, whether they were eaten raw, cooked, or processed, should always be regarded as prime suspects.

Mushroom toxins can, with difficulty, be recovered from poisonous fungi, cooking water, stomach contents, serum, and urine. Procedures for extraction and quantitation are generally elaborate and time-consuming, and, in most cases, the patient will have recovered by the time an analysis is made on the basis of toxin chemistry. The exact chemical nature of most of the toxins that produce milder symptoms are unknown.

Aflatoxins

The aflatoxins (AFs) are mycotoxins produced by certain fungi and can cause serious illness in animals and humans. The four major aflatoxins are AFB1, AFB2, AFG1, and AFG2. In adverse weather or under poor storage conditions, these toxins are produced mainly by certain strains of *Aspergillus flavus* and *A. parasiticus* in a broad range of agricultural commodities, such as corn and nuts. The name "aflatoxin" reflects the fact that this compound was first recognized in damaged peanuts contaminated with *Aspergillus flavus*. The aflatoxins then were described according to other mechanisms (i.e., on the basis of their blue or green fluorescence under UV light and relative chromatographic mobility after thin-layer chromatographic separation). Another aflatoxin, aflatoxin M1 (AFM1), is produced by mammals after consumption of feed (or food) contaminated by AFB1. Cows are able to convert AFB1 into AFM1 and transmit it through their milk. Although AFM1 in milk is, by far, not as hazardous as the parent compound, a limit of 0.5 parts per billion is applied, largely because milk tends to constitute a large part of the diet of infants and children. In the United States, strict regulations in place since 1971, as well as FDA monitoring of the food supply and the population's consumption of a diverse diet, have prevented human health problems. However, aflatoxin-induced chronic and acute disease is common in children and adults in some developing countries.

Other Pathogenic Agents

Prions and transmissible spongiform encephalopathies

Transmissible spongiform encephalopathies (TSEs) are neurodegenerative diseases. There are examples of these diseases in both humans and animals. The spongiform portion of their name is derived from the fact that microscopic analysis of the affected brain tissue shows the presence of numerous holes, which gives the brain a "sponge-like" appearance. The disease-causing entity that elicits all TSEs is neither a cellular organism (i.e., a bacterium or parasite) nor a virus. Rather, it is the prion protein, a normal mammalian cell protein that causes these diseases. Under normal physiologic conditions, the prion protein is found on the surface of a wide variety of cells within the body, most notably in the nervous tissue, such as nerve cells and brain tissue. While our understanding of the precise function of this protein is still evolving (and somewhat controversial), current evidence suggests that prions have a role in long-term memory and/or maintaining normal nerve-cell physiology. Prion diseases are initiated when normal cellular prions come in contact with a disease-causing prion. The disease-causing prion is a misfolded form of the normal prion. Once it is misfolded, it can induce other, normally folded prion proteins to become misfolded. This folding/misfolding process is responsible for the amplification of the disease (Halliday et al., 2014).

There are several naturally occurring human TSEs, for instance, Kuru, fatal familial Insomnia, Gerstmann Straussler-Scheinker Syndrome, and Creutzfeldt-Jakob Disease (CJD). Kuru and CJD are the only human-specific TSEs that can be transmitted between people. There are three different types of classic CJD—spontaneous, familial, and iatrogenic. Only iatrogenic, or acquired, CJD is transmissible. This form of CJD is transmitted through unintended exposure to infected tissue during medical events (for example, from *dura mater* grafts or from prion-contaminated human growth hormone). Spontaneous CJD accounts for approximately 85% of all CJD cases and occurs in people with no obvious risk factors. Familial, or hereditary CJD is a disease passed from parent(s) to offspring and comprises approximately 10% of all CJD cases.

Only the variant, Creutzfeldt-Jakob disease (vCJD), is transmitted through food. vCJD and the cattle disease bovine spongiform encephalopathy (BSE), also known as "mad cow" disease, appear to be caused by the same agent. Other TSE diseases exist in animals, but there is no known transmission of these TSEs to humans. Included among these are chronic wasting disease (CWD) of deer and elk, and scrapie, the oldest known animal TSE, which occurs in sheep and goats. vCJD is always fatal. There is no known cure for this disease. Cases of vCJD usually present with psychiatric problems such as depression. The traditional route of entry into humans of the BSE-causing agent is oral, through consumption of meat or meat products derived from BSE-infected animals. Development of vCJD is believed to be the result of eating meat or meat products from cattle infected with BSE. The available scientific information strongly supports the supposition that the infectious agent that causes BSE in cattle is the same agent that causes vCJD in humans.

The first case of vCJD was discovered in 1996, in Great Britain. Since then, a total of 217 patients worldwide have been diagnosed with vCJD (until August 2010). A total of 170 patients have been diagnosed in Great Britain, with 25 cases in France, 5 in Spain, 4 in Ireland, 3 each in the U.S. and the Netherlands, 2 each in Portugal and Italy, and one each in Canada, Japan, and Saudi Arabia. Two of the three patients in the U.S. contracted vCJD while living in Great Britain, while the third patient most likely contracted this disease while living in Saudi Arabia. Three patients from Great Britain contracted vCJD following a blood transfusion from a single, asymptomatic vCJD blood donor. The peak number of new cases occurred in 2000, and the number of new cases has continued to decline in the subsequent years. The most effective means of preventing vCJD is to prevent the spread and dissemination of BSE in cattle. The prohibitions on feeding rendered cattle by-products to cattle have been very effective in helping to reduce the number of new cases of BSE-infected cattle, worldwide. Latest on April 24, 2012, the USDA confirmed a BSE case in a dairy cow in California. This cow was tested as part of the USDA targeted BSE surveillance at rendering facilities in the United States. The cow was 10 years and 7 months old and was classified as having the L-type BSE strain (USDA, 2012).

Conclusion

Salmonella, Campylobacter and strains of *E. coli* are well-established examples of common foodborne pathogens. They are markedly different in terms of epidemiology, physiology, ecology, host association and virulence properties, but together enable some generic conclusions to be drawn on the overall persistence of foodborne bacterial disease over the last 20 years. Although these are the major bacterial pathogens monitored, many others are also transmitted through food. At any time such relatively minor foodborne pathogens, like *L. monocytogenes*, can also become major problems. Investigating the reasons for such shifts in patterns of foodborne disease provides valuable information for future risk management strategies.

From the examples cited above, for which data can be assessed over relatively long periods, it would seem that there is little evidence that we are winning significant ground in the battle against foodborne illnesses caused overall by bacterial pathogens. In fact, even the most successful interventions, such as vaccination of chickens against *Salmonella*, have done little more than reduce the pathogen load in the food chain. It can be speculated that such reductions in exposure could have some adverse effects, for example, by altering the immune status of the population. There are already disturbing trends such as shifts away from illnesses, such as campylobacteriosis and listeriosis, in the young towards the increasingly growing older population. In addition, these pathogens are constantly evolving and adapting, enabling the exploitation of novel opportunities, for example, new vehicles created by modern processing techniques, new retailing fads or new food consumption habits. This highlights the need for multidisciplinary research, and especially the inclusion of social sciences, to investigate changing trends in foodborne disease.

Despite the substantial investments made by governments and industry alike, these bacterial pathogens still feature as major public health problems. However, a combination of biosecurity and vaccination has largely eliminated *S. enteritidis* PT 4 from the breeder and layer flocks in many European countries, while legislation and retailer education have sufficiently improved hygiene to reduce *E. coli* O157 in cooked meat. Clearly we need to share and implement those strategies throughout the world that are effective, while at the same time maintaining constant vigilance against the ability of such organisms to adapt to changing environments and to exploit the opportunities that arise to occupy novel niches.

References

Amir Attaran, MacDonald Noni, Matthew B. Stanbrook, Sibbald Barbara and Flegel Ken. October 7, 2008. Canadian Medical Association Journal (CMAJ), 179(8): 739–740. doi: 10.1503/cmaj.081477.

Barbee, G., C. Berry-Caban, J. Barry et al. 2009. Analysis of mushroom exposures in Texas requiring hospitalization, 2005–2006. J. Med. Toxicol., 2: 59–62.

Bryden, W.L. 2012. Mycotoxin contamination of the feed supply chain: implications for animal productivity and feed security. Anim. Feed Sci. Technol., 173: 134–158.

Caroline, A.R., M.K. Nickels, M.S. Nancy, T. Hargrett-Bean, M.E. Potter, T. Endo, L. Mayer, C.W. Langkop, C. Gibson, R.C. McDonald, R.T. Kenney, N.D. Puhr, P.J. McDonnell, R.J. Martin, M.L. Cohen and P.A. Blake. 1987. Massive Outbreak of Antimicrobial-Resistant Salmonellosis Traced to Pasteurized Milk. JAMA, 258(22): 3269–3274. doi:10.1001/jama.1987.03400220069039.

CDC (Centers for Disease Control and Prevention). 2002. Outbreak of Listeriosis—Northeastern United States, Morbidity and Mortality Weekly Report, 51(42): 950–951.

CDC. 2003. Hepatitis A outbreak associated with green onions at a restaurant—Monaca, Pennsylvania, MMWR, 52(47): 1155–1157.

CDC. 2009. Multistate outbreak of Salmonella infections associated with peanut butter and peanut butter–Containing Products—United States, MMWR, 58: 85–90.

CDC. 2011. Estimates of Foodborne Illness in the United States. Retrieved from http://www.cdc.gov/foodborneburden/index.html. Accessed on June 26, 2014.

Chitambar, S., V. Gopalkrishna, P. Chhabra, P. Patil, H. Verma, A. Lahon, R. Arora, V. Tatte, S. Ranshing, D. Hale G., R. Kolhapure, S. Tikute, J. Kulkarni, R. Bhardwaj, S. Akarte and S. Pawar. 2012. Diversity in the enteric viruses detected in outbreaks of gastroenteritis from Mumbai, Western India. Int. J. Environ. Res. Public Health, 9: 895–915.

Davis, M. 1993. Update: Multistate outbreak of *Escherichia coli* O157:H7 infections from Hamburgers—Western United States, 1992–1993. MMWR, 42(14): 258–263.

Diaz, J.H. February 2005. Evolving global epidemiology, syndromic classification, general management, and prevention of unknown mushroom poisonings. Crit. Care Med., 33(2): 419–426.

Diaz, J.H. Feb. 2005. Syndromic diagnosis and management of confirmed mushroom poisonings. Crit. Care Med., 33(2): 427–436.

Dorny, P., N. Praet, N. Deckers and S. Gabriel. 2009. Emerging foodborne parasites. Vet. Parasitol., 163: 196–206.

Dutton, M.F. 2009. The African Fusarium/maize disease. Mycotoxin Res., 25: 29–39.

Dutta, A., B.C. Kalita and A.K. Pegu. 2013. A study of clinical profile and treatment outcome of mushroom poisoning—A Hospital Based Study. Assam Journal of Internal Medicine, 3(2): 13–17.

Fleury, M.D., J. Stratton, C. Tinga, D.F. Charron and J. Aramini. 2008. A descriptive analysis of hospitalization due to acute gastrointestinal illness in Canada, 1995–2004. Canadian Journal of Public Health, 99(6): 489–493.

Frank, C., D. Werber, J.P. Cramer, M. Askar, M. Faber and M. der Heiden. 2011. Epidemic profile of Shiga-toxin-producing *Escherichia coli* O104:H4 outbreak in Germany. N. Engl. J. Med., 365: 1771–1780. doi: 10.1056/NEJMoa1106483.

Grant, J., A.M. Wendelboe, A. Wendel, B. Jepson, P. Torres and C. Smelser. 2008. Spinach-associated *Escherichia coli* O157:H7, Utah and New Mexico, 2006. Emerg. Infect. Dis. 14(10): 1633–6. Available from http://wwwnc.cdc.gov/eid/article/14/10/07-1341.

Guyader, F.S. and R.L. Atmar. 2008. Binding and inactivation of viruses on and in food, with a focus on the role of matrix. pp. 189–208. *In*: M.P.G. Koopmans, D.O. Clive and A. Bosch (eds.). Foodborne Viruses: Progress and Challenges. Washington DC: ASM press.

Halliday, M., H. Radford and G.R. Mallucci. 2014. Prions: generation and spread versus neurotoxicity. J. Biol. Chem. 2014 Jul. 18, 289(29): 19862–19868.

Hatten, B.W., N.J. McKeown, R.G. Hendrickson and B.Z. Horowitz. The epidemiology of mushroom ingestion calls to US Poison Control Centers: 2001–2011 [abstract]. Clin. Toxicol. Aug. 2012, 50(7): 274–720.

Helms, M., P. Vastrup, P. Gerner-Smidt and K. Mølbak. 2003. Short- and long-term mortality associated with foodborne bacterial gastrointestinal infections: registry-based study. BMJ (Clinical Research Ed.), 326(7385): 357.

HPSWR (Health Protection Scotland Weekly Report). 2014. Listeria outbreak associated with deli meats, Denmark. HPSWR, 48(35): 434–436. Available at: http://www.hps.scot.nhs.uk/documents/ewr/pdf2014/1435.pdf and accessed on November 15, 2014.

Johnston, R.W., J. Feldman and R. Sullivan. 1963. Botulism from canned tuna fish. Public Health Rep., 78(7): 561–564.

Khan, A., C. Jordan, C. Muccioli, A.L. Vallochi, L.V. Rizzo, R. Belfort, Jr., R.W. Vitor, C. Silveira and L.D. Sibley. 2007. Genetic Divergence of *Toxoplasma gondii* strains associated with ocular toxoplasmosis, Brazil. Emerg. Infect. Dis., 12: 942–949.

Koopmans, M. and E. Duizer. 2004. Foodborne viruses: an emerging problem. Int. J. Food Microbiol., 90: 23–41.

Linscott, A.J. 2011. Foodborne illnesses. Clin. Microbiol. Newsletter, 33(6): 41–46.

Mead, P.S., E.F. Dunne, L. Gravis, M. Wiedman, M. Patrick, S. Hunter, E. Saleh, F. Mostashar, A. Craig, P.M. Shar, T. Bannerman, B.D. Sauders, P. Hayes, W. Dewitt, P. Sparling, P. Griffin, D. Morse,

L. Slutsker and B. Swaminathan. 2006. Epidemiol. Infect., 134: 744–751. doi: 10.1017/S0950268805005376 Printed in the United Kingdom.

Michino, H., K. Araki, S. Minami, S. Takaya, N. Sakai, M. Miyazaki, A. Ono and H. Yanagawa. 1999. Massive outbreak of *Escherichia coli* O157:H7 infection in schoolchildren in Sakai City, Japan, associated with consumption of white radish sprouts. Am. J. Epidemiol., 150(8): 787–96.

Newell, D.G., M. Koopmans, L. Verhoef, E. Duizer, A.A.-Kane, H. Sprong, M. Opsteegh, M. Langelaar, J. Threfall, F. Scheutz, J. van der Giessen and H. Kruse. 2010. Food-borne diseases—The challenges of 20 years ago still persist while new ones continue to emerge. International Journal of Food Microbiology, 139: S3–S15.

Nyenje, M.E. and R.N. Ndip. 2013. The challenges of foodborne pathogens and antimicrobial chemotherapy: a global perspective. African Journal of Microbiology Research, 7(14): 1158–1172.

Nyenje, M.E., C.O. Odjadjare, N.F. Tanih, E. Green and R.N. Ndip. 2012b. Foodborne pathogens recovered from ready-to-eat foods from roadside cafeterias and retail outlets in Alice, Eastern Cape Province, South Africa: public health implications. Int. J. Environ. Res. Public Health, 9: 2608–2619.

Nyenje, M.E., N.F. Tanih, E. Green and R.N. Ndip. 2012a. Current status on antibiogram of *Listeria ivanovii* and Enterobacter cloacae isolated from ready-to-eat foods in Alice, South Africa: a cause for concern. Int. J. Environ. Res. Public Health, 9: 3101–3114.

NYT (New York Times). 1971. Botulism Death in Westchester Brings Hunt for Soup. New York Times. July 2, 1971. Available at http://thelede.blogs.nytimes.com/2007/10/05/in-a-beef-packagers-demise-a-whiff-of-vichyssoise/#more-845 and accessed on 15 November 2014.

Pennington, T.H. 2009. The Public Inquiry into the September 2005 outbreak of *E. coli* O157 in South Wales. Available from: http://wales.gov.uk/ecolidocs/3008707/reporten.pdf?lang=en and accessed on November 15, 2014.

Sánchez-Hervás, M., J.V. Gil, F. Bisbal, D. Ramón and P.V. Martínez-Culebras. 2008. Mycobiota and mycotoxin producing fungi from cocoa beans. Int. J. Food Microbiol., 125: 336–340.

Schelin, J., N. Wallin-Carlquist, M.T. Cohn, R. Lindqvist, G.C. Barker and P. Rådström. 2011. The formation of *Staphylococcus aureus* enterotoxin in food environments and advances in risk assessment. Virulence, 2: 580–592.

Schmidt, R.H., R.M. Goodrich, D.L. Archer and K.R. Schneider. 2009. General overview of the causative agents of foodborne illness. Food Science and Human Nutrition Department, Florida Cooperative Extension Service, IFAS, University of Florida. Publication: FSHN033, pp. 1–5.

Sean Patrick Nordt and Anthony Manoguerra. 2000. 5-Year analysis of mushroom exposures in California. West J. Med., 173(5): 314–317.

Teplitski, M., A.C. Wrigh and G. Lorca. 2009. Biological approaches for controlling shellfish-associated pathogens. Curr. Opin. Biotechnol., 20: 185–190.

USDA (United States Department of Agriculture). 2012. Final report of USDA/APHIS on California Bovine Spongiform Encephalopathy Case Investigation. Retrieved from http://www.aphis.usda.gov/animal_health/animal_diseases/bse/downloads/BSE_Summary_Report.pdf. Accessed on 20/10/2014.

Vaillant, V., H. De Valk, E. Baron, T. Ancelle, P. Colin, M.C. Delmas, B. Dufour, R. Pouillot, Y. Le Strat, P. Weinbreck, E. Jougla and J.C. Desenclos. 2005. Foodborne pathogens and disease. Fall, 2: 221–232.

Vasickova, P., L. Dvorska, A. Lorencova and I. Pavlik. 2005. Viruses as a cause of foodborne diseases. Vet. Med. Czech., 50: 89–104.

Wild, C.P. and Y.Y. Gong. 2010. Mycotoxins and human disease: a largely ignored global health issue. Carcinogenesis, 31: 71–82.

Wagacha, J.M. and J.W. Muthomi. 2008. Mycotoxin problem in Africa: current status, implications to food safety and health and possible management strategies. Int. J. Food Microbiol., 10: 124: 1–12.

World Health Organization [WHO]. 2011a. Enterohaemorrhagic *Escherichia coli* (EHEC). Fact sheet no.125 http://www.who.int/mediacentre/factsheets/fs125/en/. Accessed on 9/10/2014.

World Health Organization [WHO]. 2011b. Initiative to estimate the Global Burden of Foodborne Diseases. Retrieved June 26, 2011, from http://www.who.int/foodsafety/foodborne_disease/ferg/en/index1.html. Accessed on 9/10/2014.

3

Facts on Foodborne Pathogens[#]

Abdul Khaleque[1],* and Md. Latiful Bari[2]

Introduction

Foodborne illness is a preventable public health challenge that causes illnesses and deaths every year worldwide. It is an illness that comes from eating contaminated food. Everyone is at risk for getting a foodborne illness. However, some people are at greater risk for experiencing a more serious illness or even death should they contract a foodborne illness. Thousands of types of bacteria are naturally present in our environment. Microorganisms that cause disease are called "pathogens". When certain pathogens enter the food supply, they can cause foodborne illness. Not all bacteria cause disease in humans. For example, some bacteria are used beneficially in making cheese and yogurt. Foods, including safely cooked and ready-to-eat foods, can become cross-contaminated with pathogens transferred from raw food products, other contaminated products, or from food handlers with poor personal hygiene. Most cases of foodborne illness can be prevented with proper cooking or processing of food to destroy pathogens. Foodborne disease is caused by consuming contaminated foods or beverages. Many different disease-causing microbes, or pathogens, can contaminate foods, so there are many different foodborne infections. In this chapter, the microbial agents' including bacteria, virus and protozoa that causes foodborne illness has been described briefly. The information provided here is the summary of information available in the literature. It should be used as a guide only as many variables impact on the survival and inactivation of pathogens in foods. The absence of certain information does not mean it does not exist.

[1] Dept. of Biology & Chemistry, North South University, Dhaka 1229, Bangladesh.
[2] Food Analysis and Research Laboratory, Center for Advanced Research in Sciences, University of Dhaka, Dhaka-1000, Bangladesh.
* Corresponding author: abdul.khaleque@northsouth.edu

[#] The information provided here is a summary of the information available in the literature. It should be used as a guide only as many variables impact on the survival and inactivation of pathogens in foods. The absence of certain information, e.g., on impact of a specific preservative, process, etc. does not mean it does not exist.

Aeromonas and Plesiomonas

Important species: *Aeromonas hydrophila, Plesiomonas shigelloides*

Why is it important?

A. hydrophila may cause gastroenteritis in healthy individuals or septicemia in individuals with impaired immune systems or various malignancies. *Plesiomonas* can cause gastroenteritis. While there is still controversy as to the enteropathogenicity of these bacteria, they may be considered emerging pathogens. Certainly, not all strains cause human illness. Plesiomonas can also grow at low refrigeration temperatures so it could be a potential problem in chilled foods.

Sources

A. hydrophila is found in brackish, fresh, estuarine, marine, chlorinated and unchlorinated water supplies worldwide. *P. shigelloides* mainly occurs in freshwater and freshwater fish and shellfish but also many animals.

Foods at risk

A. hydrophila is frequently found in fish and shellfish but sometimes it is associated with red meat and poultry. *P. shigelloides* is mainly associated with water and raw freshwater fish and shellfish.

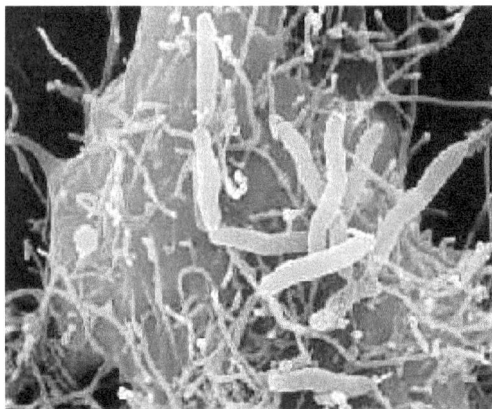

Characteristics and growth conditions

- Gram-negative.
- Rod shaped.
- Facultatively anaerobic.
- Motile (polar flagella).
- Nonspore former.
- Produces a toxin which is preformed in the food.

- Usually encapsulated.
- Unlikely to grow at salt levels greater than 3–3.5%.

Growth	*A. hydrophila*	*P. shigelloides*
T	5–45°C (28°C)	0–45°C (38–39°C)
pH	5.5–10	5–8

Transmission: Eating of contaminated food

Illness

Two types of gastroenteritis have been associated with *A. hydrophila*: a cholera-like illness with watery diarrhea and a dysenteric illness characterized by loose stools containing blood and mucus. Septicemia has been observed in individuals with underlying illness.

P. shigelloides has been implicated in cases of gastroenteritis. *P. shigelloides* infection may cause diarrhea of 1–2 days' duration in healthy adults. However, there may be high fever and chills and protracted dysenteric symptoms in infants and children under 15 years of age. Complications (septicemia and death) may occur in people who are immunocompromised or have an underlying illness. Most cases reported have occurred in people living in tropical and subtropical areas.

All people may be susceptible to infection. Infants, children and chronically ill people are more likely to experience protracted illness and complications.

Dose

The infectious dose of *A. hydrophila* is unknown, but it is thought to be high, ranging from 10^4 to 10^9. For *P. shigelloides* the infective dose is also presumed to be quite high – > 10^6 cells.

Issues relating to control

Aeromonas contribute to spoilage of many foods but can reach high numbers (10^8) in some foods such as milk without detectable organoleptic changes.

Survival

- Quite resistant to freezing.
- Some strains of both *Aeromonas* and *Plesiomonas* can grow at refrigeration temperatures.
- While *Aeromonas* can grow in modified atmospheres, *Plesiomonas* are very susceptible to modified atmosphere storage.

Prevention

- Research on methods of destruction is relatively limited but appears to be readily killed by heat or radiation.
- *Aeromonas* and *Palestinians* spp. are sensitive to low pH and > 4% NaCl.
- Susceptible to disinfectants including chlorine.

However, recovery of *Aeromonas* from chlorinated water has commonly been reported (post-treatment contamination, inactivation of chlorine by organic matter, high initial cell numbers, "viable but nonculturable" state.

Testing

- Moeller decarboxylase and dihydrolase reactions are used to identify aeromonads to species level.
- *Plesiomonas* can be distinguished from *Shigella* in diarrheal stools by an oxidase test.
- *Plesiomonas* is oxidase-positive and *Shigella* is oxidase-negative. *Plesiomonas* is negative for DNAse; this and other biochemical tests distinguish it from *Aeromonas* sp.
- Commercial kits are also available for both the organisms.

In general

- *Cook food thoroughly and rise at a high temperature of 70°C to 80°C.*

Bacillus

Important species: *Bacillus cereus*

Why is it important?

It causes two types of food poisoning as a result of toxin production: an emetic toxin (preformed in the food) that causes vomiting; and an enterotoxin (produced in the intestine) that causes diarrhea. This toxin-producing bacteria is often associated with reheated foods.

Sources: Ubiquitous in nature. Normal inhabitant of soil, dust, air, water and decaying organic matter.

Foods at risk

Found in a wide range of foods including meat, milk, dairy products, vegetables, fish, rice dishes, sauces, pastas, dried mixes, spices, ready-to-eat foods.

Characteristics and growth conditions

- Gram-positive rod-shaped bacterium.
- Aerobic, but also grows well anaerobically lower toxin production under anaerobic conditions.
- Spore former (central spores)—heat-resistant spores are an important factor in foodborne illnesses.
- Motile.
- Produces two types of toxins: an emetic toxin in foods which is highly stable and survives high temperatures, exposure to enzymes (trypsin, pepsin) and pH extremes; and an enterotoxin which is produced in the intestine; this toxin is acidic and heat labile.

- Growth temperature: 4° to 48°C (optimal 28–35°C).
- Growth pH: 4.9–9.3 (optimum 6.8–7.2).
- Growth $a_w \geq 0.91$.

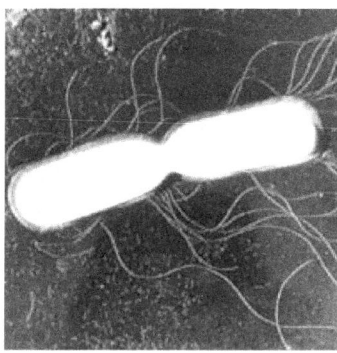

Transmission: Eating contaminated food.

Illness

Rapid-onset **emetic syndrome** is characterized by nausea and vomiting, which begin one to five hours after contaminated food is eaten. Slow-onset **diarrheal syndrome is characterized by** diarrhea and abdominal pain which occurs 8 to 16 hours after consumption of contaminated food. Recovery from both is usually rapid and all people are thought to be susceptible. Symptoms may vary among individuals.

Dose

Large numbers ($> 10^5$/g of food) required to produce toxin or cause infection. Small numbers of *B. cereus* in foods are not a direct hazard to health.

Issues relating to control

Survival

- Vegetative cells readily killed by heat (D value at 60°C = ~1 min).
- Cereus cells die in yogurt when pH reaches 4.5.
- Spores are moderately heat-resistant and can survive pasteurization (D value at 100°C = 2.7–3.1 mins). High fat/oil and low a_w increase heat resistance.
- Spores survive for very long periods in dry foods.
- Spores activated by a variety of treatments—e.g., heat shock (exposure to several hours of elevated but sub-lethal temperatures).
- Heat activation is reversible if spores then return to lower temperature (i.e., unsuitable for growth).
- Spores are hydrophobic—difficult to remove during cleaning—Can allow contamination of food during processing.
- Emetic toxins are very to heat and extremes of pH. It can survive a pH of 2–11 and for 90 min at 126°C.
- Enterotoxin is less resistant and is inactivated after 5 min at 56°C.

Prevention

- Effective prevention and control measures depend on inhibiting spore germination and preventing the growth of vegetative cells in cooked, ready-to-eat foods.
- Effect of temperature on spores varies according to strain (D_{85} = 33.8–106 min) and environment (D_{95} = 1.5–36.2 in distilled water and 1.8–19.1 min in milk).
- Inactivated by 0.1 M acetic, formic and lactic acids in broth.
- Growth inhibited by > 7.5% salt.
- Modified atmospheres can be used to control growth of *B. cereus.*
- Nisin is commonly used in dairy products to prevent outgrowth of spores.
- Growth is inhibited by sorbic acid (0.26% at pH 5.5), potassium sorbate (0.39% at pH 6.6), benzoate, EDTA and polyphosphates.
- 0.2% calcium propionate prevents spore germination in bread.
- Sensitive to most chemical disinfectants used in food industry.
- Spores are more resistant to radiation than vegetative cells.

Note: Some strains capable of growing at refrigerated temperatures have recently been reported.

Testing

- PEMBA agar used for identification; the bacteria colonies retain a turquoise-blue color after 24 hours of incubation at 37°C.
- Commercial kits also available.

In general

- *Cook food thoroughly and cool rapidly to prevent spore germination.*
- *Keep hot foods above 60°C and cold foods below 4°C.*
- *Reheat cooked food at 74°C.*

Bird Flu

Important species: Influenza A virus (H5N1)

Why it is important?

The highly pathogenic influenza A virus subtype H5N1 is an emerging avian influenza virus that has been causing global concern as a potential pandemic threat. It is often referred to simply as "bird flu" or "avian influenza", even though it is only one subtype of avian influenza-causing virus. H5N1 has killed millions of poultry in a growing number of countries throughout Asia, Europe and Africa. Health experts are concerned that the coexistence of human flu viruses and avian flu viruses (especially H5N1) will provide an opportunity for genetic material to be exchanged between species-specific viruses, possibly creating a new virulent influenza strain that is easily transmissible and lethal to humans. The mortality rate for humans with H5N1 is 60%.

Ecological niche

The avian flu virus (H5N1) has been shown to survive in the environment for long periods of time. Infection may be spread simply by touching contaminated surfaces. Birds who were infected with this flu can continue to release the virus in their faeces and saliva for as long as 10 days.

Food with which it is most frequently associated: Raw or undercooked poultry meat, eggs.

Characteristics and survival

Influenza A virus (H5N1) is a single stranded RNA, enveloped virus belonging to the Orthomyxoviridae family. Influenza A virus is transmitted directly from person to person or indirectly via contaminated food or having come into contact with an infected bird. Influenza A virus is sensitive to treatment with heat, lipid solvents, nonionic detergents, formaldehyde, oxidizing agents; its infectivity is reduced after exposure to radiation.

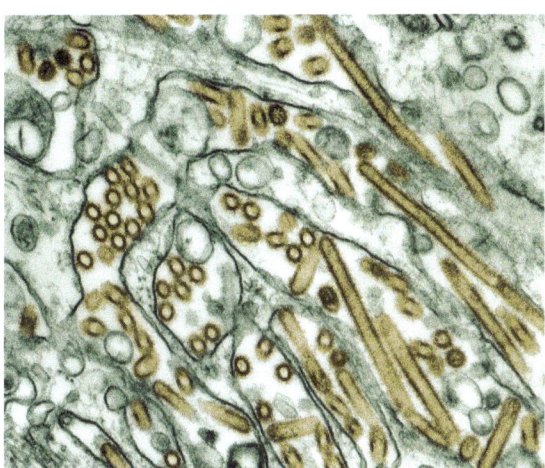

At-risk populations

- Farmers and those who work with poultry.
- Travelers visiting the affected countries.
- Those who touch an infected bird.
- Those who consume raw or undercooked poultry meat, eggs, or blood from infected birds.
- Healthcare workers and household members of patients with avian flu may also be at an increased risk of contracting bird flu.

Illness

The symptoms of influenza A virus illness usually include cough (dry or productive), diarrhea, difficulty in breathing, fever higher than 100.4°F (38°C), headache, malaise, muscle ache, runny nose, sore throat. Median time from onset of illness to acute respiratory distress syndrome is 6 days.

Route of transmission

Influenza A virus is usually transmitted via air droplets, and subsequently contaminates the mucosa of the respiratory tract. Health care workers and household contacts of patients with avian influenza may also be at an increased risk of the bird flu. The avian flu virus (H5N1) has been shown to survive in the environment for long periods of time. Infection may be spread simply by touching contaminated surfaces. Birds who were infected with this flu can continue to release the virus in their faeces and saliva for as long as 10 days.

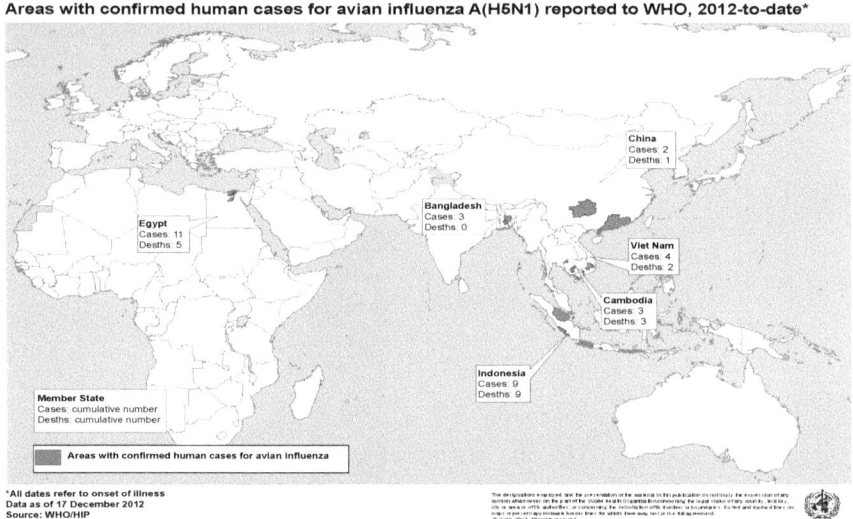

Areas with confirmed human cases for avian influenza A(H5N1) reported to WHO, 2012-to-date*

Signs and tests

If a person has been exposed to avian influenza A virus, he should call the health care provider before visiting. This will give the staff a chance to take proper precautions that will protect them and other patients during office visit.

Tests to identify the avian flu exist but are not widely available. A test for diagnosing strains of bird flu in people suspected of having the virus gives preliminary results within 4 hours. Older tests took 2 to 3 days.

Doctor might also perform the following tests:

- Auscultation (to detect abnormal breath sounds).
- Chest x-ray.
- Nasopharyngeal culture.
- White blood cell differential.
- Other tests may be done to look at the functions of your heart, kidneys, and liver.

Treatment

Different types of avian flu virus may cause different symptoms. Therefore, treatment may vary.

- In general, treatment with the antiviral medication oseltamivir (Tamiflu) or zanamivir (Relenza) may make the disease less severe if the medicine is started within 48 hours after symptoms start.
- Oseltamivir may also be prescribed for persons who live in the same house as those diagnosed with avian flu.

Control and prevention strategies

At this time, the US Center for Disease Control and Prevention (CDC) has no recommendations against travel to the countries affected by H5N1.

- However; travelers should avoid visits to bird markets in areas with an avian flu outbreak.
- People who work with birds who might be infected should use protective clothing and special breathing masks.
- Avoiding undercooked or uncooked meat reduces the risk of exposure to avian flu and other foodborne diseases.

Selected outbreaks

The earliest infections of humans by H5N1 coincided with an epizootic (an epidemic in nonhumans) of H5N1 influenza in Hong Kong's poultry population. This panzootic (a disease affecting animals of many species, especially over a wide area) outbreak was halted by the killing of the entire domestic poultry population within the territory. However, the disease has continued to spread. On December 21, 2009 WHO announced a total of 447 cases which resulted in the deaths of 263.

Campylobacter

Important species

Campylobacter jejuni, *Campylobacter coli* are most often associated with disease but *Arcobacter* spp. can also be important. Both pathogenic and non-pathogenic strains exist and it is often difficult to differentiate them.

Why is it important?

Becoming one of the main causes of foodborne disease. Usually causes gastrointestinal illness—Campylobacteriosis—(diarrhea) in humans. Can also cause systemic illness, reactive arthritis, and Guillain-Barré syndrome (GBS). Death rarely occurs. Some species cause spontaneous abortions in animals. Children and young adults are most frequently affected.

Sources

Intestines of wild animals, farm animals (cattle, sheep, pigs), birds (chickens), Domestic pets, sewage, mushroom. Seasonal variation in occurrence. Can also be found in water supply.

Foods at risk: Raw chicken, raw milk, non-chlorinated water.

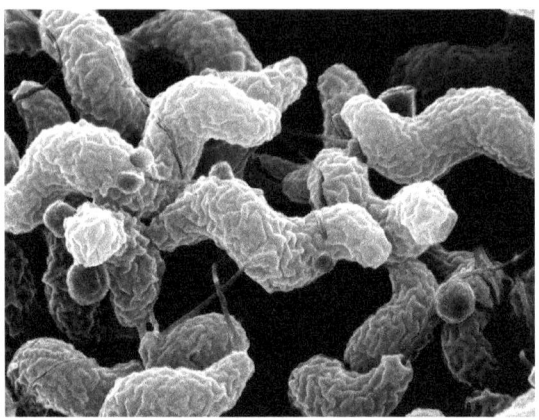

Characteristics and growth conditions

- Gram-negative, S-curved spiral rods.
- Exists as vegetative cells only—it is not a spore former.
- Motile—polar flagella.
- Reported to produce some toxins.
- Microaerophillic sensitive to high oxygen conditions (*C. jejuni*: 3–5% O_2 and 2–10% CO_2).
- Growth temperature > 30–45°C (optimum 42°C).
- Growth pH 4.9–9 (optimum 6.5–7.5).
- Growth a_w: ≥ 0.987. Optimum a_w is 0.997. Will grow in 1.5% salt (NaCl) but not > 2%.
- Is slow growing compared to other bacteria (generation time of 1 hour under optimum conditions).

Transmission

Food is a vehicle for transmission. Direct transmission person-to-person or from contact with infected animals.

Illness

Campylobacteriosis occurs 1–10 days after ingestion of the bacteria. Symptoms include muscle pain, headache and fever followed by watery diarrhea, abdominal pain and nausea. It can last up to one week and is usually self-limiting. Can affect any age group but most often found in infants and young adults. Infection is occasionally followed by arthritis or GBS (~1% of cases). On rare occasions it may cause non-enteric disease such as invasion of bloodstream.

Dose

Infective dose thought to be small and some studies suggest that 400–500 bacteria may cause illness in some individuals. But usually 1000 to 10000 cells are needed to cause illness.

Issues relating to control

Survival

- Relatively sensitive to environmental stresses (e.g., 21% oxygen, drying, heating, disinfectants, acidic conditions).
- Can survive on hands and moist surfaces for up to an hour.
- Numbers decline slowly at normal freezing temperatures, but freezing does not instantly inactivate cells.
- Can survive in faeces, milk, water, urine for 3–5 weeks at 4°C.
- Survives well in modified atmosphere and vacuum packaging but poorly at atmospheric oxygen concentrations.
- Survives better in refrigerators than at room temperature.

Prevention

- Easily inactivated by heat (D_{55} = ~1 min; D_{60} = 0.2–0.3 min).
- Susceptible to low pH–Die-off in foods < pH 4.0.
- Sensitive to oxygen.
- Appears to be sensitive to drying but under some refrigeration conditions can remain viable for weeks.
- Reduced by freeze-thawing.
- Inactivated by frozen storage < −15°C over a period of time.
- Sensitive to NaCl concentrations above 1% and death occurs > 2%.
- Sensitive to ascorbic acid and some spices.
- Susceptible to disinfectants such as chlorine.
- Sensitive to gamma radiation and ultraviolet radiation as used in water treatment units.

Note: Despite ease of heat inactivation and oxygen sensitive *Campylobacter* appears to be able to survive environmental stresses as it is an increasingly important cause of foodborne illness. Under adverse conditions *Campylobacter* is said to undergo a transition to a "Viable but non-culturable" state.

Testing

Incubation media (FBP) containing blood, pyruvate, ferrous salts, charcoal and metabisulfite.

In general

- *Avoid cross-contamination from raw foods to cooked foods.*
- *Reheat cooked food at 74°C.*

Clostridium

Important species: *Clostridium botulinum, Clostridium perfringens*

Why is it important?

Clostridium botulinum causes botulism, recognized as a foodborne disease since the late 1800s. There are 4 forms of disease (foodborne, infant, wound and animal botulism) caused by 7 types of botulism toxin. While the incidence is low the disease is very severe and can be fatal if not treated immediately and properly. Two groups of *C. botulinum* are important in food: Group I-Types A, B, F (proteolytic strains) and Group II-Types B, E, F (non-proteolytic strains). Because they are proteolytic Group I organisms they generally cause spoilage of contaminated food.

C. perfringens* causes two types of food poisoning—a common form known as type A (diarrhea and abdominal cramps) and a rarer form called necrotic enteritis which can be fatal. *C. perfringens* poisoning occurs much more frequently than botulism.

Sources

C. botulinum (esp. type A) is widely found in soils faeces marine sediments (the type varies from country to country). Types B, C, D and E appear to be animal parasites. *C. perfringens* is considered ubiquitous in the natural environment.

Foods at risk

Canned corn, honey, pepper, green beans, soups, beet, asparagus, mushrooms, ripe olives, spinach, tuna fish, chicken and chicken liver and liver pate, and luncheon meats, ham, sausage, stuffed eggplant, honey, lobster, and smoked and salted fish. *C. botulinum* is often associated with home canned foods. Meats, meat products, and gravy are the foods most frequently implicated with *C. perfringens* poisoning but it can occur in any prepared foods that are temperature abused.

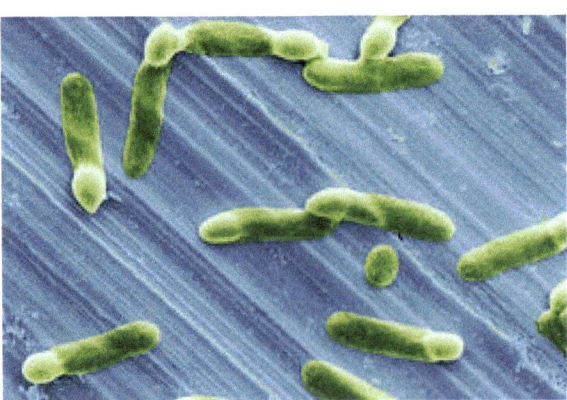

Characteristics and growth conditions

- Gram-positive rods.
- Anaerobic (but can grow in the presence of low levels of oxygen).
- Sporeformers (spores are heat-resistant).
- Produce toxins. *C. botulinum* produces a potent neurotoxin. *C. perfringens* produces an enterotoxin.
- *C. botulinum* is motile.
- *C. perfringens* is non-motile and encapsulated.

Growth	*C. botulinum* Group I	*C. botulinum* Group II
T	12–50°C (varies with strain type) optimum 35–40°C	3.3–45°C optimum 18–25°C
pH	4.6–8.9 (varies with strain type)	
a_w	≥ 0.95	

- *C. botulinum* is produced at pH down to 5.2 and at a lower pH in certain foods, e.g., pH 4.85 in potato.

Growth	*C. perfringens*
T	12–50°C (optimum 43–45°C)
pH	5.5–9 (optimum 6–7.5)
a_w	≥ 0.95

- *C. perfringens* does not readily from spores in food. These are mostly formed in the intestine. It sporulates well at pH 6–8.
- The enterotoxin is produced during spore formation. Occasionally the toxin is produced in food but large numbers of cells are required and so food containing the toxin is usually spoiled.

Transmission: Foods contaminated with spores or vegetative cells.

Illnesses

C. botulinum. Classic symptoms of foodborne botulism include double vision, blurred vision, drooping eyelids, slurred speech, difficulty swallowing, dry mouth, and muscle weakness. These are all symptoms of the muscle paralysis caused by the toxin. If untreated, these symptoms may progress to cause paralysis of the arms, legs, trunk and respiratory muscles. In foodborne botulism, symptoms generally begin 18 to 36 hours after eating a contaminated food, but they can occur as early as 6 hours or as late as 10 days. It is thought that all people are susceptible to foodborne toxicity. Infants with botulism appear lethargic, feed poorly, are constipated, and have a weak cry and poor muscle tone. Muscle weakness and loss of head control can reach a point where the infant appears "floppy". Infant botulism occurs 3–30 days following ingestion of spores which germinate and produce toxins in the intestine.

- *C. perfringens.* Symptoms occur 6–24 hours after eating contaminated food and include profuse watery diarrhea and abdominal pain. Recovery is usually rapid, within 24 hours. All people are thought to be susceptible but the severity of symptoms may vary among individuals.

Dose

C. botulinum. Botulism toxin is very potent and even low concentrations can cause sickness. It is estimated that the dose required to kill humans ranges from 0.1–1.0 µg. In cases of infant botulism implicated honey samples have contained 10^4–10^5 spores/kg *C. perfringens.* Large numbers are required to cause illness, at least 10^6/g food.

Issues relating to control

C. botulinum

Survival

- Spores are resistant to freezing (not defined for vegetative cells).
- The vegetative cells are killed by only a few minutes' exposure to 60°C heat.
- The spores survive drying and are very resistant to heat.
- The toxin is stable at low pH but inactivates quickly at pH 11.
- Toxins may be slightly more heat stable at lower pH values and are resistant to freezing.

Prevention

- For spores the D_{100} for Group I *C. botulinum* is 25 min and for Group II is < 0.1 min, D_{121} for Group I is 0.1–0.2 min and for Group II is < 0.001 min.
- A 12 D process, controlling group I spores, has been adopted for the canning of low-acid (pH > 4.6) foods. This is the equivalent of heating to 121°C for 3 min.
- Thermal death of spores is accelerated at extremes of pH (< 5.0 and > 9.0).
- The toxin is inactivated by treatment at 85°C heat for 1 min, 80°C for 6 min or 65°C for 1.5 hours.
- Nitrite is important to control *C. botulinum.*
- Lactic acid bacteria used in starter cultures inhibit *C. botulinum* in meat products.
- Nisin is widely used in dairy products.
- Liquid smoke appears to be effective for fish but not meat.
- A range of commonly used preservatives (e.g., sorbates, parabens, nisin, phenolic antioxidants, polyphosphates, ascorbates) can be useful in the control of *C. botulinum* as part of a hurdle approach. However the interactions of various preservatives when used in hurdle technology are complex and combinations for food use need to be validated.
- Spores are inactivated by ozone and chlorine (more effective at low pH), hydrogen peroxide and iodophors. Normally chlorinated water should inactivate the toxin.
- *Botulinum* spores are the most resistant bacterial spores to radiation and doses used in food preservation do not effectively eliminate the toxin.
- The toxin is not inactivated by irradiation.

C. perfringens

Survival

- Vegetative cells are readily killed by heating, are very susceptible to freezing and decline slowly under refrigeration.
- Vegetative cells are not very tolerant of low water activity.
- Spores are very heat-resistant and some spores survive boiling for 1 hour.
- Spores very resistant to freezing, refrigeration and desiccation.

Inactivation

- D_{60} for vegetative cells = 5.4–14.5 min.
- D_{100} for spores varies between strains from 0.31 min and > 38 min.
- Heating food to between 70 and 80°C followed by cooling will induce germination of spores.
- *perfringens* enterotoxin is inactivated by heating for 5 min at 60°C.
- *perfringens* will not grow at < 12°C (a very important means of control).
- Cells will die after several days below pH 5.0 and above pH 8.3.
- 6–8% NaCl inhibits growth.
- While *C. perfringens* growth is inhibited by sodium nitrite and sodium nitrate, the levels required exceed those permitted in foods.
- The application of several hurdles enables control of growth of *C. perfringens.*
- Susceptibility of spores to irradiation varies according to strain.

Testing

- Horse-blood or egg-yolk agar and incubated in anaerobic condition for 3 days, colonies show lipolytic activity **(C. botulinum).**
- TSC and OPSP media containing antibiotics and incubated in anaerobic condition for 24 hours at 37°C **(C. perfringens).**
- Commercial kits are available.

In general

- *Store foods at ≤ 3.3°C or ≥ 60°C and cool cooked foods under refrigeration (C. botulinum).*
- *Cook food thoroughly and cool rapidly to prevent spores germination and keep hot foods above 60°C and cold foods below 4°C (C. perfringens).*

Cryptosporidium

Important species: *Cryptosporidium hominis* and *Cryptosporidium parvum*

Why is it important?

Cryptosporidium is one of the most frequent causes of waterborne diseases (drinking water and recreational water) among humans. Many species of *Cryptosporidium* exist that infect humans and a wide range of animals. The parasite is protected by an outer shell that allows it to survive outside the body for long periods of time and makes it very resistant to chlorine disinfection.

Ecological niche

Cryptosporidium may be found in soil, food, water, or surfaces that have been contaminated with the faeces from infected humans or animals.

Foods with which it is most frequently associated

Cryptosporidium sp. could occur, theoretically, on any food touched by a contaminated food handler. Flies can also act as vectors.

Characteristics and survival

Cryptosporidium belongs to the phylum Apicomplexa, class sporozoa, subclass Coccidiasina. *Cryptosporidium* oocysts and measures between 4 and 6 μm in diameter. *Cryptosporidium* can cause mild or severe symptoms depending on the oocyst ingested, the virulence of the strain of *C. parvum*, and the immuno-competence of the affected individuals. Oocysts can survive in fresh, brackish and salt water at 15 and 30°C for several months and are readily filtered out of the water and taken up by clams, oysters and other shellfish. *C. parvum* has been detected in flies, and these may serve as a vector by transferring the oocysts from the faeces to food. The cryptosporidium oocyst is resistant to many disfectants, including 3% hypochlorite solution, iodophore, cresylic acid, benzalkonium chloride, and 5% formaldehyde.

Illness

Symptoms of cryptosporidiosis generally begin 2 to 10 days (average 7 days) after becoming infected with the parasite.

The most common symptom of cryptosporidiosis is watery diarrhea. Other symptoms include stomach cramps or pain, dehydration, nausea, vomiting, fever and weight loss. Some people with crypto will have no symptoms at all. Symptoms usually last about 1 to 2 weeks (with a range of a few days to 4 or more weeks) in persons with a healthy immune system. Occasionally, people may experience a recurrence of symptoms after a brief period of recovery before the illness ends. Symptoms can come and go for up to 30 days. Once infected, people with decreased immunity are most at risk for a more severe infection. The risk of developing severe infection may differ depending on each person's degree of immune suppression.

Selected outbreaks

Location	Year	Vehicle of infection	Cases	Death
Australia	2001	Unpasteurized milk	8	0
Ireland	2000	Drinking water	576	0
England	2000	Drinking water	58	0
Washington DC (US)	1998	Raw produce; Food handler	88	0
Spokane, WA (US)	1997	Green onions? Food handler?	51	0
Minnesota (US)	1995	Chicken salad	50	0
Maine (US)	1993	Cider	213	0
Milwaukee, WI (US)	1993	Drinking water	> 400,000	69

Dose

Less than 10 organisms and, presumably, one organism can initiate an infection. The mechanism of the illness is not known; however, the intracellular stages of the parasite can cause severe tissue alteration.

Control and intervention strategies

- High temperatures are known to be lethal to these protozoa and therefore boiled water and adequately heat-processed foods should be safe to consume.
- Recent experiments evaluating the efficacy of high-temperature–short-time pasteurization treatments demonstrated that heating to 73°C for 1 min was sufficient to destroy the infectivity of *Cryptosporidium* oocysts suspended in water or milk.
- However, oocysts are more resistant to cold and freezing temperatures.
- Oocysts suspended in water retained their infectivity after 168 hours storage at +5°C and at –10°C. At colder temperatures, infectivity was destroyed: at –15, –20, and –70°C, no infective cells remained after 168, 24, and 1 hour of storage, respectively.
- Treatment with 1000 µg/ml for 1 min of chlorine and chlorine dioxide resulted in 0.5 and 2.0 log reductions, respectively. More than 90% reduction was achieved when treated with ozone at 1 µg/ml for 5 min, chlorine dioxide at 1.3 µg/ml for 1 h, and chlorine or monochloramin at 80 µg/ml for 90 min.
- Washing hands is the most effective means of preventing cryptosporidiosis.

To avoid becoming infected, the following precautions should be taken:

- Practice good hygiene (hand washing).
- Avoid water that might be contaminated.
- Avoid food that might be contaminated.

Cyclospora

Important species: *Cyclospora cayetanensis*

Why is it important?

Cyclospora is a microscopic parasite that can affect the intestinal tract and cause diarrhea. *Cyclospora* infections can occur in people of all ages and spread by eating or drinking contaminated food or water. Although the route of transmission is undocumented, faecal-oral route or via water and food, is the probable route of transmission.

Ecological niche

Cyclospora is found in warm climates, mainly the tropics, and sub-tropics. This parasite infects fruits, vegetables, and water that has been contaminated with unsporolated oocysts from a former host's faeces.

Foods with which it is most frequently associated

Any food contaminated with faecal material. Infected rodents, insects, snakes, flies and humans can act as a vehicle of contamination for food and water. Fresh fruits and vegetables, and water is the most common food associated with cyclosporiasis.

Characteristics and survival

Cyclospora cayetanensis is an apicomplexan, cyst-forming coccidian protozoan that causes a self-limiting diarrhea. Morphologically, *C. cayetanensis* has round oocysts that are between 7.5 and 10 μm in diameter with 50 nm thick walls with an outer threadlike coat. The incubation period between acquisition of infection and onset of symptoms averages approximately 1 week (ranges from 2 to 14 or more days). Little information is available regarding animal hosts and environmental survival time. Oocysts, however, tend to be resistant to adverse conditions and have been known to survive for long periods of time if kept moist.

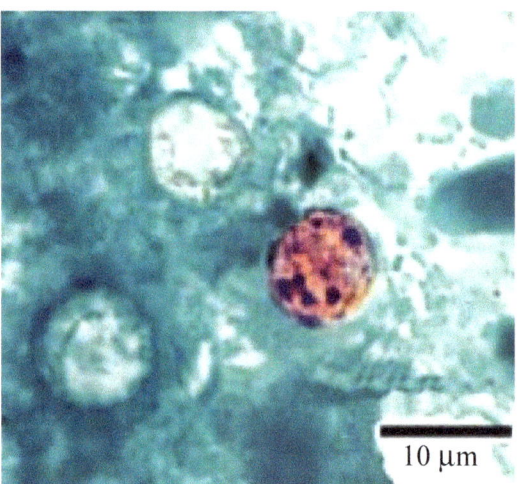

10 μm

Illness

Cyclospora infects the small intestine (bowel) and usually causes watery diarrhea, with frequent bowel movements. Other symptoms include loss of appetite, weight loss, bloating, increased gas, stomach cramps, nausea, vomiting, tiredness, muscle aches, and low-grade fever. Other infectious organisms can cause similar illness. Some persons infected with *Cyclospora* do not develop any symptoms. The time between becoming infected and developing symptoms is usually several days to a week. If not treated, the illness may last for a few days to a month or longer. It may also return one or more times.

Selected outbreaks

Location	Year	Vehicle of infection	Cases
Pennsylvania (US)	2004	Snow Peas	96
Germany	2000	Salad	34
Pennsylvania (US)	2000	Imported raspberries	54
Ontario and Florida (US)	1999	Berries	198
Missouri (US)	1999	Basil (Imported and Domestic)	64
Ontario, Canada	1998	Imported raspberries	315
US (13 states), Canada (1 province)	1997	Imported raspberries	1012
US (3 states)	1997	Basil	341
Florida (US)	1997	Mesclun lettuce	67
US (20 states), Canada (2 provinces)	1996	Imported raspberries	1465

Dose: Minimum infective dose is unknown, but suspected to be low.

Control and intervention strategies

- Like many parasites, *Cyclospora* can be killed by thorough washing and/or cooking of foodstuffs.
- The best way to avoid exposure to *Cyclospora* is to avoid food from unsafe sources.
- Any food to be eaten raw, should be thoroughly washed with potable water before consumption. This will decrease, *but will not eliminate* the risk of *Cyclospora* transmission.

In general, cyclosporiasis can be prevented through the following precautions:

1) *Practice good hygiene*: Thoroughly wash hands with soap after using the restroom and also after assisting a patient with cyclosporiases and handling food later.
2) Clean the food preparation work surfaces, equipment, and utensils with soap and water before, during, and after food preparation.
3) *Separate raw and cooked foods*: Avoid cross-contamination by separating produce, ready-to-eat foods, and cooked foods. Use separate equipment and utensils to handle raw foods. Eat safe foods and drink safe water (Contaminated foods may look and smell normal).
4) *Unpasteurized dairy products or juices*: Wash all produce before cooking or eating raw. Use treated water for washing food, cooking, and drinking. Avoid swallowing untreated water.

Encephalitis

Important species: Nipah virus (NiV)

Why it is important?

Nipah virus causes severe illness characterized by inflammation of the brain (encephalitis) or respiratory diseases. Nipah virus can be transmitted to humans from animals, and can also be transmitted directly from human-to-human; in Bangladesh, half of reported cases between 2001 and 2008 were due to human-to-human transmission. Nipah virus can cause severe disease in domestic animals such as pigs. There is no treatment or vaccine available for either people or animals.

Foods with which it is most frequently associated

Fruits or fruit products (e.g., raw date palm juice) contaminated with urine or saliva from infected fruit bats was the most likely source of infection.

Characteristics

Nipah virus genome is non-segmented, single-stranded-negative-sense RNA belonging to the Paramyxoviridae family.

Illness

Human infections range from asymptomatic infection to fatal encephalitis. Infected people initially develop influenza-like symptoms including, fever, headache, myalgia (muscle pain), vomiting, sore throat, dizziness, drowsiness. Severe respiratory problems, including acute respiratory distress, etc. The incubation period (interval from infection to onset of symptoms) varies from four to 45 days.

Route of transmission

Nipah is presumably derived from a fruit bat virus that spread to swine and adapted over a few years in closely confined swine herds. Once established within the swine population, the virus had ample opportunity to adapt to human hosts. Close contact with infected animal secretions and/or tissues may transmit the virus between pigs. The mode of disease spread between farms has not been established.

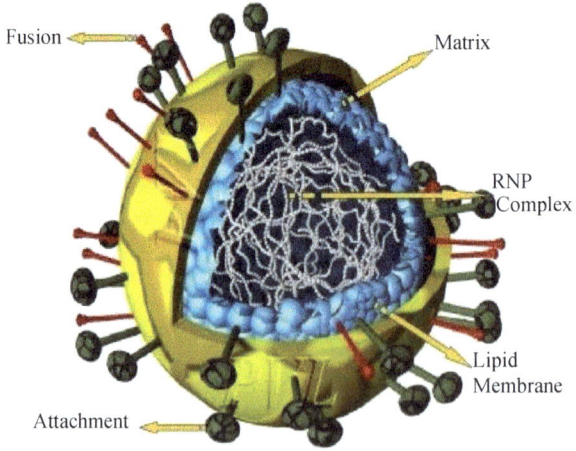

Clinical Presentation of Nipah Virus by Route of Exposure	
Route of exposure	Characteristics
Oral inoculation	Incubation period 14–16 days Mild clinical signs and gross pathology
Parenteral inoculation	Narrow study of two pigs revealed a more severe disease, closer resembling natural exposure Incubation period about 7–10 days
In-contact pigs	Rapid infection Neutralizing antibodies detected at day 14

Signs and tests

No standard protocol exists for detecting Nipah virus at this time. However, several methods have been used to confirm viral infection:

- History/clinical signs.
- Virus isolation by cell culture (kidney, liver, cerebrospinal fluid).
- IgG-, IgM-capture enzyme linked immunosorbent assay (ELISA).
- Virus neutralization.
- Immunofluorescence assay.
- Reverse transcriptase polymerase chain reaction (RT-PCR).

Treatment

Effective treatment has not yet been developed for Nipah infection.

- Ribavirin, an antiviral drug, may reduce mortality among patients with encephalitis caused by Nipah virus, although further study is needed.
- Intensive supportive care is required for infected human.

Control and prevention strategies

In the absence of a vaccine, the only way to reduce infection in people is by raising awareness of the risk factors and educating people about the measures they can take to reduce exposure to the virus.

- Reducing the risk of bat-to-human transmission. Efforts to prevent transmission should first focus on decreasing bat access to date palm sap. Freshly collected date palm juice should also be boiled and fruits should be thoroughly washed and peeled before consumption.
- Reducing the risk of human-to-human transmission. Close physical contact with Nipah virus-infected people should be avoided. Gloves and protective equipment should be worn when taking care of ill people. Hands should be thoroughly washed after caring for or visiting sick people.
- Reducing the risk of animal-to-human transmission. Gloves and other protective clothing should be worn while handling sick animals or their tissues, and during slaughtering and culling procedures.

Selected outbreaks

Year	Location	Cases	Death
2001	Siliguri	66	49
2001	Meherpur	13	9
2003	Naogaon	12	8
2004	Rajbari	31	23
2004	Faridpur	36	27
2005	Tangail	12	11
2007	Thakurgaon	7	3
2007	Kushtia	8	5
2007	Nadia	5	5
2008	Manikgonj	4	4
2008	Rajbari	6	5
2009	Rangpur, Gaibandha, Rajbari, Nilphamari	4	1
2010	Faridpur, Rajbari, Gopalganj, Kurigram	17	15
2011	Lalmonirhat, Dinajpur, Comilla, Nilpahmari, Faridpur, Rajbari	28	28
Total		253	194

Enterohemorrhagic *Escherichia coli*

Important species: *Escherichia coli* O157:H7 and non-O157 shiga toxin producing *Escherichia coli* (STEC).

Why is it important?

E. coli are part of the normal microflora of man and many strains are non-pathogenic. However, some are pathogenic and these can be divided into 4 groups:

- Enteropathogenic *E. coli* (EPEC);
- Enteroinvasive *E. coli* (EIEC);
- Enterotoxigenic *E. coli* (ETEC); and
- Enterohemorrhagic *E. coli* (EHEC).

In recent times the toxin-producing forms of *E. coli* have become an important cause of foodborne illness. They can cause severe illness such as hemorrhagic colitis and haemolytic ureic syndrome, which can be fatal, particularly in young children and the elderly. Non-O157 shiga toxin producing *Escherichia coli* (STEC) are a diverse group of organisms with varying pathogenic potential. By definition all STEC produce 1 or 2 toxins but might not possess other factors critical for pathogenicity.

Sources: Intestinal tracts of domesticated and wild animals and their faeces, also soil and water.

Foods at risk

Beef (esp. ground beef), milk, leafy green and salad vegetables, poultry foods, potatoes. May also be transmitted via faecal-oral route from person to person or food handling.

Characteristics and growth conditions

- Gram-negative
- Rod shaped
- Facultative—Can grow in presence or absence of oxygen
- Produce toxins
- Most strains are motile
- Growth temperature: 8–45°C (optimum 37°C).
- Growth pH 4.4–9.0 (optimum 6–7)
- A_w 0.955 (optimum 0.995)
- Indicator of faecal contamination and the possible presence of enteric pathogens (ex: *Salmonella typhi*)

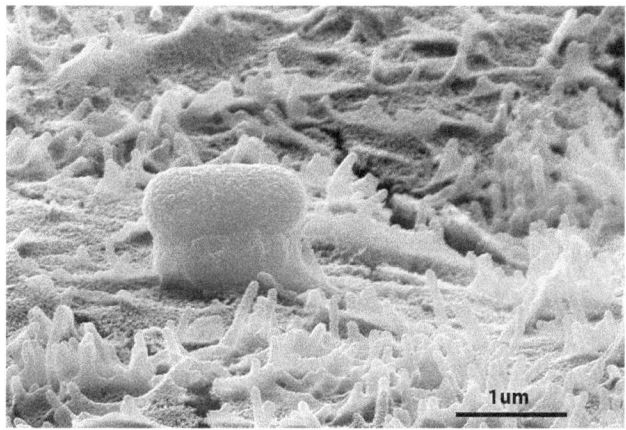

Transmission: Faecal contamination of water supplies and contaminated food due to poor hygiene practices of food handlers.

Illness

EPEC, which is rarely foodborne, causes infant diarrhea 1–3 days after ingestion. EIEC causes a dysentery-like syndrome, 8–24 hours after ingestion. ETEC is a cause of travellers' diarrhea. It is caused by the production of a cholera-like toxin and symptoms occur 8–44 h after ingestion. EHEC (STEC and *E. coli* O157:H7) invades the gut and then produces a toxin which can lead to a bloody-diarrhea syndrome, kidney disease and even death. Resulting conditions include haemorrhagic colitis (HC), Haemolytic Uraemic Syndrome (HUS). Any age group can be affected but disease most often occurs in young children and the elderly.

Dose

Foods with very low numbers of cells—0.3–0.4 EHEC cells/g—have been implicated in disease but it has been estimated that ingestion of 10^5 cells gives a 50% probability of disease. For other pathogenic *E. coli* the minimal infective dose for adults is probably 10^6/g.

Issues relating to control

Survival

- *Coli* O157:H7 survives well in chilled and frozen foods. At low temperatures cells enter a viable but non-culturable state.
- It is more acid-resistant than other *E. coli*. Prior exposure to acidic conditions can increase acid tolerance further. Has also been shown to survive stomach pH (1.5) for longer than 3 hours.
- Can survive for weeks under dry conditions and in dried food products, e.g., dried meat.

Prevention

- Rapidly inactivated by heating to 71°C, so pasteurization is effective.
- D value at 60°C is 30–45 seconds.
- Inactivated by proper cooking.
- Thermal resistance is higher in foods with a high fat content.
- Freeze thawing can reduce numbers but this effect is strain dependent.
- Inactivation by decreasing pH. Usually more effective at warmer temperatures.
- Sensitive to UV and gamma radiation.
- While commonly used disinfectants such as chlorine are effective, their efficacy may be reduced in the presence of solids or organic matter.

Testing

- Tests for *E. coli* are not sensitive to *E. coli* O157:H7 in selective medium.
- Latex agglutination kits commercially available.
- Serological tests.

In general

- *Cook food thoroughly and cool rapidly.*
- *Keep hot foods above 60°C and cold foods below 4°C.*
- *Reheat cooked food at 74°C.*

Giardia

Important species: *Giardia lamblia* and *Giardia intestinalis*

Why is it important?

Giardia food poisoning is caused by a parasite known as *Giardia lamblia* or *Giardia intestinalis*. *Giardia* is also a common cause of waterborne disease.

Ecological niche

Giardia lives in the intestines of infected people or animals, and is excreted through the faeces. Infection, therefore, is the faecal to oral route. *Giardia* cysts are present in surface waters, even in areas remote from significant human activity and apparently originate from wild animals.

Foods with which it is most frequently associated

Giardiasis is most frequently associated with the consumption of contaminated water. *Giardia* cysts detected in some ground water, and supply water. Infected or infested food handlers have traced five foodborne outbreaks. Prepared food such as sandwiches, salads, canned salmon, and ice also have been implicated in foodborne outbreaks.

Characteristics and survival

Giardia infection occurs after cysts are ingested. This marks the beginning of the life-cycle. After ingestion, mature cysts in the small intestine release trophozoites through a process called excystation. Cysts are able to survive exposure to gastric acid which may actually trigger excystation. The cysts are approximately 7–10 μm in length and are oval in shape. The mature cyst contains four nuclei. They are environmentally resistant and may remain viable for several months in cool, moist conditions, and freezing for a few days. They are also able to survive standard concentrations of chlorine used in water purification systems. *Giardia* cysts may be released in the stool for weeks and even months and may be present in concentrations as high as 10^7/gram.

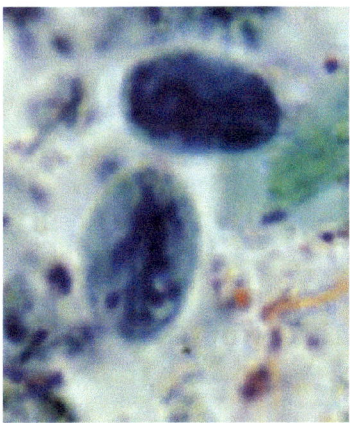

Illness

Giardia infection occurs when an individual eats food, drinks water, or comes into contact with surfaces or objects contaminated by the parasite or its cysts (a dormant stage in which the parasite is resistant to many adverse environmental conditions allowing it to survive outside the body).

Symptoms of *Giardia* infection, or Giardiasis, may include stomach cramps, diarrhea, gas, belching, bloating, greasy stools, nausea or upset stomach.

Symptoms may occur 1 to 2 weeks after infection, and usually last from 1 to 3 weeks. Weight loss and dehydration may also occur as a result of these symptoms. Some individuals with Giardiasis are asymptomatic, meaning that they do not exhibit symptoms.

Selected outbreaks

Giardiasis associated with drinking water reported 39 outbreaks to the CDC in 1999–2000, of them 6 outbreaks, affecting 52 people, were traced to the presence of *G. intestinalis* in inadequately treated well or river water or in water contaminated by cross-connections with pipes containing sewage or drinking water for animals.

Dose: Infections may result from the ingestion of 10 or fewer *Giardia* cysts.

Control and intervention strategies

Boiling is a very effective method in inactivating *Giardia* cysts. They are also susceptible to inactivation by ozone and halogen; however, inactivation by chlorine requires prolonged contact time and higher concentration. Filtration can be used for purifying water.

In addition, various public health and prevention strategies should be taken to decrease risk of infection.

1) Avoid contaminated water or treat all water with halogenated tablets or solutions.
2) Avoid foods washed in contaminated water or which cannot be cooked or peeled, which is especially important for travelers.
3) Avoiding eating raw fruits and vegetables.
4) Wash hands with soap frequently, especially before eating and after using the bathroom for at least 15 seconds.
5) Avoid swallowing water in swimming pools and spas.
6) Ultimately, practice good hygiene including avoiding contact with the faeces of an infected person. Being aware of this helps to prevent spread of the infection.

Hepatitis A Virus

Important species: Hepatitis A virus

Why it is important?

Hepatitis A virus is an acute infectious disease of the liver caused by the hepatitis A virus (HAV), and is the most prominent foodborne virus and may cause hemolysis, acalculous cholecystitis, acute renal failure, acute reactive arthritis, and pancreatitis to human. The mortality rate for HAV infection is approximately 0.01%. Because of their low infective dose, this virus can be transmitted via infected food handlers. HAV is primarily transmitted by the faecal-oral route, either by person-to-person contact or by ingestion of food or water contaminated with human faeces.

Ecological niche

HAV is found in the faeces of infected persons. It also can be found in fresh produce, food, restaurant, water and the environment.

Food with which it is most frequently associated

Pre- and post-harvest fresh produce (green onions, blueberries), ready to eat foods, Molluscan shellfish, fresh and seawater. Cold cuts and sandwiches, fruits and fruit juices, milk and milk products, vegetables, salads, and iced drinks are commonly implicated in outbreaks. Water, shellfish, and salads are the most frequent sources. Contamination of foods by infected workers in food processing plants and restaurants is common.

Characteristics and survival

The Hepatitis virus (HAV) is non-enveloped and contains a single-stranded RNA packaged in a protein shell. There is only one type of the virus. The virus is resistant to detergents, acid (pH 1), solvents (e.g., ether, chloroform), drying, and temperatures up to 60°C. It can survive for months in fresh and salt water. Many factors influence the stability of HAV in the environment, the most important of which are relative humidity, temperature, degree of inoculum drying, type of suspending medium (faecal or otherwise), and the type of surface contaminated. For example, half life of HAV ranges more than 7 days at low relative humidity and 5°C but only 2 h at high relative humidity and 35°C.

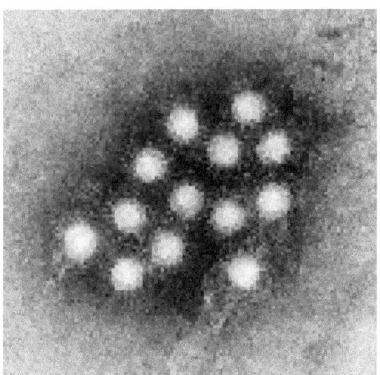

Illness

Hepatitis A infection begins with ingestion of the virus, which reaches the gastrointestinal tract in the bile and replicates in the hepatocytes. Approximately 10^4 virions/ml of blood can be detected from infected patients. The incubation period of HAV is 2–7 weeks, with an average of 28 days. Infection of hepatitis consists of four phases. First phase, viral replication but lack of symptoms. Second phase lasts for 5–7 days, patients may experience anorexia, nausea, vomiting, fatigue, urticaria and pruritus. Third phase lasts 7–28 days and patients experience darkening of urine, pale-colored stool, jaundice and hepatomegaly. In the last phase symptoms resolve and liver enzymes return to normal. Infection is common in children in developing countries, reaching 100% incidence, but following infection there is life-long immunity.

Selected outbreaks

Viral outbreaks associated with contaminated green onions occurred in 2003 with more than 900 cases. Raw green onions imported from Mexico have been implicated in this Tennessee and Georgia outbreaks. The contamination of the produce may occur either at the pre-harvest stage (water, infected pickers) or at the post-harvest phase (food handlers).

Dose

The infectious dose of HAV is unknown, but it is thought to be low, ranging from 10^1–10^4 virions/ml of blood.

Control and prevention strategies

- Hepatitis A can be prevented by vaccination and good hygiene and sanitation practices. Boiling at 100°C could inactivate rotavirus. Irradiation of < 3.1 KG could inactivate 90% or rotavirus in shellfish.
- Other treatments include, chlorine treatment (drinking water), formalin (0.35%, 37°C, 72 hours), peracetic acid (2%, 4 hours), and beta-propiolactone (0.25%, 1 hour). UV radiation (2 µW/cm^2/min) could reduce to undetectable levels.
- Cleaning equipment and surfaces with 70% ethanol or sodium hypoclorite (0.125%, 10 min) or sodium chlorite (30%, 10 min) could inactivate HAV.

Listeria

Important species

Listeria monocytogenes is the most important pathogenic strain. Not all strains of *Listeria* are pathogenic.

Why is it important?

Listeriosis is a relatively rare but potentially fatal disease. Invasive listeriosis is a life-threatening systemic infection. Those at risk include pregnant women and their foetuses, neonates, the elderly and immunocompromised people. There is also a non-invasive form—gastroenteritis—but there is little available information about the occurrence of this milder form. Most healthy people do not seem to be affected by *Listeria*.

Sources

It is widely distributed in the environment and has been isolated from a variety of sources, including soil, vegetation, silage, faecal material, sewage and water. It is ubiquitous in the natural environment, i.e., soil and water. It is prevalent in intensive animal/bird farming practices. Also present in cracks of tiles, ventilation systems and filters.

Foods at risk

Found in a wide range of foods. Milk, semi-soft and soft mold-ripened cheeses, smoked fish, modified atmosphere packaged vegetables, hot dogs, pork tongue in jelly, processed meats, pâté, salami, butter, cooked shrimp, salads, raw vegetables and coleslaw and ready-to-eat food products.

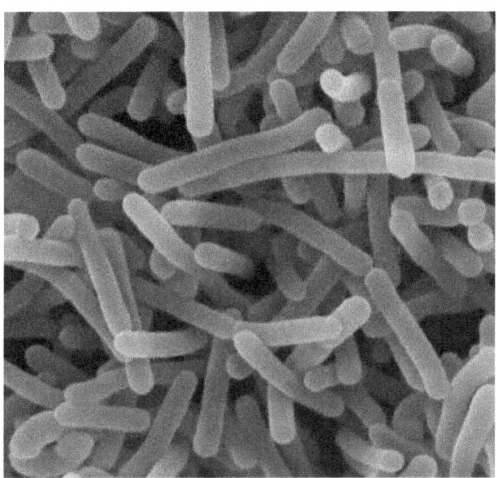

Characteristics and growth conditions

- Gram-positive.
- Facultative anaerobe: Growth is optimal under micro-aerophilic conditions.
- Non-spore-forming rod.
- A typical tumbling motility at 20–25°C, but not at 35°C.
- Psychrotrophic and grows over a temperature range of 0° to 45°C (optimum 37°C). Some reports indicate a minimum temperature of –1.5°C.
- Growth pH 4.4–9.4.
- Growth $a_w \geq 0.92$.
- Salt tolerant, e.g., can grow in 10% sodium chloride.

Transmission

- Contaminated food, ready-to-eat foods with long shelf life (e.g., in hot dogs and delimeats occur after cooking but before packaging).
- Transmitted from person to person.

Illness

The invasive disease is normally associated with people with weakened immune systems and can occur 1–90 days after ingestion of cells. Symptoms include "flu"-like symptoms, diarrhea, vomiting, septicemia, meningitis and abortion. In most cases patients require hospitalization and the fatality rate is 20–30%. There may be long-term effects such as neurological problems. Non-invasive listeriosis can occur in anyone consuming a high number of cells. Symptoms occur 12 h to 7 days after ingestion and include diarrhea, fever, muscle pain, headache, cramps and vomiting.

Dose

The infective dose is not really known but it seems that in most cases of invasive listeriosis 100 to 1000 cells are required. In cases of non-invasive listeriosis, a higher does has been implicated, at least $> 10^5$ and more often around 10^{11} cells.

Issues relating to control

Survival

- Resistant to various environmental conditions, such as high salinity or acidity that allows it to survive longer under adverse conditions than most other non-spore forming bacteria.
- These foodborne pathogens are widely distributed in nature.
- One of the most thermally resistant vegetative cells it has caused some concern in terms of effectiveness of pasteurization.
- *Listeria* can survive in wet conditions for many months and up to 2 years in dry soil or dust.

Prevention

- *Listeria* is killed by heat treatments in most normal processing operations but its heat resistance varies according to strain type and food matrix.
- The commonly reported $D_{60°C}$ is 2.6 min but it may be greater, e.g., $D_{60°C}$ of strain Scott A is 5.29 min in chicken slurry, 8.32 min in beef slurry and 5.02 min in carrot slurry.
- Can grow at low temperatures (3°C) and survive down to 0°C or just below.
- *Listeria* can survive and grow at a pH down to 4.4 but it seems that the type of acid and the storage temperature influence the effect of pH.
- Sensitive to commonly used disinfectants in the absence of organic matter. In presence of organic matter many disinfectants are ineffective.
- Similar resistance as other gram-positive bacteria to gamma radiation but more sensitive than other gram-positive bacteria to UV radiation.

Testing

Selective agar containing lithium chloride, phynylethanol, glycine anhydride and antibiotics. CAMP tests (Production of a characteristic zone of hymolysis when grow in proximity to *S. aureus*).

In general

- *Avoid post-cooking contamination of ready-to-eat foods with a long shelf life.*
- *Reheat cooked food at 74°C.*

Noroviruses

Important species: Norwalk virus

Why is it important?

Noroviruses are a group of viruses that cause the "stomach flu" or gastroenteritis, in people. Noroviruses are very infectious and can survive on practically any surface including door handles, sinks, railings and glassware. They occur throughout the year but are more common in winter and affect all age groups.

Food with which it is most frequently associated

Food items implicated in norovirus outbreaks include consumption of cold foods, including various salads, sandwiches, and bakery products. Liquid items (e.g., salad dressing or cake icing) that allow virus to mix evenly are often implicated as a cause of outbreaks. Molluscan shellfish and ready-to-eat foods that become contaminated by human handling, such as fruit salad, raspberries, cake icing, box lunch, potato salads and deli meat.

Characteristics and survival

Noroviruses (NoV) are a genetically diverse group of single stranded RNA, nonenveloped viruses belonging to the Caliciviridae family. Noroviruses can genetically be classified into five different genogroups (GI, GII, GIII, GIV, and GV) which can be further divided into different genetic clusters or genotypes. The most common genotype identified in hospitalized children was GII.4. Virus (10^7 particles per gram) may shed in the stool for a week or two after symptoms resolved. Noroviruses are transmitted directly via person to person or indirectly via contaminated water and food. The incubation period ranges from 10–51 h after virus ingestion and symptoms generally persist for 24–48 h. Noroviruses are relatively resistant to environmental challenge: they are able to survive freezing, temperatures as high as 60°C, and have even been associated with illness after being steamed in shellfish. Moreover, noroviruses can survive in up to 10 ppm chlorine, well in excess of levels routinely present in public water systems.

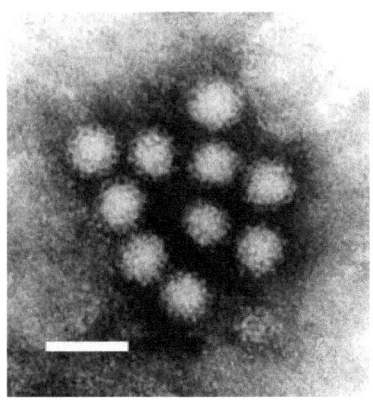

Norovirus is rapidly inactivated by chlorine-based disinfectants, but because the virus particle does not have a lipid envelope, it is less susceptible to alcohols and detergents.

Illness

The symptoms of norovirus illness usually include nausea, vomiting, diarrhea, and some stomach cramping. Sometimes people additionally have a low-grade fever, chills, headache, muscle aches, and a general sense of tiredness. The illness often begins suddenly, and the infected person may feel very sick. In most people the illness is self-limiting with symptoms lasting for about 1 or 2 days. In general, children experience more vomiting than adults.

Selected outbreaks

Most foodborne outbreaks of norovirus illness occurs though direct contamination of food by a food handler immediately before its consumption.

In 2007, 10 NoV foodborne outbreaks were reported affecting 392 persons in Belgium. The major implicated foods were sandwiches (4/10), where food handlers reported a history of gastroenteritis in two outbreaks. Forty foodborne and waterborne outbreak events due to NoV, epidemiological and/or laboratory confirmed, from 2000 to 2007 revealed that in 42.5% of the cases the food handler was responsible for the outbreak, followed by water (27.5%), bivalve shellfish (17.5%) and raspberries (10.0%).

Dose

Noroviruses are highly contagious and as few as 10 viral particles are sufficient to infect an individual.

Control and intervention strategies

Norovirus is extremely infectious, and can be controlled by the following preventive steps:

1) Frequently washing hands, especially after visiting the toilet and before eating or preparing food.
2) Carefully washing fruits and vegetables, and steaming oysters before consumption.
3) Thoroughly cleaning and disinfecting contaminated surfaces by using a bleach-based household cleaner.
4) Immediately removing and washing clothes or linen that may be contaminated with the virus after an episode of illness (use hot water and soap for washing).
5) Flush or discard any vomitus and/or stool in the toilet and make sure that the surrounding area is kept clean.

Rota virus

Important species: Rotavirus A, B and C most commonly infect humans.

Why is it important?

Rotavirus A accounts for more than 90% of rotavirus gastroenteritis in humans. It is the leading single cause of severe diarrhea among infants and young children. Rotavirus is transmitted by the faecal-oral route, via contact with contaminated hands, surfaces and objects, and possibly by the respiratory route.

Ecological niche: Rotaviruses have been found in water, sewage and in the environment.

Foods with which it is most frequently associated

Infected food handlers may contaminate foods that require handling and no further cooking, such as salads, fruits, and hors d'oeuvres. Rotaviruses are quite stable in the environment and have been found in estuaries. In temperate areas, it occurs primarily in the winter, but in the tropics it occurs throughout the year. The number attributable to food contamination is unknown.

Characteristics and survival

Rotavirus is a genus of double-stranded RNA virus in the family Reoviridae. The genome of rotavirus consists of 11 unique double helix molecules of RNA which are 18,555 nucleoside base-pairs in total. Rotaviruses are ubiquitous pathogens present in faeces at levels approximately 10^9–10^{10} particles/gram. It is transmitted by the faecal-oral route. They are acid labile and can survive at pH 2.0 of the stomach and the digestive enzymes in the gut. Rotavirus can survive in the atmosphere up to 9 days at 20°C and may persist up to 60 days with reduced infectivity on paper, cotton, cloth, aluminium, tiles, latex and polystyrene.

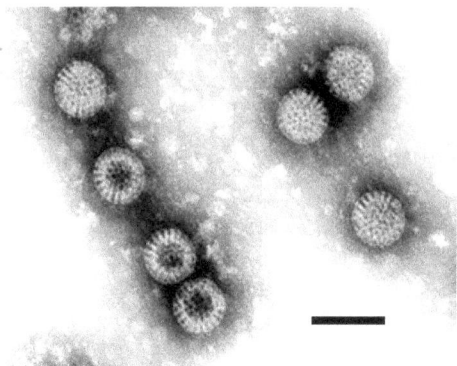

Illness

Rotaviruses cause acute gastroenteritis, infantile diarrhea, anorexia, dehydration, depression and occasional vomiting. Rotavirus gastroenteritis is a self-limiting, mild to severe disease characterized by vomiting, watery diarrhea, and low-grade fever. The incubation period ranges from 11 h to 6 days. Vomiting lasts 1–2 days and diarrhea persists for about 5 days and may be accompanied by several days of fever.

However, severe diarrhea without fluid and electrolyte replacement may result in severe dehydration and death.

Group B rotavirus, also called adult diarrhea rotavirus or ADRV, has caused major epidemics of severe diarrhea affecting thousands of persons of all ages in China.

Group C rotavirus has been associated with rare and sporadic cases of diarrhea in children in many countries. However, the first outbreaks were reported from Japan and England.

Selected outbreaks

In March 2000, 85 students at a university in Washington, D.C., were affected by an epidemic of gastroenteritis. The illness was associated with eating deli sandwiches at the dining hall. The cause of the outbreak was identified as Group A rotavirus.

Dose: The infective dose is presumed to be 10–100 infectious viral particles.

Control and intervention strategies

- Rotavirus was found readily inactivated by chlorine dioxide at alkaline pH and by chlorine and ozone at neutral pH.
- The virus also appeared to be more sensitive to the action of ether and chloroform.
- The infectivity of viruses was lost after UV irradiation for 15 min and after treatment with 8% formaldehyde for 5 min, 70% (vol/vol) ethanol for 30 min, and 2% lysol, 2% phenol, and 1% H_2O_2 for 1 h each.
- Boiling at 100°C could inactivate rotavirus. Irradiation of < 3.1 KG could inactivate 90% or rotavirus in shellfish.
- Other chemical disinfectants include, chloramin-T, chlorhexidine gluconate, glutaraldehyde, hydrochloric acid, isopropyl alcohol, peracetic acid, providone iodine, quaternary ammonium compound and sodium-o-benzyl-p-cholorophenate, has reduction capability of rotavirus.

Salmonella

Important species

Salmonella enterica—many different serotypes that can be divided into two groups (i) non-typhoid and (ii) typhoid

Most foodborne *Salmonella* are non-typhoid strains. Names usually written as *Species* Serotype. The "*enterica*" is often left out, e.g., *Salmonella* Enteritidis

Why is it important?

Salmonellosis, caused by non-typhoid *S. enterica* is one of the most frequently reported foodborne disease in the world. *S. enterica* comprises a large number of different serotypes although a few dominate in terms of causing disease, e.g., *S. enteritidis* and *S. typhimurium.* The typhoid fever causing serotypes are *S. typhi* and *S. paratyphi* which produce typhoid (enteric) fever or typhoid-like fever in humans, which is a much more serious form of *salmonella* infection.

Sources

Non-typhoid *Salmonella* is ubiquitous in the natural environment, i.e., soil and water. It is prevalent in intensive animal/bird farming practices. Man is the reservoir for *S. typhi* and *S. paratyphi.* Water may become contaminated via human faecal waste.

Foods at risk

Non-typhoid salmonellae are found in a wide range of foods as they are prevalent in many sectors of the food chain. Associated in particular with foods of animal origin, e.g., poultry and egg products, meat and meat products, cakes and ice cream, milk and dairy products, fish and shrimp, peanut butter, cocoa, and chocolate.

Contamination of foods with *S. typhi* and *S. paratyphi* is usually via food handlers or faecally contaminated water.

Characteristics and growth conditions

- Gram-negative rod.
- Non-spore forming.
- Motile–peritrichous flagella.
- Facultative anaerobic (i.e., can grow in the presence or absence of air).
- Growth temperature 8–45°C (optimum 37°C).
- Growth pH 4–8 (optimum 6.5–7.5).
- Growth $a_w \geq 0.93$.

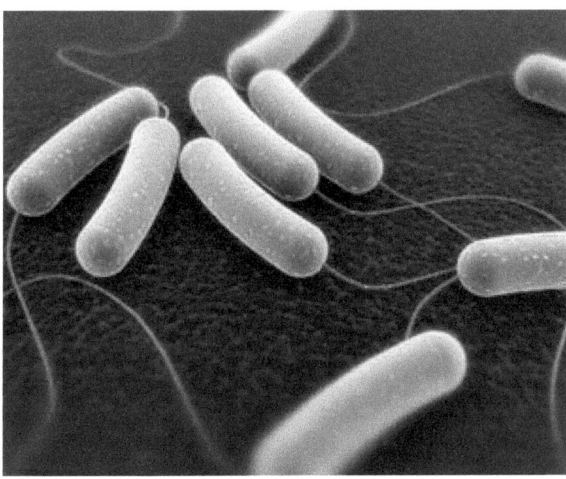

Transmission

- Ingestion of contaminated food.
- Disseminated via animal and human faeces to soil and water.

Illness

- Non-typhoid *salmonella* causes salmonellosis—a gastrointestinal illness which occurs 6–48 h after ingestion.

- Symptoms include nausea, vomiting, cramps, diarrhea, fever and abdominal pain and last 1–7 days.
- Can have long-term effects (e.g., septicemia may occur) and chronic consequences (e.g., arthritic symptoms in 3–4 weeks).
- All age groups are susceptible, but symptoms are most severe in the elderly, infants, and the immunocompromised.
- Typhoid occurs 7–28 days after ingestion of *S. typhi* cells. Symptoms include fever, malaise, anorexia, diarrhea or constipation, delirium.
- Recovery is slow and may take up to 2 months.
- Hospitalization is often required. *S. typhi* is not a problem in industrialized countries. Those at risk include people living in conditions of poor sanitation and international travellers.

Dose

The dose required to cause non-typhoid salmonellosis varies with many factors. While in general 10^5–10^6 cells are required to cause illness, doses as low as 4–45 cells have been reported to cause illness. As with non-typhoid salmonellosis the infectious dose to cause typhoid fever also appears to be variable.

Issues relating to control

Survival

- Can survive well in foods and on surfaces for long periods at room temperature or below and for long periods in animal sewage applied to land (up to 8 months).
- Can survive in dried foods.
- Has survived in broth with 12% salt (22 days at 20°C and 55 days at 5°C).

Prevention

- It is one of the most heat-resistant vegetative cells. D value at 60°C = 2–6 min.
- Varying heat resistance—Increased heat resistance following heat shock and also at a lower a_w values (e.g., milk concentrate and chocolate).
- Freezing usually kills *Salmonella* but some foods, e.g., meat, appear to be protective of *Salmonella* so freezing does not ensure inactivation.
- Inactivated by radiation (D value around 0.5 kGy, up to 0.8) but effectiveness depends on food type, e.g., D times are higher in drier foods such as desiccated coconut.
- Effect of pH is related to other factors such as acid type and temperature, e.g., inactivation is more rapid in commercial mayonnaise at 20°C than it is at 4°C.
- Sensitive to disinfectants used in food industry.
- Avoid direct handling of food by infected employees.

Testing

- Impedence-conductance techniques using TMAO culture Medium.
- ELISA kits are also commercially available.

In general

- *Cook food thoroughly and cool rapidly.*
- *Keep hot foods above 60°C and cold foods below 4°C.*
- *Reheat cooked foods at 74°C.*

Shigella

Important species: *Shigella dysenteriae, Shigella flexneri, Shigella boydii* and *Shigella sonnei*

Why is it important?

Causes bacillary dysentery or shigellosis. It is painful and incapacitating and may require hospitalization. Not usually fatal but can be in undernourished children, immunocompromised individuals and the elderly.

Sources: The primary reservoir is humans and other primates. Also found in water.

Foods at risk

Not associated with any particular foods but outbreaks have been associated with salads, raw vegetables, milk, dairy products, chicken and shellfish. Contamination of foods is usually via the faecal-oral route. Can be introduced into food by an infected food handler. Often regarded as waterborne.

Characteristics and growth conditions

- Gram-negative rod.
- Non-spore forming.
- Non-motile.
- Can grow in the presence or absence of oxygen.
- Growth temperature 6–47°C.
- Growth pH 4.8–9.3.
- $A_w > 0.95$.
- 5.2% NaCl.
- Some strains produce enterotoxin and Shiga toxin.

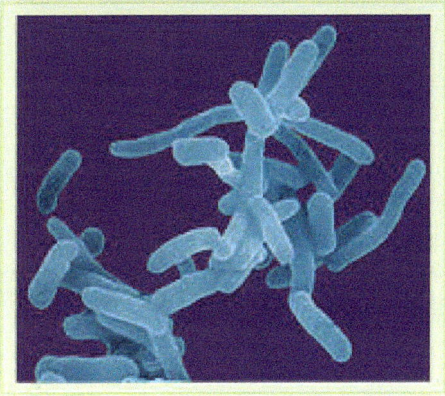

Transmission

Ingestion of contaminated foods through faecal-oral route, faecally contaminated water, unsanitary food handling and flies carrying sewage contamination.

Illness

Bacillary dysentery or shigellosis can occur within 12 hours to 4 days of ingestion of *Shigella*. Symptoms include abdominal pain, profuse watery or bloody diarrhea, fever and malaise. It can last from 3 to 14 days. Everyone is susceptible to some degree of it but some groups in the population are more at risk such as young children in day-care centres, people in nursing homes, prisons, and those living in places with poor sanitary conditions. Infants, the elderly, and the immunocompromised are susceptible to the severest form of the illness. Shigellosis is common among AIDS and non-AIDS homosexual men. In severe cases of diarrhea/dehydration hospitalization may be required. Sequelae may occur such as Reiter's disease, reactive arthritis, haemolytic uremic syndrome.

Dose: The dose required to cause illness is low with 10–200 cells capable of causing illness.

Issues relating to control

Survival

- *Shigella* spp. have not been widely studied in foods and very little is known about the growth and survival of the organism in food.
- *S. sonnei* is more robust than *S. flexneri* and it has been shown to be able to grow on foods (e.g., parsley).
- Usually introduced by a food handler and so is often difficult to control.
- Survive well at low temperatures, e.g., up to 100 days on butter.
- Survives heating at 63°C for 2–3 min.
- Among the most acid-resistant foodborne pathogens.
- Can survive up to 15% salt and survival is better in low moisture foods.

Inactivation

- Rapidly inactivated in temperatures above 65°C.

Prevention

- Prevent at pH values < 4 but die-off can be slow.
- Decline slowly at 6% NaCl.
- Susceptibility to preservatives can vary among strains, e.g., *S. flexneri* is inhibited by 450 ppm nitrite at pH 5.5 while *S. sonnei* requires 700 ppm for inhibition.
- Inactivated by chlorine.
- Sensitive to gamma radiation.
- Good personal hygiene of food handlers is critical for control as in some cases carriers of *Shigella* may not exhibit any symptoms.

Testing

- Gram negative broth and selenite broth.
- Rapid techniques based on immunoassays.
- Polymerase chain reaction (PCR).

In general

- *Cook food thoroughly and cool rapidly.*
- *Keep hot foods above 60°C and cold foods below 4°C.*

Staphylococcus

Important species: *Staphylococcus aureus*

Why is it important?

S. aureus produces a toxin which causes food poisoning. Growth and toxin production are greatest in aerobic conditions but it can also grow anaerobically. Although total number of staphylococcal poisoning cases are difficult to estimate it is thought to be relatively common, especially in areas where there are deficiencies in food handling.

Sources

Ubiquitous in man and animals (skin, nose, throat) which are the primary reservoirs. Carried by a large proportion of the healthy population. Also exists in air, dust, sewage, water, milk, and food or on food equipment, environmental surfaces.

Foods at risk

Fish, meat, poultry and egg products, salads, cream, confectionary products, milk and cheese. Any food contaminated by a food handler.

Characteristics and growth conditions

- Gram-positive cocci (characteristic grape like clusters of cells).
- Aerobe but capable of growing anaerobically.
- Non-motile.
- Non-spore forming.
- Some strains produce toxins (5 types of toxins produces by *S. aureus*).
- Growth temperature 6.5–50°C (optimum 37°C). Toxin production 10–48°C (optimum 35–40°C).
- Growth pH 4–9.3 (6–7). Toxin Production 4.8–9 (5.3–7).
- A_w 0.83–0.99 (0.98). Toxin production 0.85–0.99 (0.98) (Ability to grow at such low water activity gives it a competitive advantage in foods with low water activity/high salt (grows at up to 25% NaCl).
- In general it does not compete well with other bacteria.
- Toxin production is greater in the presence of oxygen.

Transmission: Food contaminated by food handlers from skin lesions, by coughing and sneezing over food.

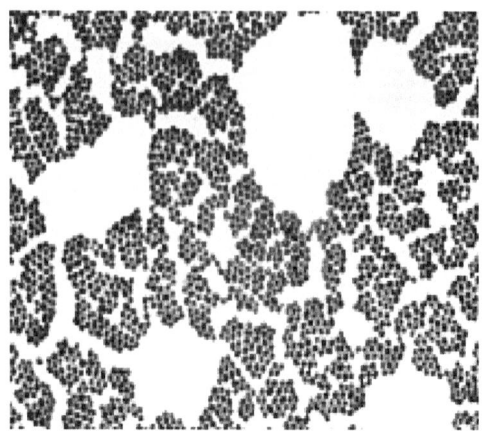

Illness

Staphylococcal food poisoning is acute with symptoms occurring within 30 mins to 7 hours after eating contaminated foods. They include nausea, vomiting, abdominal cramps, sometimes followed by diarrhea. Recovery is usually rapid, within 2 days. All people are thought to be susceptible but severity of symptoms varies among individuals.

Dose

Need high numbers of cells before sufficient toxin to cause illness is produced. Toxin production begins when there are 10^5 to 10^6 cells. Sufficient toxin to cause illness is produced when there are 5×10^6. A toxin dose of about 1.0 mg in contaminated food will produce symptoms of staphylococcal poisoning. Compared to other exotoxins a large amount is needed to cause illness—potent but not as potent as other exotoxins.

Issues relating to control

Survival

- Relatively susceptible to high temperatures.
- Survives frozen storage.
- Resistant to drying and can grow and produce toxin in foods of low A_w.
- Can grow in up to 25% NaCl.
- Can survive in foods at pH down to 4.2 but this is dependent on type of acid.
- Toxins are very resistant to heat (e.g., D value of enterotoxin B at 149°C is 100 min at an A_w of 0.99).

Prevention

- Usually readily killed at cooking and pasteurization temperatures. D_{60} = ~2 min
- Heat resistance is increased in dry, high-fat and high-salt foods.
- High resistance is reduced at high and low pH.
- Lemon and lime juice and lactic acid bacteria inhibit *S. aureus*.

- Toxin production is inhibited at low pH (lactic acid) and in the absence of oxygen.
- Commonly used preservatives, with the exception of NaCl, are effective against *S. aureus*.
- The effectiveness of some preservatives towards *S. aureus* increases with decreasing pH.
- Sensitive to most chemical disinfectants routinely used in the food industry.
- Relatively resistant to ionizing radiation but not UV radiation compared to other non-spore formers.

Testing

- Baird-Parker agar shows the characteristic shiny, jet-black colonies surrounded by a clearing zone.
- Enterotoxin identification from contaminated food identified by ELISA or reverse passive latex agglutination tests.

In general

- *Keep hot foods above 60°C and cold foods below 4°C.*
- *Reheat cooked food to 74°C.*

Vibrio

Important species: *Vibrio parahaemolyticus, Vibrio vulnificus, Vibrio cholerae*

Why is it important?

V. parahaemolyticus and *V. cholerae* cause gastrointestinal illness while *V. vulnificus* causes non-intestinal illnesses such as septicemia. Important cause of gastroenteritis where large amounts of undercooked or raw seafood is consumed. Toxigenic *V. cholerae* O1 and O139 cause water- and food-borne cholera outbreaks with epidemic and pandemic potential.

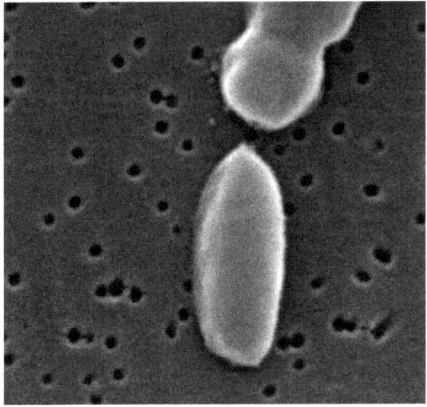

Sources

Marine, coastal and estuarine waters and marine organisms. Pathogenic vibrios sometimes found in fresh water reaches of estuaries.

Foods at risk

Fish and seafood. Can be a big problem when seafood products are eaten raw (sushi, sashimi). *V. cholerae* in particular is associated with inadequate sanitary conditions and faecal contamination of water and foods.

Characteristics and growth conditions

- Gram-negative curved rods.
- Facultatively anaerobic.
- Non-spore forming.
- Curved rod shape.
- Motile.
- Mildly halophilic (except *V. cholerae*—does not require presence of salt to grow).
- *V. cholerae* produces cholera toxin in the small intestine.

Growth	*V. parahaemolyticus*	*V. cholerae*	*V. vulnificus*
T	5–45°C (37°C)	10–43°C (37°C)	8–43°C (37°C)
pH	4.8–11 (7.8–8.6)	5–9.6 (7.6)	5–10 (7.8)
a_w	> 0.94 (0.98)	> 0.97 (0.984)	> 0.96 (0.98)
NaCl	0.5–10% (3%)	0.1–4%	0.5–5% (2.5%)

Transmission: waterborne infection, ingestion of food that was in contact with contaminated water.

Illness

V. cholerae: Cholera, caused by serotypes O1 and O139 which produce a toxin in the small intestine, occurs within 12–72 hours of ingesting cells. Symptoms include initially mild diarrhea progressing to "rice water" stools, nausea, abdominal cramps and low blood pressure. Dehydration, if not treated, may cause death. Otherwise recovery occurs within 1 week. Children up to 9 years in endemic countries are most at risk. More prevalent in the developing world. There have been up to 8 cholera pandemics.

Non-O1/O139 types (do not produce toxin) can also cause a milder illness with diarrhea, cramps and fever. Everyone seems to be susceptible.

V. parahaemolyticus: Symptoms occur 4–72 hours after ingestion and include abdominal cramps, watery diarrhea, nausea, vomiting and fever. It can last for up to a week but is usually self-limiting. Produces a haemolysin and shiga-like toxin. Sequelae such as reactive arthritis have been reported. All people appear to be susceptible.

V. vulnificus: Illness can occur within 12 hours to 3 days after consumption. Illness can take the form of gastroenteritis (vomiting, diarrhea, abdominal pain) or more often primary septicemia. Initial symptoms include fever, malaise and chills and death can occur rapidly in 50% of cases. There is a high fatality rate. The "at risk" people are those with impaired liver function and those who are immunosuppressed.

Dose

V. cholera: Around 106 cells when ingested in food by healthy adults. Lower numbers will cause illness in those taking antacids.

V. parahaemolyticus: 2×10^5 to 10^7 cells required to cause illness in healthy adults but also may be lower in those taking antacids.

V. vulnificus: Dose for healthy adults unknown but could be less than 100 for "at risk" people.

Issues relating to control

V. cholerae (Survival)

- Survives better under refrigeration than ambient temperature and can survive freezing.
- Variation in survival during refrigerated and frozen storage (up to 3 weeks at 7°C and up to 5 months of frozen storage).
- Highly sensitive to acidic environments and desiccation.
- *V. cholerae* has been shown to undergo a transition into a "Viable but non-culturable" state.

V. cholerae (Prevention)

- D_{60} = 2.7 min D_{71} = 0.30 min.
- Rapidly inactivated at pH values < 4.5 at room temperature.
- Freshly squeezed lemon juice inactivates organism after 5 minutes exposure.
- Depuration is not a reliable way to remove vibrios from shellfish.
- Sensitive to disinfectants routinely used in the food industry.
- Sensitive to irradiation.

V. parahaemolyticus (Survival)

- Can survive refrigeration temperatures for several weeks and freezing for up to 7 weeks.
- Normally starts to die off at < 5°C.

V. parahaemolyticus (Inactivation)

- Sensitive to heating $D_{65°C}$ = < 1 min and $D_{55°C}$ = 2.5 min.
- A low temperature pasteurization (50°C) for 10 min is effective in inactivating the organism in oysters (outbreaks are often associated with oysters).
- Very sensitive to drying.
- Fresh water inactivates the organism.
- Growth inhibited by 0/1% acetic acid (pH 5.1).

- Very sensitive to preservatives.
- Very sensitive to chlorine and iodophors.
- Depuration is not effective for removal from shellfish.

V. vulnificus (Survival)

- More sensitive to environmental conditions than other *Vibrio* spp.
- Can survive well under refrigeration by entering "Viable but non culturable" (VBNC) state.
- Survives quite well in oysters at 0–4°C.
- Survives to pH 5 at low salinity.

V. vulnificus (Inactivation)

- Inactivated by mild heat treatment and low pH. $D_{45°C}$ = 50 min and $D_{51°C}$ = 10 min.
- A low temperature pasteurization (50°C) for 10 min is effective in inactivating the organism in oysters.
- Freezing reduces the organism in oysters by 95–99% but survivors remain fairly stable during frozen storage.
- Inactivation at low pH and is slower at lower temperatures.
- Inhibited by some but not all preservatives. Depuration is ineffective.
- Relaying shellfish to higher salinity environments can reduce pathogen numbers.
- Irradiation is effective and the irradiation dose can be reduced if used at higher temperatures, e.g., increasing from 25 to 40°C means that approximately half the dose has the same killing effect.

Testing: TCBS agar incubated for 18–24 hour at 35°C aerobically produces yellow colonies.

In general

- *Keep hot foods above 60°C and cold foods below 4°C.*

Yersinia

Important species

Yersinia enterocolitica. There are two other pathogenic species *Y. pseudotuberculosis* and *Y. pestis* (the causative agent of "the plague") but these are not foodborne.

Why is it important?

It causes yersiniosis which is frequently characterized by gastroenteritis. *Yersinia* infections mimic appendicitis and is often a major "complication" in the performance of unnecessary appendectomies. Yersinia is capable of growing at refrigeration temperatures.

Sources: Intestinal tracts of many mammalian species and birds. Soil, vegetation, lakes, rivers, wells, streams.

Foods at risk

Meats (pork, beef, lamb, etc.), oysters, fish, and raw milk. Mostly associated with pig and pig meat products. Poor sanitation and improper sterilization techniques by food handlers, including improper storage, can contribute to contamination of food.

Characteristics and growth conditions

- Gram-negative rod shaped bacterium.
- Aerobe but capable of growing anaerobically.
- Non-spore forming.
- Non-toxin producing.
- Growth temperature 0–44°C (optimum 28–29°C).
- Growth pH 4.2–10 (7.2–7.4).
- A_w: 0.945 (5% NaCl).
- Not all strains are pathogenic.
- Geographical variation of pathogenic strains worldwide.
- A good competitor with other bacteria.
- Tolerate 6% of salt.

Transmission: Person-to-person transmission, by ingestion of contaminated food and waterborne transmission.

Illness

Occurs between 1 and 11 days after ingestion. Symptoms include abdominal pain (sometimes confused with appendicitis), headache, fever, diarrhea, nausea and vomiting. The disease is self-limiting with recovery within about 1 week. In rare cases this bacterium can be transmitted from person to person. It usually affects the very young and very old as well as those who are immunocompromised and debilitated. In some cases enterocolitica may persist for several months.

Dose: Unknown.

Issues relating to control

Survival

- Can readily withstand freezing and refrigeration.
- Survived for 64 weeks in water stored at 4°C.
- At low pH (below pH value for growth) survival is better at lower temperatures.
- Survives well in water and moist soil but declines as soil dries out.

Prevention

- Easily inactivated by heat. D value at 60°C = ~30 sec at 65°C D value = 2 sec or so of pasteurization is an effective heat treatment.
- Impact of pH varies according to acidulant.
- Susceptible to drying.
- 5–7% NaCl inhibits growth.
- Its presence often indicates a failure in the processing system.
- Preservatives retard growth to varying degrees, e.g., sodium nitrite is more effective than sodium nitrate.
- It is thought to be somewhat resistant to chlorine.
- Sensitive to radiation and commercial UV water radiation systems are considered to be effective.

Testing

CIN agar incubated at 28°C for 24 hours produces dark-red centre surrounded by a transparent border colonies.

In general

- *Keep hot foods above 60°C and cold foods below 4°C.*
- *Reheat cooked food to 74°C.*

References and Resources

Book

Basic Food Microbiology. 2nd Edition. Banwart, G.J. 1989. Chapman and Hall, New York.

Food Microbiology: Fundamentals and Frontiers. 1997. Doyle, M.P., Beuchat, L.R. and Montville, T.J. eds. ASM Press, Washington DC.

Foodborne Pathogens: Microbiology and Molecular biology. 2005. Pina M Fratamico, Arun K. Bhunia and James L. Smith eds. Caister Academic Press, UK.

Microbial Hazard Identification in Fresh fruits and vegetables. 2006. Jennylynd James ed. Wiley-interscience, a John Wiley and Sons, Inc., Publication.

Microbiology of fresh produce-emerging issues in food safety. 2006. Karl R. Matthews ed. ASM Press, Washington DC.

Viruses in Food. 2006. Sagar M. Goyal ed. Springer Science business media LLC, New York, USA.

Principles and Practices for the safe Processing of Food. 1998. Shapton, D. A. and Shapton, N. F. Woodhead Publishing Limited Cambridge.

Listeria, Listeriosis and Food Safety, 2nd Edition, Ryser, E.T. and Marth, W.H. 1999. Marcel Dekker Inc. New York.

Risk assessments of *Salmonella* in eggs and broiler chickens. 2002. FAO/WHO. Microbiological Risk Assessment Series 2.

Risk assessment of *Listeria monocytogenes* in ready-to-eat foods: Technical Report, 2004. FAO/WHO. Microbiological Risk Assessment Series 4.

Vibrio cholerae and cholera: Molecular to Global Perspectives. 1994. Wachsmuth, I.K., Blake, P.E. and Olsevik, O. ASM Press, Washington DC.

Review article

Marion Koopmans and Erwin Duizer. 2004. Foodborne viruses: an emerging problem. International Journal of Food Microbiology, 90: 23–41.

FAO. 2008. Viruses in Food: Scientific advice to support risk management activities. Microbiological Risk Assessment Series. 13, 2008.

Foodborne Parasites-A review of the scientific literature by Ellin Doyle, Food Research Institute, UW-Madison, October 2003.

Marion Koopmans and Erwin Duizer. 2004. Foodborne viruses: an emerging problem. International Journal of Food Microbiology, 90: 23–41.

FAO. 2008. Viruses in Food: Scientific advice to support risk management activities. Microbiological Risk Assessment Series. 13, 2008.

Foodborne Parasites-A review of the scientific literature by Ellin Doyle, Food Research Institute, UW-Madison, October 2003.

Weblink: (All Accessed October 2014)

Todar's Online Textbook of Bacteriology

http://www.textbookofbacteriology.net/

U.S. Food & Drug Administration, Center for Food Safety & Applied Nutrition Foodborne Pathogenic Microorganisms and Natural Toxins Handbook. The *"Bad Bug Book"*

http://vm.cfsan.fda.gov/~mow/intro.html

CDC disease information

http://www.cdc.gov/ncidod/dbmd/diseaseinfo/

Infectious diseases; eMedicine Instant access to the minds of Medicine

http://www.emedicine.com/med/INFECTIOUS_DISEASES.htm

Microbial Pathogen Data Sheets, New Zealand food Safety Authority. 2001.

http://www.nzfsa.govt.nz/science-technology/data-sheets/

Aeromonas

http://www.who.int/water_sanitation_health/dwq/en/admicrob2.pdf

Foodborne Outbreak of Group A Rotavirus Gastroenteritis Among College Students District of Columbia, March-April 2000. MMWR *December 22, 2000/Vol.49/RR-16.*

http://www.cdc.gov/od/oc/media/mmwrnews/n2k1222.htm#mmwr2

CDC—Norovirus: Technical Fact Sheet www.cdc.gov/ncidod/dvrd/revb/gastro/norovirus-factsheet.htm

4

Survival, Injury and Inactivation of Human Bacterial Pathogens in Foods: Effect of Non-Thermal Treatments#

Dike O. Ukuku,[1,]* *Md. Latiful Bari,*[2] *Lamin S. Kassama,*[3]
Sudarsan Mukhopadhyay[1] *and Modesto Olanya*[1]

Introduction

For health reasons, people are consuming foods with or without minimal processing, thereby, exposing themselves to the risk of foodborne illness if such foods are contaminated with bacteria. Food manufacturers and researchers are responding to consumers' demand for fresh food that are safe by proposing and developing non-thermal process intervention treatments that can keep food fresh or near fresh without altering the sensorial characteristic. In this chapter, the efficacy of *high hydrostatic pressure,* antimicrobials, radio frequency electric field and pulse electric field, pulsed ultraviolet light, and ultrasonic *treatment on* bacterial injury, leakage of intracellular

[1] Eastern Regional Research Center, Agricultural Research Service, U.S. Department of Agriculture, 600 E. Mermaid Lane, Wyndmoor, PA 19038, USA.
[2] Center for Advanced Research in Sciences, University of Dhaka, Dhaka-1000, Bangladesh.
[3] Department of Food and Animal Sciences, Alabama A & M University, P.O. Box 1628, Normal, Alabama 35762.
* Corresponding author: dike.ukuku@ars.usda.gov

Mention of trade names or commercial products in this article is solely for the purpose of providing specific information and does not imply recommendation or endorsement by the U.S. Department of Agriculture. "USDA is an equal opportunity provider and employer".

bacterial substances leading to inactivation is evaluated. The information presented in this chapter will expand the knowledge base regarding processing variables associated with these non-thermal technologies in achieving optimum efficacy for bacterial inactivation in food while maintaining quality attributes of heat sensitive foods.

Non-thermal Treatments

Thermal processing is among the oldest method of food preservation and this process includes pasteurization, sterilization and blanching. Thermal processing, although effective in microbial reduction, often results in significant degradation in product quality (Besser et al., 1993; Linton et al., 1999). These concerns have motivated research efforts in the application of non-thermal processing techniques to ensure microbial food safety while preserving the natural flavor, taste and heat labile nutritional components. For health reasons, people are consuming foods with or without minimal processing, thereby, exposing themselves to the risk of foodborne illness if such foods are contaminated with bacteria. Contamination of juices with pathogenic microorganisms has caused numerous illnesses and some fatalities. From 1923 to 2000, consumption of contaminated fruit juices has been implicated in at least 28 foodborne illness outbreaks (FAO, 2001). Eleven out of 28 (almost 40%) outbreaks were associated with *Salmonella* spp. Eight out of 28 (close to 30%) outbreaks were caused by enteropathogenic *Escherichia coli* especially *E. coli* O157:H7. Although most bacteria cannot grow at low pH, *E. coli*, *Shigella* and *Salmonella* species can survive for several days or weeks in acidic foods (Conner and Kotrola, 1995; Garcia-Graells et al., 1998; Leyer et al., 1995; Weagant et al., 1994). Foodborne outbreaks involving *E. coli* O157:H7 in apple and orange juices (CDC, 1996a; CDC, 1996b; CDC, 1999; Cody et al., 1999) have raised concerns about the safety of consuming unpasteurized fruit juices. Alternative, non-thermal processes must be developed in order to make further improvements on microbial reductions in fruit juices (Mazzotta, 2001).

Food manufacturers and researchers are responding to consumers' demand for fresh food that are safe by proposing and developing non-thermal process intervention treatments that can keep food fresh or near fresh without altering the sensorial characteristics. Non-thermal food processing techniques are applied to control microbial proliferation in food. The ultimate goal is to ensure food safety while satisfying the organoleptic and nutritional requirement of the consumer. Non-thermal treatment for inactivation of bacteria in food systems includes but is not limited to the use of any processing treatments involving no or minimal heat, pressure and antimicrobial agents, and or any combination thereof. Ultrasound (US) and pulsed ultraviolet light (PUV) technologies are key technologies which have generated significant research interest in its applications to develop minimal processing that ensures food quality and safety (Ercan and Soysal, 2013; Ulusoy et al., 2007). Ultrasound technology could be applied to solid, liquids and gaseous systems for different purposes and its application in food processing for microbial destruction and enzymatic control in the past decade is increasing (Ercan and Soysal, 2013). Earnshaw et al. (1995) reported the use of ultrasound to have reduced energy, flavor and greater homogeneity in foods. Fonteles et al. (2012) reported total inactivation of peroxidase, polyphenol oxidase and ascorbate peroxidase activities in cantaloupe melon juices. Zenker et al. (2003) reported a 6 log reduction of *Escherichia coli* in fruit juices and milk and inactivation of *Pseudomonas fluorescens* and *Streptococcus thermophilus* in trypticase-soy broth and total bacteria in milk by continuous-flow ultrasonic treatment (Villamiel and De Jong, 2000; Vercet and Burgos, 1997).

Ultrasonic applications in foods can be classified into two categories depending on their applications in food process. Direct and indirect applications exist in food processing applications. The direct application is when ultrasound is directed towards microbial inactivation while surface cleaning is used for indirect applications. It is effective in surface decontamination where incoming fluid accompanying cavitation collapse near a surface and the crevices, thus provide mechanical potential for removing attached or entrapped bacteria in localized spots. One mode of generating ultrasounds is by electrostrictive transformers that cause elastic deformation of ferroelectric materials within a high frequency electrical field, thus mutual attraction of the molecules are polarized in the field (Raichel, 2000). Hence, high frequency alternating currents are transmitted via two ferroelectric electrodes to cause polarization of molecules. In a nutshell, electrical energy is converted into mechanical oscillation of sound waves through an amplifier to a sonotrode that generates vibrational energy (Knorr et al., 2004). When ultrasound energy is passed through a liquid medium it generates continuous motion of longitudinal waves. These waveforms create alternative compression and rarefaction (decompression) of particles in the medium, a phenomenon described as "cavitation" (Povey and Mason, 1998). The biochemical effect of ultrasound is the direct result from acoustic cavitation phenomenon, which is the formation, growth, and implosive collapse of gas bubbles in liquids, as shown in Fig. 1, that henceforth releases large amounts of high localized energy (Suslick, 1989). Furthermore, water molecules are broken to form free radicals thus intensifying chemical reactions, induce crosslinking of proteins (Cavalieri et al., 2008). The reaction generates local turbulence and liquid micro-circulation (acoustic streaming) (Gogate and Pandit, 2011).

Hydroxide radicals ($OH-$) and hydrogen atoms are generated as a result of dissociation of the water molecules in aqueous solutions due to the high temperature (5000 K) and pressure (1000 atm) of the collapsing gas bubbles associated with cavitation (sonolysis) (Riesz and Kondo, 1992). There are evidences of the formation of free radicals by cavitation in non-aqueous solutions and polymers. However, the application of cavitation is very effective in aqueous media compared to organic media. These mechanisms induce physicochemical effects with several potential applications in the food industry hence reveals the significance of ultrasound technology.

Other non-thermal processing technologies have been developed and some commercialized ones (FDA, 2002; Tewari et al., 1999; Geveke and Brunkhorst, 2002, 2005, 2004a,b; Ukuku et al., 2008; Barbosa-Canovas et al., 1998), including

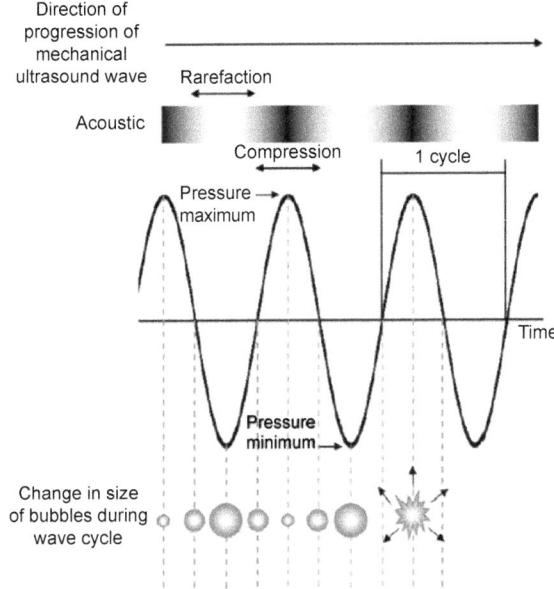

Figure 1. Ultrasonic bubble formation and cavitation (Soria and Villamiel, 2010).

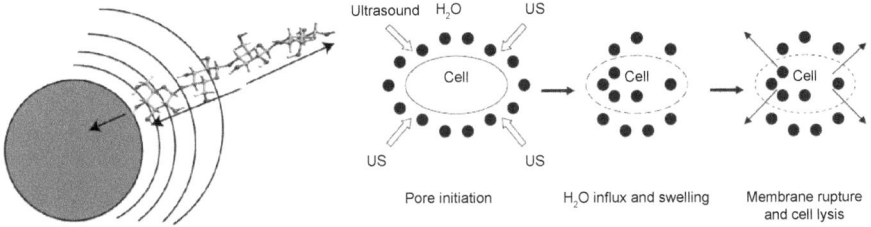

Figure 2. Degradation of biological cell during ultrasonic treatment (Chemat et al., 2011).

pulsed electric fields (PEF) (Qin et al., 1995; Selma et al., 2006). The use of PUV to inactivate vegetative cells and spores on the surface of foods, food contacting surfaces and packaging materials was developed by a US company in the 1980s (Fellow, 2009). The application of PUV can also be used during the process of fresh produces designated for fresh-cut preparations. It uses intense and short duration pulses of broad spectrum white light. The PUV spectrum involves wavelength in the UV and the near infrared region. Application of PUV treatment was approved by the Food and Drug Administration of the United States (FDA) for its use in treating food during production, processing and handling of food (Federal Register, 1999). The microbial safety of PEF and radio frequency electric field (RFEF) treated juices such as apple juice, orange juice, tomato, watermelon and melon juices has been investigated during refrigerated storage (Min et al., 2003; Mosqueda-Melgar et al., 2008; Ukuku et al., 2010; Ukuku et al., 2011). Similarly, the use of high-hydrostatic pressure (HHP) as an alternative to heat pasteurization for inactivation of bacteria (Hoover, 1993; Ukuku et al., 2009) and its ability to inactivate enzymes responsible for quality loss (Knorr, 1993) in food has been reported. The effect of HHP on the survival of bacterial populations in liquid foods or liquid buffered systems has been reported (Bari et al., 2007; Koseke and Yamamoto, 2006; Koseke and Yamamoto, 2006). In these reports, inactivation and survival/recovery of *E. coli* populations were discussed but not the details of bacterial injury that led to inactivation.

Use of biocontrol strains or biocontrol microbes for suppression or inactivation of foodborne pathogens have been documented previously, with various degrees of success. For example, the use of community and single strain inoculants have been shown to be effective for biocontrol of *Salmonella* on alfalfa sprouts (Matos and Garland, 2006). In this research, the authors demonstrated that *Salmonella* populations were reduced by approximately 2 logs CFU/g of produce. Similarly, Olanya et al. (2013) documented the biocontrol potential of *Pseudomonas fluorescens* against *Escherichia coli* O157:H7 on baby spinach. In this study, reductions of *E. coli* O157:H7 in the range of 0.5–1.5 log CFU/g of produce were achieved. Similarly, suppressive activities of resting populations of *E. coli* O157:H7 in broths when co-cultured with *P. fluorescens* strains were also documented (Olanya et al., 2014). In other studies, bell pepper disks were treated with *P. fluorescens* strain 2–79 or a strain of *Bacillus* sp. and subjected to microbial control. The reductions in the growth of *E. coli* O157:H7 in the range of 1 to 4 logs CFU/g for produce were recorded (Liao, 2009). In addition to microbial inoculants, phage-based biocontrol of foodborne pathogens have also assumed considerable importance (Goodrigde and Bisha, 2011). This has been applied in both animal systems where phages may be applied in chicken and animals reared for meat as well as in produce (Toro et al., 2005). In this case, phage therapy is used prior to the entry of chicken and meat into the processing plant as well as directly in post-harvest situations (post-processed food). Phages have shown considerable advantages not possessed by microbial inoculants. These include safety of phages, host-specificity to certain foodborne bacteria, its ability to replicate in the host-bacteria and lack of resistance development. In general, the utility of phages for biocontrol of foodborne pathogens have been greater on produce that in animals systems, perhaps due to the ease of applications on produce. It should be pointed out that both microbial inoculants and phage-based therapy are by no means confined to a single biocontrol or antagonistic microbe. More often than not, a cocktail of biocontrol agents are mixed or combined to achieve a wide array of efficacy or a broad spectrum of effectiveness (Toro et al., 2005; Sharma et al., 2009). In other cases, combinations of bacteriocins, bacteriophages and competitive bacteria may also be used simultaneously (Dykes and Morehead, 2002).

Microbial inactivation of foodborne pathogens based on biological control (biocontrol) principles and approaches may be included under the broad umbrella of antimicrobials. This may specifically refer to the suppression, reduction or inactivation of human pathogens or any organisms with another usually in the same ecological niché or closely related niché (Olanya et al., 2013). Antimicrobials may be broadly defined as microbial-derived biological compounds that have the capacity to inactivate foodborne pathogens when present in food components or matrices (Lacroix, 2011). This may be based on its application in food during pre-processing and post-processing stages or its production *in vivo* by microbial agents. Biocontrol approach implies the utilization of live organisms or its components for pathogen suppression, while antimicrobials may refer to microbial products or compounds derived from its metabolism. Antimicrobials may consist of bacteriocins, biocontrol microbes (competitive bacteria, antagonists, bacteriophages and predatory bacteria), food-grade enzymes and protective cultures of certain bacteria. While some of the antimicrobials (pathogen derivatives) are used primarily for food preservation and self-life extension (Holzapfet et al., 1995), others may be applied at pre-harvest or post-harvest to primarily inactivate human pathogens in food (Cleveland et al., 2001; Galvez et al., 2007; Garcia et al., 2010). It should be mentioned that some antimicrobial compounds may be strictly chemical in nature and known as "Generally Recognized as Safe" (GRAS). These types of antimicrobials (GRAS) have been approved for use in the food industry by the US Food and Drug Administration (FDA). The GRAS antimicrobials may consist of compounds such as Lauric-Arginate-Ester (LAE), potassium lactate (PL), sodium diacetate (SDA) and others (Sommers et al., 2010). In other studies, potassium lactate and sodium diacetate in combination with irradiation has been used to control *L. monocytogenes* on frankfurters (Knight et al., 2007). Also, several reports are documented on the use of short chain organic acids as antimicrobial agents in reducing the population of human bacterial pathogens in foods (Mass et al., 1989; Bacus and Bontenbal, 1991; Papadopoulos et al., 1991a,b; Shelef and Yang, 1991; Shelef, 1994; Ukuku et al., 2009). In many other instances, the antimicrobials are applied in food products to enhance food quality (aroma, flavor, color). In many cases, the antimicrobials are preferred due to other beneficial properties such as non-toxic to the consumers, effectiveness of pathogenic bacteria at low doses or concentrations, and they do not have adverse effects on sensory or organoleptic properties of foods.

Effect on Bacterial Cell Surfaces and Membrane Damage

As opposed to bacteriocins and endolysins, the mechanism of activity of microbial antagonists or competitive bacteria for the suppression or inactivation of foodborne pathogens is based primarily on competitive exclusion in which the biocontrol microbes usually outcompete or suppress the foodborne pathogens for nutrients and space. In other cases, increase in populations of microbial antagonists and competition for attachment sites are of importance. In the case of phages, its primary mechanism of activity on foodborne pathogens is its bacteriocidal property. Phages often replicate in host-bacteria and may result in either lysogeny or lytic processes resulting in the destruction of the host (foodborne) bacteria (Hudson et al., 2005). In general, the lytic phages have better microbial control due to its integration into the genome of the hostbacteria upon attachment to specific bacteria cell wall. Once the genome integration occurs in the host bacteria and it is expressed, new phage particles will be produced in bacterial cells and subsequently result in the rupture of bacterial cell walls based on the activity of phage-encoded enzymes. In contrast to bacteriocins, biocontrol microbes as components of antimicrobial agents for inactivation of human pathogens offer potentially different disadvantages.

Current knowledge of the mechanisms of bacterial inactivation by most of these non-thermal processes is limited. Several studies have reported that inactivation of bacteria by PEF is through cell membrane permeabilization and injury after a critical value is achieved (Babosa-Canovas et al., 2007; Aronsson et al., 2005; Lebovka and Vorobiev, 2004; Zimmerman et al., 1976; Zimmerman, 1986). More information on the exact mechanism of bacteria inactivation by these technologies is needed. The reason being that, liquid food matrix, such as vegetable juices, fruit juices, liquid egg, sauces, beer, wine and soups may require a particular technology that works best for it (Ukuku et al., 2010, 2011). For example, there is no information on the effect of PEF on bacterial cell surface charge or hydrophobicity. Bacterial cell surfaces have net negative charge due to the presence of ionized phosphoryl and carboxylate moieties on the outer envelope exposed to the extracellular environment. Therefore, if the membrane structure of the bacteria is damaged due to electrostatic charge separation in the cell membrane, the injured bacteria may lose its biological activities leading to its death.

The mode of microbial inactivation occurs via a combination of photochemical and photothermal mechanisms. When microbes are exposed to PUV, the UV-C provides greater photochemical effect that induces chemical modification of the proteins and other cellular materials and the DNA (Fig. 3). As a result of high absorption of the UV-C irradiation crosslinking occurs between the pyrimidine nucleoside based in

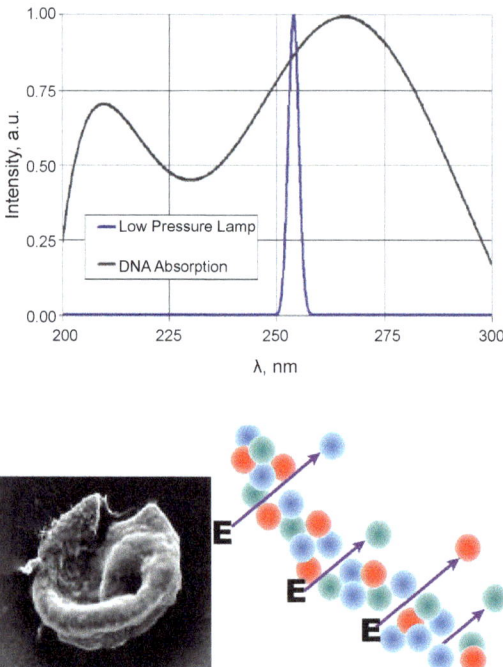

Figure 3. Interactive effects of PUV spectrum and microbial DNA (Macdonald et al., 2000, Plasma Science, IEEE Transactions on, 28(5): 1581–1587; Wang et al., 2005, Water research, 39(13): 2921–2925).

DNA, thus disrupting cellular metabolism (Fig. 3). Hence, microbial reduction is attributed to the formation of irreversible damage to DNA (Macdonald et al., 2000; Wang et al., 2005).

The energy in the visual spectrum, contributes to the photothermal effects, due to the large amount of energy transferred rapidly to the exposed surface. The effective temperature of a pulsed discharged in a flash lamp reaches 10,000 to 20,000°K (www.steribeam.com) and the instantaneous temperature change contributes to the destruction of vegetative cells. The bactericidal effect was found to be attributed to the instantaneous overheating which causes cell disruption primarily because of the heat and the difference in the UV absorption of bacteria and surrounding media (Takeshita et al., 2003).

Similarly, Kassama et al. (2005) studied the effect of energy per pulse on the inactivation of *Salmonella enteritidis* in poultry chiller water (Fig. 4). They applied frequency (0.1, 1.0 and 10 Hz) and voltage (500, 800 and 1000 V) in their treatments, hence over 6 log reduction was significantly ($p < 0.5$) achieved for all treatments except for T_1 (0.1 Hz and 500 V) as shown in Fig. 4. The effect of frequency seems to be independent of the discharged voltage for this particular application. For example, intensive heat generation especially at higher frequency (10 Hz) and discharged voltage (1000 V) treatment, tremendous synergy between photothermal and photochemical effect, contributed to the efficacy of the system.

Ukuku et al. (2007) investigated the effect of radio frequency electric field on surface membrane of *E. coli* bacteria using scanning electron microscopy (SEM).

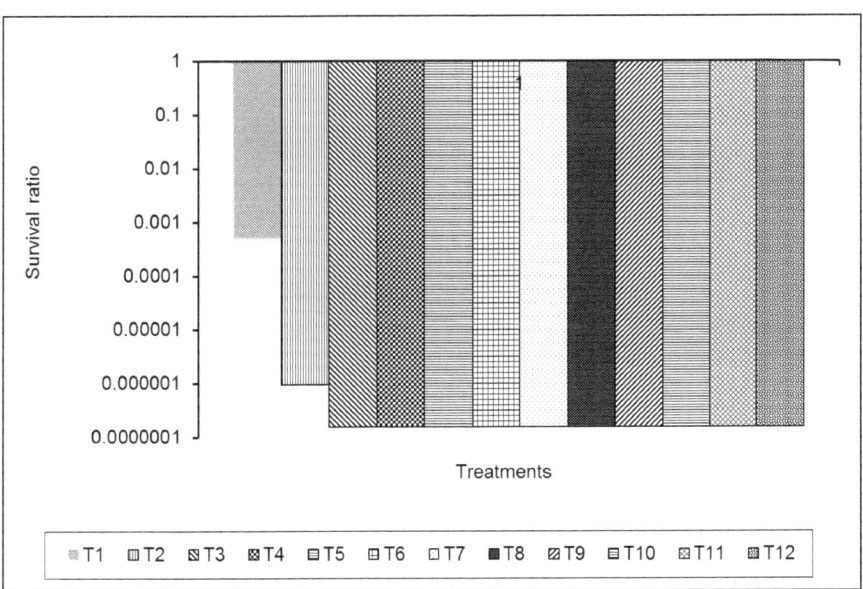

Figure 4. The effect of PUV treatment of *Salmonella enteritidis* in poultry chiller water. Where T1: 0.1 Hz x 500 V, T2: 0.1 Hz x 600 V, T3: 0.1 Hz x 800 V, T4: 0. 1 Hz x 1000 V, T5: 1 Hz x 500 V, T6: 1 Hz x 600 V, T7: 1 Hz x 800 V, T8: 1 Hz x 1000 V, T9: 10 Hz x 500 V, T10: 10 Hz x 600 V, T11: 10 Hz x 800 V, T12: 10 Hz x 1000 V (*Source:* Kassama Ngadi and Smith, 2005).

The authors clearly showed differences on treated bacterial cell surfaces as compared to untreated cells (Fig. 5). The authors reported that it is possible that the disruption of the surface structure of *E. coli* in this study was due to charge-charge interactions between the bacterial negative charge and the energy charge produced by the RFEF. Similarly, the gross effects of electrical interaction with biological cells are well known and a macroscopic intravascular electrode maintained at a constant current intensity of 1 mA was found to induce thrombosis and injury on the vascular wall, ranging from minimal lesion of endothelium to almost total necrosis of the vascular wall (Weidenbach et al., 1978).

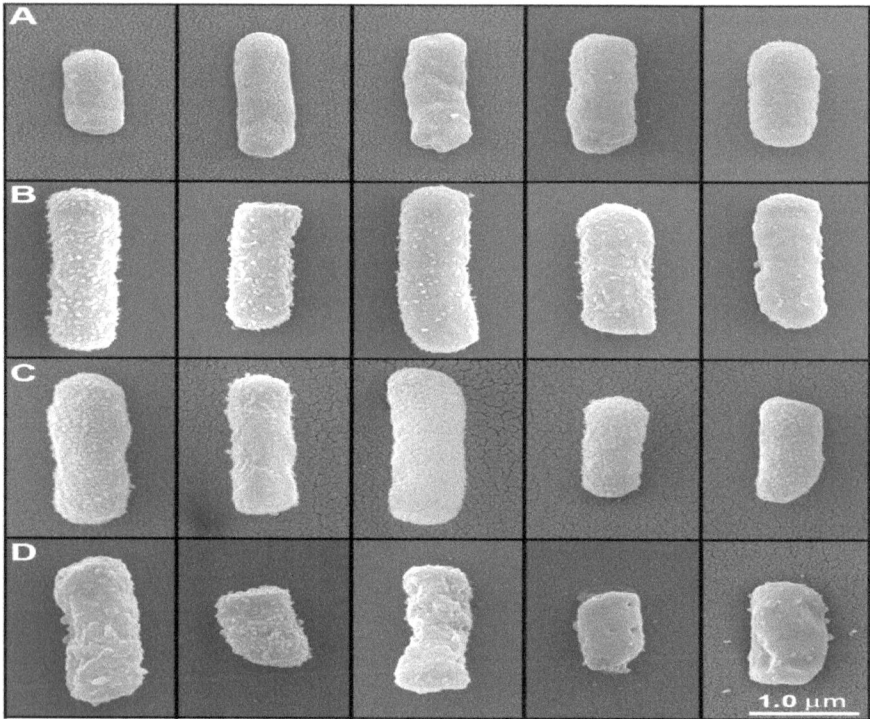

Figure 5. Surface structure of *Escherichia coli* K-12 (ATCC 23716) in apple juice treated with RFEF (30 kV/cm) at 23°C (A = control), 45°C (B), 50°C (C) and 55°C (D) as observed by SEM. Printed with permission. Ukuku et al., 2008. J. Food Prot. 2008. 71: 684–690.

In another study where the effect of RFEF and thermal treatment on internal membrane organelles of *E. coli* bacteria was investigated, the outer membranes of the regular, rod-shaped control *E. coli* bacteria were smooth; and the internal organelles densely dispersed (Fig. 6A).

Heat treatment alone at 75°C resulted in condensed and aggregated cytoplasm (Fig. 6B), but the outer membranes lacked the smooth regular, rod-shaped membrane that were typical of the control *E. coli* cells (Fig. 6A). The effect of treatment temperature at 75°C alone caused changes in the internal alignment of the cellular

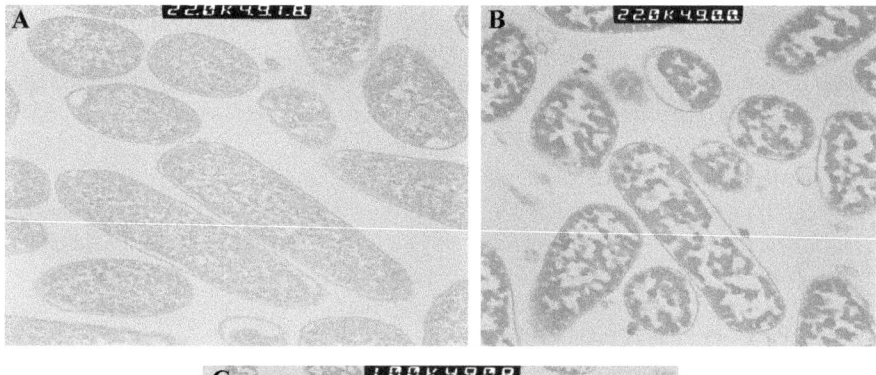

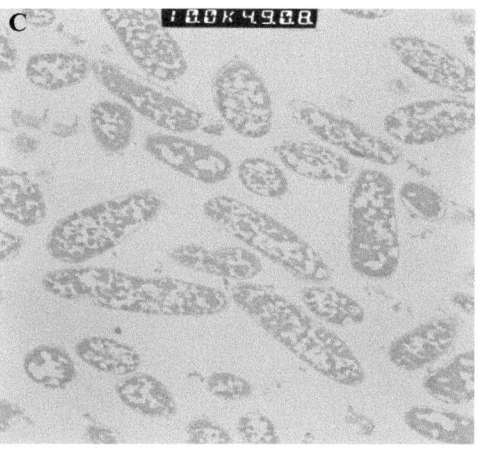

Figure 6. TEM observation of changes in RFEF (25 kV/cm) at 55°C and heat (75°C) treated *Escherichia coli* K-12 (ATCC 23716) in apple juice (A = control; B = 75°C and C = 55°C + RFEF, Bar = 0.5 µm). Values are means of three determinations ± standard deviation. Ukuku et al. (2012). J. Biochem. & Bioproces. 4(3): 76–81.

organs. The TEM observation of treated and untreated bacterial cells revealed an altered outer membrane structure; although distinctly trilaminar, the electron-density of the cytoplasm was markedly condensed and aggregated as compared to controls. The authors concluded that it is possible that the condensed and aggregated electron-density of the cytoplasm by the RFEF treatment led to the separation and disruption of cytoplasmic membranes and the formation of surface blebs or vice versa, reported earlier by Ukuku et al. (2008).

Several researchers have reported that bacterial inactivation of RFEF and PEF follows the same trend (Ukuku et al., 2010; Ukuku et al., 2010, 2011) including a combination treatment using UV-light (Ukuku and Geveke, 2010). Ukuku et al. (2010, 2011) used hydrophobic interaction chromatography (HIC) and electrostatic interaction chromatography (ESIC) to estimate changes on bacterial cell surface, net negative charge and hydrophobicity of *E. coli* cells exposed to sub-lethal and lethal PEF treatment. The electric field strength at all temperatures tested deformed the bacterial cell surface structure leading to leakage of intracellular biological materials. Also, the treatments clearly showed changes on both external cell surfaces and intracellular membranes of *E. coli* bacteria. It will be wise to state that during RFEF treatments, the outlet treatment temperatures (34.1°C, 45.9°C and 55.3°C) were higher than the inlet temperatures (Ukuku et al., 2012). Similar observations on increase in outlet treatment temperatures was noted during PEF treatments (Ukuku et al., 2012). Similarly, the membrane damage caused by the RFEF led to leakage of intracellular ATP and UV-absorbing materials of bacteria leading to the increased level of extracellular ATP (Fig. 7A) and UV-absorbing substances measured at 260 and 280 nm (Fig. 7B) in treated apple juice. The authors concluded that these changes at the bacterial cell surfaces led to injury/damage, causing efflux/leakage of intracellular ATP, protein, and/or nucleic acid of the bacteria, all of which affected the energy status and the enzymatic activity of the bacterial cell, leading to cell death (Ukuku et al., 2008).

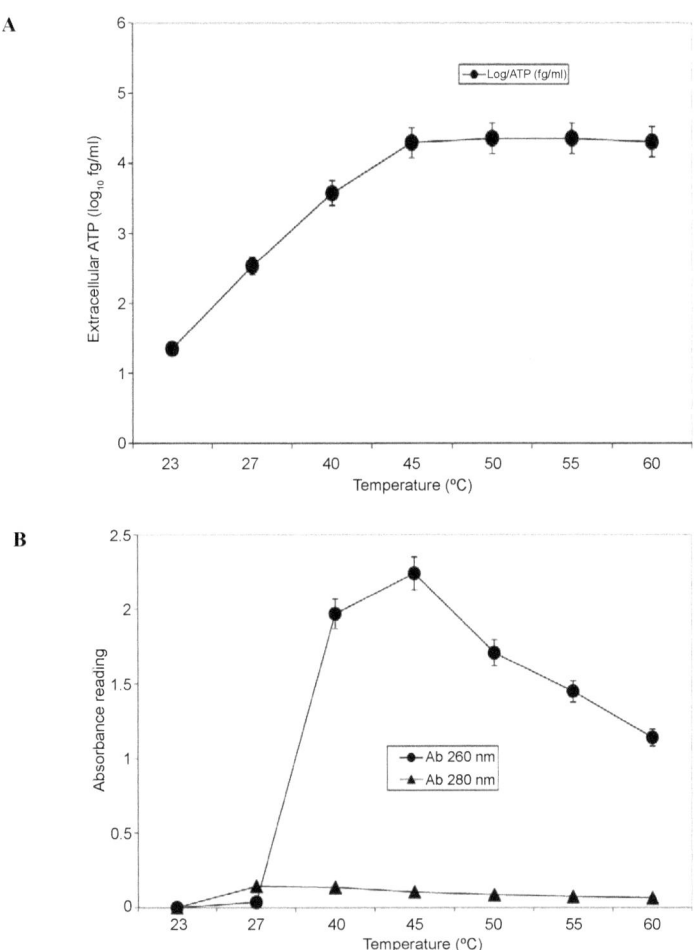

Figure 7. Leakage of intracellular UV-absorbing substances of *Escherichia coli* K-12 (ATCC 23716) in apple juice treated with RFEF (15 kV/cm) at 23, 27, 40, 45, 50 and 55°C. Values are means of three determinations ± standard deviations (With permission, Ukuku et al., 2008, J. Food Prot., Vol. 71, No. 4, pages 684–690).

In conclusion, all the non-thermal processing intervention technologies discussed here have shown promises on microbial inactivation in food without causing any adverse effect on the sensorial characteristics of the food system tested. Some have been commercialized while some are still under consideration due to specificity of bacterial inactivation or treatment cost associated with the technology. For example, one of the major constraints of biocontrol microbes in the lack of effectiveness and application technology for pathogen controls. In comparison to various other technologies for inactivation of foodborne pathogens, biocontrol microbes generally result in pathogen reductions of less than 2 logs CFU/g of produce (Olanya et al., 2013). In other cases, the production processes and culturing technology to produce these microbes are often cumbersome and costly. In other cases, the application technology to deliver biocontrol microbes onto surfaces of food produce such as leafy greens and fruits are also difficult. This results in less adequate coverage and ineffective concentrations of the biocontrol relative to foodborne bacteria. A much larger problem relates to perception of biocontrol microbes on the surfaces of produce or fruit being eaten by consumers. Use of short chain organic acid antimicrobials has been in existence for a long time especially those "generally regarded as safe (GRAS)". As demand for fresh foods by consumers continue to rise, the need to accommodate these demands will continue to encourage food manufacturers and researchers to explore more non-thermal processing intervention technologies that can improve the microbial safety of foods.

References

Aronsson, K., U. Ronner and E. Borch. 2005. Inactivation of *Escherichia coli*, *Listeria innocua* and *Saccharomyces cerevisiae* in relation to membrane permeabilization and subsequent leakage of intracellular compounds due to pulsed electric field processing. Inter. J. Food Microbiol., 99: 19–32.

Bacus, J. and E. Bontenbal. 1991. Controlling *Listeria*. Meat Poult., 37: 64–65.

Barbosa-Canovas, G.V., U.R. Pothakamury, E. Palou and B.G. Swanson. 1998. Nonthermal preservation of foods. New York: M Dekker Co., pp. 139–150.

Bari, M.L., D.O. Ukuku, M. Mori, S. Kawamoto and K. Yamamoto. 2007. Effect of hydrostatic pressure pulsing on the inactivation of *Salmonella enteritidis* in liquid whole egg. Foodborne Pathogens and Disease, 5: 175–182.

Besser, R.E., S.M. Lett, T. Webber, Jr., M.P. Doyle, T.J. Barrett, J.G. Wells and P.M. Griffin. 1993. An outbreak of diarrhea and hemolytic uremic syndrome from *Escherichia coli* O157:H7 in fresh-pressed apple cider. J. Amer. Med. Assoc., 169: 2217–2224.

Cavalieri, F., M. Ashokkumar, F. Grieser and F. Caruso. 2008. Ultrasonic synthesis of stable, functional lysozyme microbubbles. Langmuir, 24(18): 10078–10083.

Centers for Disease Control and Prevention (CDC). 1996a. Outbreaks of *Escherichia coli* O157:H7 infection and cryptosporidiosis associated with drinking unpasteurized apple cider—Connecticut and New York. Morbid. Mortal. Weekly Rep., 46: 4–8.

Centers for Disease Control and Prevention (CDC). 1996b. Outbreak of *Escherichia coli* O157:H7 infections associated with drinking unpasteurized commercial apple juice—British Columbia, California, Colorado, and Washington. Morb. Mortal. Wkly. Rep., 45: 975.

Centers for Disease Control and Prevention (CDC). 1999. Outbreak of *Salmonella* serotype Muenchen infections associated with unpasteurized orange juice—United States and Canada. Morb. Mortal. Wkly. Rep., 48: 582–585.

Chemat, F., Z.E. Huma and M.K. Khan. 2011. Applications of ultrasound in food technology: processing, preservation and extraction. Ultrasonics Sonochemistry, 18: 813–835.

Cleveland, J., T.J. Montville, I.F. Nes and M.L. Chikindas. 2001. Bacteriocins: safe, natural antimicrobials for food preservation. Inter. J. Food Microbiol., 71: 1–20.

Cody, S.H., M.K. Glynn, J.A. Farrar, K.L. Cairns, P.M. Grifin, J. Kobayashi, M. Fyfe, R. Hoffman, A.S. Arlene, J.H. Lewis, B. Swaminathan, R.G. Bryant and D.J. Vugia. 1999. An outbreak of *Escherichia coli* O157:H7 infection from unpasteurized commercial apple juice. Ann. Intern. Med. 130: 202–209.

Conner, D.E. and J.S. Kotrola. 1995. Growth and survival of *Escherichia coli* O157:H7 under acidic conditions. Appl. Environ. Microbiol., 61: 382–385.

Dykes, G.A. and S.M. Moorhead. 2002. Combined antimicrobial effect on nisin and a listeriophage against *Listeria monocytogenes* in broth but not in buffer or on raw beef. Inter. J. Food Microbiol. 73: 71–81.

Earnshaw, R.G., J. Appleyard and R.M. Hurst. 1995. Understanding physical inactivation processes: Combined preservation opportunities using heat, ultrasound and pressure. Inter. J. Food Microbiol., 28: 197–219.

Ercan, S.S. and C. Soysal. 2013. Use of ultrasound in food preservation. Natural Science, 5(8) A20: 5–13.

[FAO] Food and Agriculture Organization of the United Nations. 2001. Principles and practices of small and medium-scale of fruit juice processing. Chapter 16. Available from: http://www.fao.org/DOCREP/005/Y2515E00.HTM. Accessed September 15, 2014.

Federal Register. 1999. Pulsed light treatment of food. Federal Register, 66: 338829–338830.

FDA. 2002. Thermally-processed low-acid foods packaged in hermetically sealed containers. Federal Register, 21CFR113. US Food and Drug Administration, Rockville, MD.

Fellow, P.J. 2009. Food Processing Technology: Principles and Practice, 3rd Edn., New York: CRC Press.

Fonteles, V.T., M.G.M. Costa, A.L.T. de Jesus, M.R.A. de Miranda, F.A.N. Fernandes and S. Rodrigues. 2012. Power ultrasound processing of cantaloupe melon juice: effects on quality parameters. Food Res. Inter., 48: 41–48.

Galvez, A., H. Abriouel, R.L. Lopez and N.B. Omar. 2007. Bacteriocin-based strategies for food biopreservation. Inter. J. Food Microbiol., 120: 51–70.

Garcia, G., N. Gomez, P. Manas, J. Raso and R. Pagan. 2007. Pulsed electric cause bacterial envelopes permeabilization depending on the treatment intensity, the treatment medium pH and the microorganism investigated. Inter. J. Food Microbiol., 113: 219–227.

Garcia-Graells, C., K.J.A. Hauben and C.W. Michiels. 1998. High-pressure inactivation and sublethal injury of pressure-resistant *Escherichia coli* mutants in fruit juices. Appl. Environ. Microbiol., 64: 1566–1568.

Geveke, David. 2005. UV inactivation of bacteria in apple cider. J. Food Prot., 68: 1739–1742.

Geveke, D.J. and C. Brunkhorst. 2004. Inactivation of *Escherichia coli* in apple juice by radio frequency electric fields. J. Food Sci., 69: 134–138.

Geveke, D.J. and C. Brunkhorst. 2003. Inactivation of *Saccharomyces cerevisiae* with radio frequency Electric Fields. J. Food Prot., 66: 1712–1715.

Geveke, D.J. and C. Brunkhorst. 2004. RFEF pilot plant for inactivation of *Escherichia coli* in apple juice. Fruit Proc., 14: 166–170.

Gogate, P.R. and A.B. Pandit. 2011. Sonocrystallization and its application in food and bioprocessing. pp. 467–493. *In*: H. Feng, G. Barbosa-Canovas and J. Weiss (eds.). Ultrasound Technologies for Food and Bioprocessing. New York: Springer.

Goodridge, L.D. and B. Bisha. 2011. Phage-based biocontrol strategies to reduce foodborne pathogens in foods. Bacteriophage, 1: 130–137; http://dx.doi.org/10.4161/bact.1.3.17629.

Holzapfel, W.H., R. Geisen and U. Schillinger. 1995. Biological preservation of foods with reference to protective cultures, bacteriocins and food-grade enzymes. Inter. J. Food Microbiol., 24: 343–362.

Hoover, D.G. 1993. Pressure effects on biological systems. Food Technol., 47: 150–155.

Hudson, J.A., C. Billington, G. Carey-Smith and G. Greening. 2005. Bacteriophages as biocontrol agents in food. J. Food Prot., 68: 426–437.

Knight, T.D., A. Castillo, J. Maxim, J.T. Keeton and R.K. Miller. 2007. Effectiveness of potassium lactate and sodium diacetate in combination with irradiation to control *Listeria monocytogenes* on frankfurters. J. Food Sci., 72: M26–M30.

Knorr, D. 1993. Effect of high pressure processes on food safety and quality. Food Technol., 47: 156–162.

Knorr, D., M. Zenker, V. Heinz and D.U. Lee. 2004. Applications and potential of ultrasonics in food processing. Trends Food Sci. & Technol., 15: 261–266.

Koseke, S. and K. Yamamoto. 2006. Recovery of *E. coli* ATCC25922 in phosphate buffered saline after treatment with hydrostatic pressure. Int. J. Food Microbiol., 110: 108–111.

Lacroix, C. 2011. Protective Cultures, Antimicrobials Metabolites, and Bacteriophages for Food and Beverage Biopreservation. Woodhead Publishing Ltd., Philadelphia, PA, USA.

Lebovka, N.I. and E. Vorobiev. 2004. On the origin of the deviation from the first order kinetics in inactivation of microbial cells by pulsed electric fields. Inter. J. Food Microbiol., 91: 83–89.

Leyer, G.J., L.-L. Wang and E.A. Johnson. 1995. Acid adaption of *Escherichia coli* O157:H7 increases survival in acidic foods. Appl. Environ. Microbiol., 61: 3752–3755.

Liao, C.H. 2009. Control of foodborne pathogens and soft-rot bacteria on bell pepper by three strains of bacterial antagonists. J. Food Prot., 72: 85–92.

Linton, M., J.M.J. McClements and M.F. Patterson. 1999. Inactivation of *Escherichia coli* O157:H7 in orange juice using a combination of high pressure and mild heat. J. Food Prot., 62: 277–279.

Mass, M.R., K.A. Glass and M.P. Doyle. 1989. Sodium lactate delays toxin production by *Clostridium botulinum* in cook-in-bag turkey products. Appl. Environ. Microbiol., 55: 2226–2229.

Matos, A. and J.L. Garland. 2005. Effects of community versus single strain inoculants on the biocontrol of Salmonella and microbial community dynamics in alfalfa sprouts. J. Food Prot., 68: 40–48.

Mazzotta, A.S. 2001. Thermal inactivation of stationary-phase and acid adapted *Escherichia coli* O157:H7, *Salmonella*, and *Listeria monocytogenes* in fruit juices. J. Food Prot., 64: 315–320.

McDonald, K.F., R.D. Curry, T.E. Clevenger, K. Unklesbay, A. Eisenstark, J. Golden and R.D. Morgan. 2000. A comparison of pulsed and continuous ultraviolet light sources for the decontamination of surfaces. Plasma Science, IEEE Transactions on, 28(5): 1581–1587.

Min, S., Z.T. Jin and Q.H. Zhang. 2003. Commercial scale pulsed electric field processing of tomato juice. J. Agric. and Food Chem., 51: 3338–3344.

Mosqueda-Melgar, J., R.M. Raybaudi-Massillia and O. Martin-Belloso. 2008. Combination of high-intensity pulsed electric fields with natural antimicrobials to inactivate pathogenic microorganisms and extend the shelf of melon and watermelon juices. Food Microbiol., 25: 479–491.

Olanya, O.M., D.O. Ukuku and B.A. Niemira. 2014. Effect of temperature and storage time on resting populations of *Escherichia coli* O157:H7 and *Pseudomonas fluorescens in vitro*. Food Control, 39: 128–134.

Olanya, M.O., D.O. Ukuku, B.A. Annous, B.A. Niemira and C.H. Sommers. 2013. Efficacy of *Pseudomonas fluorescens* for biocontrol of *Escherichia coli* O157:H7 on spinach. J. Food, Agric. & Environ., 11: 86–91.

Olanya, O.M., B.A. Annous, B.A. Niemira, D.O. Ukuku and C.H. Sommers. 2012. Effects of media on recovery of *Escherichia coli* O157:H7 and *Pseudomonas fluorescens* from spinach. J. Food Safety, 32: 492–501.

Papadopoulos, L., R. Miller, G. Acuff, L. Lucia, C. Vanderzant and H. Cross. 1991a. Consumer and trained sensory comparisons of cooked beef top rounds treated with sodium lactate. J. Food Sci., 56: 1141–1146.

Papadopoulos, L., R. Miller, G. Acuff, L. Lucia, C. Vanderzant and H. Cross. 1991b. Effect of sodium lactate on microbial and chemical composition of cooked beef during storage. J. Food Sci., 56: 341–347.

Povey, J.W. and T. Mason. 1998. Ultrasound in Food Processing. New York: Blackie Academic & Professional.

Qin, B.L., F.-J. Chang, G.V. Barbosa-Canovas and B.G. Swanson. 1995. Nonthermal inactivation of *Saccharomyces cerevisiae* in apple juice using pulsed electric fields. Lebensmittel-Wisenschaft und-Teechnologie., 28: 564–568.

Raichel, D.R. 2000. The Science and Applications of Acoustics. Springer: New York.

Riesz, P. and T. Kondo. 1992. Free radical formation induced by ultrasound and its biological implications. Free Radical Biology and Medicine, 13(3): 247–270.

Sharma, M., J.R. Patel, W.S. Conway, S. Ferguson and A. Sulakvelidze. 2009. Effectiveness of bacteriophages in reducing *Escherichia coli* O157:H7 on fresh-cut cantaloupes and lettuce. J. Food Prot., 72: 1481–1485.

Selma, M.V., M.C. Salmeron, M. Valero and P.S. Fernandez. 2006. Efficacy of pulsed electric fields for *Listeria monocytogenes* inactivation and control in horchata. J. Food Safety, 26: 137–149.

Shelef, L.A. 1994. Antimicrobial effects of lactates: a review. J. Food Prot., 57: 445–450.

Shelef, L.A. and Q. Yang. 1991. Growth suppression of *Listeria monocytogenes* by lactates in broth, chicken, and beef. J. Food Protect., 54: 283–287.

Sommers, C.H., O.J. Scullen and J.E. Sites. 2010. Inactivation of foodborne pathogens of frankfurters using ultraviolet light and Gras antimicrobials. J. Food Safety, 30: 666–678.

Soria, A.C. and M. Villamiel. 2010. Effect of ultrasound on the technological properties and bioactivity of food: a review. Trends Food Sci. & Technol., 21(7): 323–331.

Suslick, K.S. 1989. The chemical effects of ultrasound. Scientific American, 260(2): 80–86.

Takeshita, K., J. Shibato, T. Sameshima, S. Fukunaga, S. Isobe, K. Arihara and M. Itoh. 2003. Damage of yeast cells induced by pulsed light irradiation. Inter. J. Food Microbiol., 85(1): 151–158.

Tewari, G., D.S. Jayas and R.A. Holley. 1999. High pressure processing of foods: an overview. Science des Aliments, 19: 619–661.

Toro, H., S.B. Price, A.S. McKee, F.J. Hoerr, J. Krehling, M. Perdue et al. 2005. Use of bacteriophages in combination with competitive exclusion to reduce *Salmonella* from infected chickens. Avian Dis 2005; 49:118–24; PMID:15839424; http://dx.doi.org/10.1637/7286- 100404R.

Ukuku, D.O., D.J. Geveke, H.Q. Zhang and P.H. Cooke. 2008. Membrane damage and viability loss of *E. coli* K-12 in apple juice treated with radio frequency electric fields treatment. J. Food Prot., 71: 684–690.

Ukuku, D.O., D.J. Geveke and P. Cooke. 2012. Effect of thermal and radio frequency electric fields treatments on *Escherichia coli* bacteria in Apple Juice. J. Biochem. & Bioproces., 4(3): 76–81.

Ukuku, D.O., Yuk Hyung-Gyun and Zhang Howard. 2010. Behavior of Pulsed Electric fields injured *E. coli* O157:H7 cells in apple juice amended with pyruvate and catalase. J. Microbial. & Biochem. Technol., 2(6): 134–138.

Ukuku, D.O., H.Q. Zhang, M.L. Bari, K. Yamamoto and S. Kawamoto. 2009. Leakage of UV-materials of High Hydrostatic Pressure Injured *E. coli* O157:H7 in Tomato Juice. J. Food Prot., 72: 2407–2412.

Ukuku, D.O., Yuk Hyung-Gyun and Zhang Howard. 2011. Hydrophobic and electrostatic interaction chromatography for estimating changes in cell surface charge of *Escherichia coli* cells treated with pulsed electric fields. Foodborne Pathogens and Disease, 8(10): 1103–1109.

Ukuku, D.O., H.Q. Zhang and Huang Lihan. 2009. Growth parameters of *Escherichia coli*, *Salmonella enteritidis* and *Listeria monocytogenes*, and Aerobic Mesophilic Bacteria of Apple cider amended with nisin-EDTA. Foodborne Pathogens and Disease, 6: 487–494.

Ukuku, D.O. and D.J.A. Geveke. 2010. Combined treatment of UV-light and radio frequency electric field for the inactivation of *Escherichia coli* K-12 in apple juice. Inter. J. Food Microbiol., 138: 50–55.

Ulusoy, H.B., H. Colak and H. Hampikyan. 2007. The use of ultrasonic waves in food technology. Res. J. Biol. Sci., 2(4): 491–497.

Vercet, A., P. Lopez and J. Burgos. 1997. Inactivation of heat-resistant lipase and protease from *Pseudomonas fluorescens* by manothermosonication. J. Dairy Sci., 80: 29–36.

Villamiel, M. and P. de Jong. 2000. Inactivation of *Pseudomonas fluorescens* and *Streptococcus thermophilus* in Trypticase-soy broth and total bacteria in milk by continuous-flow ultrasonic treatment and conventional heating. J. Food Eng., 45: 171–179.

Wang, T., S.J. MacGregor, J.G. Anderson and G.A. Woolsey. 2005. Pulsed ultra-violet inactivation spectrum of *Escherichia coli*. Water Research, 39(13): 2921–2925.

Weagant, D.S., J.L. Bryant and D.H. Bark. 1994. Survival of *Escherichia coli* O157:H7 in mayonnaise and mayonnaise-based sauces at room and refrigeration temperatures. J. Food Prot., 57: 629–631.

Weidenbach, H., K. Sedlarik, G. Reimann and J. Wilde. 1978. Light and electron microscopy studies on experimental arterial thrombosis in dwarf swine. Z. Exp. Chir., 11: 230–237.

Zenker, M., V. Heinz and D. Knorr. 2003. Application of ultrasound assisted thermal processing for preservation and quality retention of liquid foods. J. Food Prot., 66(9): 1642–1649.

Zimmerman, U., G. Pilwat and F. Rieman. 1976. Dielectric break-down of cell membrane. Biophys. J., 44: 881–889.

Zimmerman, U. 1986. Electric break-down, electropearmeabilization and electrofussion. Rev. Phyiol. Biochem. Pharmacol., 105: 175–256.

5

Emerging and Re-emerging Foodborne Diseases: Threats to Human Health and Global Stability

Shinnichi Kawamoto[1] *and Md. Latiful Bari*[2,*]

Introduction

The inevitable, but unpredictable, appearance of new diseases has been recognized for millennia, well before the discovery of causative infectious agents. Today, however, despite extraordinary advances in the development of countermeasures (diagnostics, therapeutics, and vaccines), the ease of world travel and increased global interdependence have added layers of complexity to containing these infectious diseases that affect not only the health but the economic stability of societies. One example of disease emergence includes SARS, which emerged from bats and spread into humans first by person-to-person transmission in confined spaces, then within hospitals, and finally by human movement between international air hubs. Nipah virus also emerged from bats and caused an epizootic in herds of intensively bred pigs, which in turn served as the animal reservoir from which the virus was passed on to humans. In, 2009 H1N1 pandemic influenza virus emerged from pigs as well, but only after complex exchanges of human, swine, and avian influenza genes (Morens et al., 2009). H5N1 influenza emerged from wild birds to cause epizootics that amplified

[1] National Food Research Institute (NFRI), National Agriculture and Food Research Organization (NARO), 2-1-12, Kannondai, Tsukuba-shi, Ibaraki 305-8645, Japan.
[2] Center for Advanced Research in Sciences, University of Dhaka, Dhaka-1000, Bangladesh.
* Corresponding author: latiful@du.ac.bd

virus transmission in domestic poultry, precipitating dead-end viral transmission to poultry-exposed humans. On the contrary a good example of re-emerging diseases is cholera, which has repeatedly re-emerged over more than two centuries in association with global travel, changing seasons, war, natural disasters, and conditions that lead to inadequate sanitation, poverty, and social disruption. Each of these diseases has caused global societal and economic impact related to unexpected illnesses and deaths, as well as interference with travel, business, and many normal life activities.

Historical information as well as microbial sequencing and phylogenetic constructions make it clear that infectious diseases have been emerging and re-emerging over millennia and that such emergences are driven by numerous factors. Notably, 60 to 80% of new human infections most likely originated in animals, disproportionately in rodents and bats, as shown by the examples of hantavirus pulmonary syndrome, Lassa fever, and Nipah virus encephalitis. Most other emerging/re-emerging diseases result from human-adapted infectious agents that genetically acquire heightened transmission and/or pathogenic characteristics. Examples of such diseases include multidrug-resistant and extensively drug-resistant (MDR and XDR) tuberculosis, toxin-producing *Staphylococcus aureus* causing toxic shock syndrome, and pandemic influenza. However, the variables associated with emergences are unique for each and typically complex.

Definition and Concepts

Two major categories of emerging infections—newly emerging and re-emerging infectious diseases—can be defined, respectively, as diseases that are recognized in the human host for the first time and diseases that historically have infected humans, but continue to appear in new locations or in drug-resistant forms, or which reappear after apparent control or elimination (Morens and Fauci, 2013). Emerging/re-emerging infections may exhibit successive stages of emergence. These stages include adaptation to a new host, an epidemic/pathogenic stage, an endemic stage, and a fully adapted stage in which the organism may become non-pathogenic and potentially even beneficial to the new host (e.g., the human gut microbiome) or stably integrated into the host genome (e.g., as endogenous retroviruses) (Parrish et al., 2008). Although these successive stages characterize the evolution of certain microbial agents more than others, they nevertheless can provide a useful framework for understanding many of the dynamic relationships between microorganisms, human hosts, and the environment.

Basic definitions of emerging and (re) emerging zoonoses

Newly Emerging: Diseases that are recognized in the human host for the first time

Re-emerging: Diseases that historically have infected humans, but continue to appear in new locations or in drug-resistant forms, or that reappear after apparent control or elimination

Disease Emergence

Emergence of new zoonotic pathogens seems to be accelerating for the following several reasons: global human and livestock animal populations have continued to grow, bringing increasingly larger numbers of people and animals into close contact; transportation has advanced, making it possible to circumnavigate the globe in less than the incubation period of most infectious agents; ecologic and environmental changes brought about by human activity are massive; and bioterrorist activities, supported by rogue governments as well as organized amateurs, are increasing, and in most instances the infectious agents of choice seem to be zoonotic (Esmaeil Zowghi, 2008). In general, there is no way to predict when or where the next important new zoonotic pathogen will emerge or what its ultimate importance might be. A pathogen might emerge as the cause of a geographically limited curiosity, intermittent disease outbreaks, or a new epidemic. The most prominent emerging problems stem from bacteria, viruses and protozoa. Potential emerging food-related diseases include hepatitis caused by the hepatitis E virus, intestinal spirochetosis caused by *Brachyspira pilosicoli*, gnathostomiasis caused by nematodes belonging to the genus *Gnathostoma*, and anisakidosis caused by fish nematodes. Other potential emerging pathogens include non-gastric *Helicobacter* spp., *Enterobacter sakazakii*, non-*jejuni/coli* species of *Campylobacter*, and non-O157 Shiga toxin-producing *Escherichia coli*.

An increased awareness of emerging pathogens, consumer education, changes in food production and handling practices from farm to table, and improvements in microbiological detection methods will be needed to prevent the spread of emerging foodborne diseases.

Why are Zoonoses Important?

Zoonoses are defined by the World Health Organization as "diseases and infections which are transmitted naturally between vertebrate animals and man" (WHO, 1959). They are a heterogenous group of infections with a varied epidemiology, clinical features and control measures. The causative organism may be viral, bacterial, fungal, protozoan, or parasitic (HPA, 2009). Any disease or infection that is naturally transmissible from vertebrate animals to humans and vice-versa is classified as zoonoses according to the PAHO publication Zoonoses and Communicable Diseases Common to Man and Animals. Over 200 zoonoses have been described and they have been known for many centuries. They are caused by all types of agents: bacteria, parasites, fungi, viruses and unconventional agents' (WHO, 2010). Humans have enjoyed a long and intimate relationship with other animals. Some animals are reared for food, milk or clothing, some for recreational purposes and others to keep as pets or to act as guards. Most often those interactions are decidedly for human benefit. However, there are occasional disadvantages to humans including transmission of infections. Such infections are usually called "zoonoses". In the last 20 years, 73% of all emerging human infections have been zoonotic. Economic consequences of these foodborne infections are losses in production, losses in import/export, increased sick leave, and death. Many of these diseases have the potential to spread through various means over long distances and to become global problems, e.g. (Influenza,

SARS, HIV). All major zoonotic diseases prevent the efficient production of food of animal origin, particularly of much-needed proteins, and create obstacles to international trade in animals and animal products. They are thus an impediment to overall socioeconomic development. In contrast, possible pathogens often zoonotic could be used in biological warfare/terror (e.g., Anthrax), therefore understanding why pathogens emerge or re-emerge is fundamental to effective management. The zoonoses must be considered seriously as possible future human communicable diseases, and that ignoring them will pose a threat to public health. Secondly, many zoonoses are able to cause very significant human morbidity and mortality. Amongst these are brucellosis, leptospirosis, salmonellosis, tuberculosis and echinococcosis, and a large number of other bacterial, viral and parasitic infections. In order to prevent zoonoses from occurring, it is important to identify which animals and foodstuffs are the main sources of infections. Success in the prevention and control of major zoonoses depends on the capability to mobilize resources in different sectors and on coordination and intersectoral approaches, especially between national (or international) regulatory bodies. Understanding and developing the multiple links between animal and human health fields, is essential for establishing effective surveillance, preparedness, and response strategies and for developing appropriate, government-wide mechanisms for risk assessment and management of the zoonotic diseases.

Why Do Pathogens Emerge?

An emerging infectious disease (EID) can be defined as "an infectious disease whose incidence is increasing following its first introduction into a new host population or whose incidence is increasing in an existing host population as a result of long-term changes in its underlying epidemiology" (Engering et al., 2013). EID events may also be caused by a pathogen expanding into an area in which it has not previously been reported, or which has significantly changed its pathological or clinical presentation (Jones et al., 2008). Mostly, infectious disease emergence in humans is caused by pathogens of animal origin. Likewise, cross-over events may occur between non-human species including between domestic animals and wildlife, and such events also involve transmission from a reservoir population into a novel host population (spill-over). Today, however, despite extraordinary advances in development of countermeasures (diagnostics, therapeutics, and vaccines), the ease of world travel and increased global interdependence have added layers of complexity to containing these infectious diseases that affect not only the health but the economic stability of societies. HIV/AIDS, severe acute respiratory syndrome (SARS), and the most recent 2009 pandemic H1N1 influenza are only a few of many examples of emerging infectious diseases in the modern world (Woolhouse and Dye, 2001); each of these diseases has caused global societal and economic impact related to unexpected illnesses and deaths, as well as interference with travel, business, and many normal life activities. Other emerging infections are less catastrophic than these examples; however, they nonetheless may take a significant human toll as well as cause public fear, economic loss, and other adverse outcomes.

Factors Responsible for Emergence of Pathogens

Many factors involved in the emergence or re-emergence of pathogens associated with foodborne illness worldwide. These factors may contribute to the emergence of a new zoonotic disease, including microbial/parasitical/virologic determinants such as mutation, natural selection, and evolutionary progression; individual host determinants such as acquired immunity and physiologic factors; host population determinants such as host behavioral characteristics and social, commercial, and iatrogenic factors; and environmental determinants such as ecologic and climatic influences (Esmaeil Zowghi, 2008). The emergence of pathogens consists of four broad domains of factors that influence pathogen-host interactions and the resulting emergence of new diseases. They are:

(1) Genetic and biologic factors; (2) Physical and environmental factors; (3) Ecologic factors; (4) Social, political, and economic factors (Fig. 1).

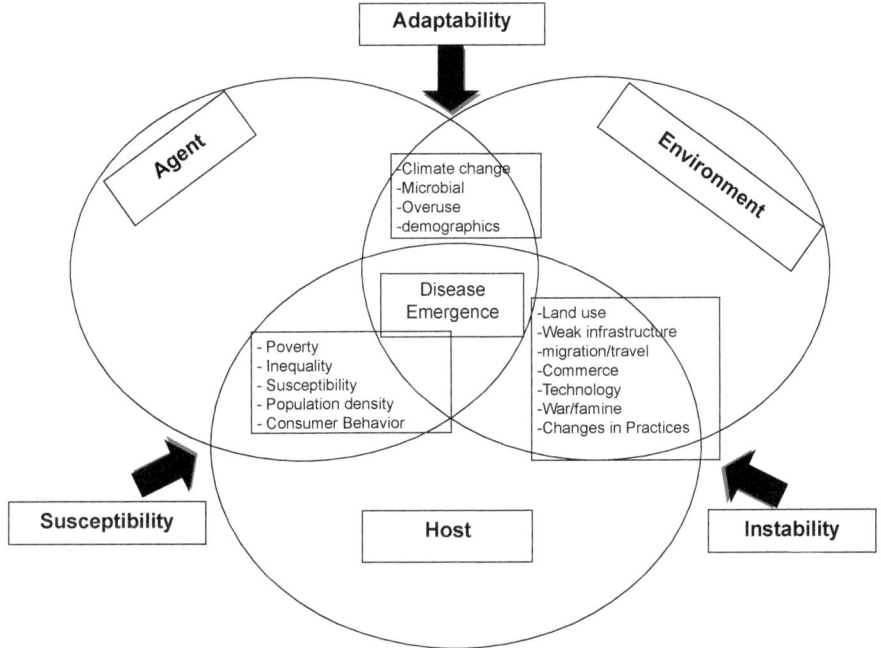

Figure 1. Factors Contributing to foodborne Disease Emergence.

Within these four broad domains, there are several specific factors that belong to one or more of these domains which contributed to the emergence of diseases. These specific factors are: (1) Microbial adaptation and change; (2) Human susceptibility to infection; (3) Climate and weather; (4) Physical events; (5) Changing ecosystems; (6) Economic development and land use; (7) Human demographics and behavior; (8) Technology and industry; (9) International travel and commerce; (10) Breakdown of public health measures; (11) Poverty and social inequity; (12) War and famine; (13) Lack of political will; and (14) Intent to harm.

The following factors play a major role in the epidemiology of emerging foodborne problems:

Microbial Adaptation/Changes in the Pathogens

Microbial adaptation through environmental and ecological changes (Natural selection) is a key process in the emergence of pathogens. Pathogens' adaptation to extreme temperatures and pH-levels, develops antimicrobial resistance or obtains new genetic characteristics through mutation, recombination or adaptation to new environment or species: Examples—Avian influenza, *Escherichia coli* O157:H7 (2005), Nipah-virus (2010), and SARS (2003). The therapeutic use of an antimicrobial agent in human or animal populations creates a selective pressure that favours survival of bacterial strains resistant to the agent. Antimicrobial resistance including Vancomycin-resistant enterococci (VRE), *Salmonella* Typhimurium DT 104, *Campylobacter jejuni* and *C. coli, S. aureus* (30–40% MRSA), *Mycobacterium tuberculosis* (15% MDR) has been detected in the US.

Figure 2. Examples of Avian influenza in chicken.

Technological Advances in Detection and Identification

Advances in science and technology allow for better methods of detection and identification of foodborne illness and the pathogen that may be causing the illness. Previously "unknown" foodborne illnesses have been identified. Technological advances, e.g., improved detection, molecular methods, enabling us to recognize new pathogens. Examples: *Y. enterocolitica, Listeria, Campylobacter*. Improved microbial media for pathogens detection; new techniques to visualization of organisms, e.g., Electron microscopy, fluorescence microscopy, antibody techniques, cell culture techniques for mammalian viruses and genetic analysis using PCR, etc. In 1972–1973, caliciviruses and rotaviruses were identified as the causative agents of diarrhea, now known to be a significant cause of mortality amongst children.

Human Susceptibility to Infections

Advances in medical technology (e.g., organ transplants and cancer therapy) have extended the life expectancy of persons with chronic diseases, thus increasing the proportion of the population with heightened susceptibility to severe foodborne illness. Number of factors related to the human host has a major impact on the occurrence and severity of foodborne disease. The host's age, gender, place of residence, ethnicity, underlying health status, and knowledge, attitudes, and practices related to health and diet all have important bearing on foodborne illness. The health of the host affects the individual's susceptibility to infection and illness, and the host's dietary and hygiene practices affect exposure to pathogens. The proportion of susceptible population to foodborne problems is increasing worldwide. In developed countries, such as the United Kingdom and the United States, between 15% and 20% of the population show greater susceptibility than the general population to foodborne disease. This proportion includes people with primary immunodeficiency, patients treated with radiation or with immunosuppressive drugs for cancer and diseases of the immune system, those with acquired immune-deficiency syndrome and diabetics, people suffering from liver or kidney disease or with excessive iron in the blood, pregnant women, infants, and the elderly. In more affluent domains, life expectancy is increasing, while elsewhere a very high birth rate often goes hand-in-hand with poverty and malnutrition. Malnutrition and the use of antacids, particularly proton-pump inhibitors, also increase susceptibility (Lund, 2011). People with primary immunodeficiencies are prone to foodborne infections (Dropulic and Lederman, 2009). For example, recurrent or chronic diarrhea was reported in 118/252 patients (47%) with common variable immunodeficiency with hypogammaglobulinemia in France; the pathogens detected most frequently were *Giardia, Campylobacter* and *Salmonella* (Oksenhendler et al., 2008). It has been predicted that by the year 2025, more than one billion of the world's population will be over 60 years of age, two-thirds of whom will live in developing countries. As a result, in some countries, one person in four faces a higher risk of contracting a foodborne disease.

Economic Development and Land Use

Economic development and land use have introduced new production systems or environmental changes and increase access to certain foods. The food chain has become longer and more complex, thus increasing opportunities for contamination. The number of farms is decreasing but the size of the farms is increasing. As a result animals are housed more densely, which increases the spread of microorganisms between animals? Likewise manure management becomes more of a concern. Manure is re-applied to fields but because of the amount that is generated in a small area, the potential for runoff and ground water contamination is increased. Areas in which foods were originally harvested may no longer support a large enough biomass to support economic harvesting. As a result production and harvesting areas may change. An example is oyster harvesting. Because of the declining harvesting in the Chesapeake Bay, the majority of oysters are harvested in the Gulf of Mexico. There has been an increase in the number of *Vibrio vulnificus* infections as prevalence of this organism

Figure 3. Urban agriculture is being practiced in or around a village, town, or city.

in oysters increases as the water temperature increases. Because oyster harvests have been low in the Chesapeake Bay, oysters are being harvested all year round from the gulf. In the past oysters were not harvested from the gulf during the warmer months of the year because of the increased prevalence of *Vibrio*. Lack of knowledge and negligence on the part of food handlers, together with an increase in mass catering, are important factors in foodborne illnesses.

Changes in Animal Husbandry

Modern intensive animal husbandry practices have been used to maximize production. This has resulted in the emergence and increased prevalence of several human pathogens, like *Salmonella* and *Campylobacter*, in flocks or herds of all the most important production animals (poultry, cattle, pigs) (Fig. 4). Crowding of animals has led to the increased use of antibiotics on so-called "factory farms" which in turn has been linked to the emergence of new strains of antibiotic-resistant bacteria.

Changes in Agronomic Process

Agricultural practices have contributed to the increased risks associated with fresh fruit and vegetables, such as the use of manure, chemical fertilizers, untreated sewage, or irrigation water containing pathogens (Fig. 5). Outbreaks linked to fruits and vegetables have increased in some regions, especially where improvements in transportation and access to imported fruits and vegetables are giving consumers more fresh produce

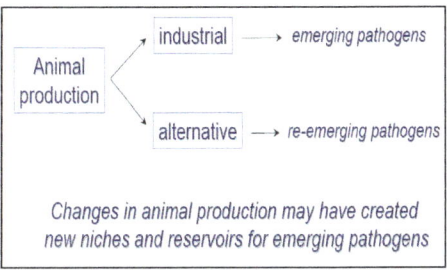

Figure 4. Chicken produced in industrial settings live an average of 49 days, far too short to study long-term safety effects of their grain-intensive diets.

Figure 5. Wastewater reusing to grow vegetables.

year round. Examples include a major *E. coli* O157:H7 outbreak in Japan linked to sprouts involving more than 9,000 cases in 1996, and several recent *Cyclospora* outbreaks associated with raspberries in North America and Canada, and lettuce in Germany (Fig. 5).

Changes in Food or Agricultural Technology

Advances in processing, preservation, packaging, shipping, and storage technologies bring new forms of foods to the market, and sometimes new hazards. For example, the increased use of refrigeration to prolong shelf-life of ready-to-eat foods has contributed to the emergence of *Listeria monocytogenes*. Consumers in many regions have expressed concern regarding the use of technologies like irradiation and genetically-engineered (GE) plants and animals (Fig. 6).

Figure 6. Production technology as concerns over the Impact of rising temperatures on crops.

Consumer Behavior

Changes in food consumption have brought to light unrecognized microbial foodborne hazards. Consumers are eating more fresh vegetables and fruits. As a result, any contamination of the surfaces of these products increases the likelihood of disease development. Dietary preferences and practices (e.g., for raw or hazardous foods) and some cultural beliefs and rituals can increase the risk of illness. Food consumption is changing as the result of a variety of factors: dietary habits may be altered by nutritional recommendations and campaigns; higher living standards have led to greater consumption of animal products; environmental changes can lead to increased access to certain foods; habits may be influenced by food policy, production systems and urban life styles; Behavioral changes leading to foodborne infections are further complicated by decreased opportunities for food safety instruction both in school and at home. Health educators in secondary schools emphasize prevention of other important health concerns (e.g., HIV infection, obesity) over consumer safety issues including food safety education (Roberts, 2003). In addition, because of two-income families and increased eating away from home, fewer opportunities may exist to pass food safety information from parent to child.

Changes in Lifestyle and Consumer Demands

Many trends impact the frequency and nature of foodborne illnesses. Consumers like to have access to seasonal foods all year; desire for foods that have a fresh taste and are minimally packaged and processed; increasing number of labor-saving food preparation equipment, i.e., food processors, microwave ovens; lack of knowledge of basic food safety principles and limited commitment to food preparation activities in the home; more interest in convenience and saving time than proper food handling and preparation; and more than one-third of meals being eaten away from home. In many developed countries, a larger share of the food budget is spent on outside food. In developing countries, there is a general rise in urban living and street food is an important component of the daily diet. As a result, outbreaks associated with food prepared outside the home are increasing in many regions.

Urbanization: Urbanization is associated with a range of health problems, including vector-borne diseases such as dengue and malaria (Rogers and Randolph 2000), diarrheal (Rose et al., 2001) and respiratory diseases (Kovats et al., 2001). Overcrowding and pollution resulting from inadequate infrastructure can trigger these conditions (Fig. 7) in this position. At present, there are an estimated four billion cases of diarrheal disease each year, causing over two million deaths. Studies have shown that water sanitation and hygiene interventions can greatly reduce water-related diseases (Moore et al., 2003). Urban sprawl as well as ecological influences may act in synergy or antagonistically with climatic factors, exacerbating or lessening the impact on infectious disease transmission of either factor acting independently. Changes in human population parameters such as increasing urbanization and crowding can cause diseases. Example: *S. aureus* (30–40% MRSA), *Mycobacterium tuberculosis*, etc.

Figure 7. Overcrowding can contributes diseases.

Breakdown of Public Health Education

Many governments are under increasing pressure to reduce staff and decentralize and privatize their health systems. The public health curriculum in primary and secondary education has changed significantly during the past 25 years with the beginning of the AIDS epidemic and the rise in teen pregnancy and drug abuse. Rapid changes and public sector austerity are having immediate, dramatic effects on health. Food safety education is being replaced by an emphasis on other important health concerns. Home science is no longer taught in many schools so proper food handling and preparation is left to home education. On the other hand, increased eating away from home have left fewer opportunities to pass food safety information from parent to child. Furthermore, many of today's meals are prepared from ready to eat packages and placed in the microwave, leaving safe food preparation to the manufacturer of the product. The emergence of foodborne diseases is a result of a set of complex issues.

By understanding these issues, one may be able to modify or alter these issues or risk factors interrupting the development of outbreaks.

Travel and Migration

International travel has increased dramatically during the 20th century. Five million international tourist arrivals were reported worldwide in 1950, and the number is expected to reach 975 million by 2015 (UNWTO, 2014). Travelers may become infected with foodborne pathogens uncommon in their nation of residence, thus complicating diagnosis and treatment when their symptoms begin after they return home. Travellers can spread disease rapidly to new and distant environments (Fig. 8), while immigrants also introduce new foods and dietary habits into new regions. Introduction and establishment of a pathogen in a new geographic area via travel, (im)migration and international trade. For example, in 1992, an outbreak of cholera caused 75 illnesses in international airline passengers; 10 persons were hospitalized, and one died (Tatem et al., 2006). Many foreign travelers visit the US for vacation, siteseeing, and visiting relatives. Often, these visitors bring foods from their country. These foods may contain pathogens resulting in disease in those consuming these foods. Soft cheeses from Mexico are frequently associated with foodborne disease as these cheeses are often made with unpasteurized milk.

Figure 8. Travelers spread disease rapidly to new places.

Human Travel Behavior

Growing human and animal population, encroachment upon natural environments, increasing interest for nature, new exotic pets and contact with wild animals can lead to infections (Fig. 9). Examples: *Borrelia*, Rabies, *Echinococcus*, *Salmonella*, etc.

Figure 9. Human Increase interest for nature.

Increased International Trade

Globalization, facilitated by the liberalization of trade, has led to an increasing number of cases where the rapid movement of food of plant and animal origin has contributed to the spread of foodborne problems to new areas (Figs. 10 and 11).

Figure 10. International foodborne disease outbreaks: Rapid spread worldwide by movement of food.

Figure 11. Globalization of Trade: *"The World on your Plate"*.

International trade allows for the rapid transfer of microorganisms from one country to another. The increased gap between processing and consumption of food leads to additional opportunities for contamination and time/temperature abuse, increasing the risk of foodborne illness.

Increasing trade also means that new and unfamiliar foodborne hazards can now more easily reach consumers who have not developed immunity to those pathogens. Foods, especially raw fruits and vegetables containing pathogens from one country, may be shipped to other countries. Moreover, growing and harvesting practices for this produce may be very different from what is accepted as a standard in the US. Recently there was an outbreak of cyclosporosis associated with raspberries from Guatemala (Calvin et al., 2002).

Changes in Industry and Technology

In the past food was produced and consumed within a community. As communities became larger and transportation more accessible, food was produced or processed at one point and shipped to another. These trends toward greater geographic distribution of products from large centralized food processors carry a risk for dispersed outbreaks. When mass-distributed food products are intermittently contaminated or contaminated at a low level, illnesses may appear sporadic rather than part of an outbreak. An example is the number of recalls on ground beef allotments. These allotments have been very large consisting of thousands of tons meat. Industry consolidation and mass distribution of foods may lead to large outbreaks of foodborne disease. Changes in animal production may have created new niches and reservoirs for emerging pathogens.

Poverty and Social Inequity

Environmental contamination, poor social conditions and lack of safe food preparation facilities are interrelated factors that lead to foodborne illnesses. While a number of related problems keep foodborne diseases at high levels within the developing countries, the root cause is poverty, which disproportionately affects women and children. Poverty exacerbates food safety problems in many ways and contributes to unsanitary conditions in rapidly growing urban centers; lack of access to clean water; unhygienic transportation and storage of foods; low education levels among consumers and food handlers, leading to reduced information on food safety (Fig. 12). Moreover, national governments lack the financial resources to enhance foodborne disease surveillance and monitoring capacities, implement food safety regulations through an efficient inspection system, invest in modern facilities and utilities, develop food safety education programmes, or conduct disaster planning and relief. Therefore, an integrated approach to combine food safety concepts with poverty reduction activities at the national level is needed to improve this situation.

Figure 12. Shanty of underprivileged children and collecting vegetable waste for consumption at a wholesale vegetable market.

Intention to Harm

Following rising incidents of terrorist attacks in many countries in recent years, concerns about intentional adulteration of food by terrorists, criminals, or other antisocial groups have risen and led to the need for new preparedness efforts. WHO (2002) states that "the key to preventing food terrorism is to enhance existing food safety programmes". Strengthening national food safety programmes requires that national policies and resources to support the infrastructure are in place and that food legislation, food monitoring and surveillance, food inspection, foodborne disease surveillance, and education and training are adequate and uptodate. In addition, the harm by bioterrorism (such as the 2001 anthrax terrorist attacks in the United States) should be added.

Lack of Political Will

Foodborne illness can have huge political implications. In Western Europe, Bovine spongiform encephalopathy (BSE) has led to more political and structural changes than any other food or agricultural issue. In Germany, the emergence of BSE in early 2001 led to the resignation of both the agriculture and health ministers and the restructuring of the agriculture ministry to become more consumer-oriented. In the United Kingdom, responsibilities for food control were transferred from the Ministry of Agriculture, Fisheries, and Food to a new, separate food authority, the Food Standards Agency. Elsewhere in Europe, similar national agencies have been created to ensure adequate regulation of food safety and restore public confidence, and a European Food Safety Authority has been established (Tirado et al., 2004). Many countries still lack the necessary surveillance capacity for foodborne disease outbreak detection and response. In addition, many foodborne disease outbreaks go undetected, in part due to lack of health infrastructure, communication between agencies and lack of political will.

War and Famine

War and famine can lead to severe shortage of food and other necessary supplies, generally affecting a large number of people. Natural causes include droughts, floods, earthquakes, insect plagues, and plant disease. Human causes include wars, civil war, sieges, deliberate crop destruction, poverty, and inefficient food distribution. The immediate consequences of famine are weight loss in adults and retarded growth in children (Fig. 13). Deaths are due in part starvation, in part to diminished ability to fight infection. One of the most dramatic large-scale consequences of famine is population migration.

In recent years, major famines have occurred in Africa, which may be susceptible to famine for the next few decades. Contributing factors have included drought, desertification, poor soils, rapid increases in population, and inadequate attention by some governments to food production. Famine in Africa has also been most severe where wars or civil unrest exist, such as in Chad, the southern Sudan, Ethiopia, Mozambique, and Somalia. On the other hand, migration and the mass movement of millions of refugees or displaced persons from one country to another—as a result

Figure 13. Nuclear war and famine: A billion people at risk: Global Impacts of Limited Nuclear War on Agriculture, Food Supplies, and Human.

of wars, civil turmoil or natural disasters—also provide fertile breeding grounds for infectious diseases and keep them on the move (IPCC, 2009).

In the early 1990s, the world was producing more than adequate food for the 5.3 billion people on the planet, and it was believed to be capable of growing enough to feed the population projected for the first part of the 21st century. The cholera outbreak in Haiti in 2011, which killed more than 3,000 people and made over 100,000 Haitians severely sick, is a stark reminder of the importance of safe food and water supplies. War in Arab countries may also provide fertile grounds for infectious diseases in the coming days. To eliminate famine and reduce malnutrition, however, attention would need to be given not only to food production, but also to food distribution and consumption and to family planning.

Climate and Weather

Environmental contamination, poor social conditions and lack of safe food preparation facilities are interrelated factors that lead to foodborne illnesses. Changes in the environment (e.g., deforestation), climate and natural disasters affect animal reservoirs and vectors (e.g., Ebola, malaria, cholera, dengue fever, leptospira); increased risk of flooding that can contaminate agricultural land and drinking water sources with chemical and microbiological hazards from industry, mining, waste disposal, etc.; increasing temperatures; spread of pests and disease vectors; increased risk for mycotoxin formation; and challenges to foresee and prepare for coming changes.

Climate change will likely modify the relationships between pathogens and hosts directly by: altering the timing of pathogen development and life histories; changing seasonal patterns of pathogen survival; changing hosts' susceptibility to pathogens (Eisenberg et al., 2012). However, ecosystem processes can influence human infectious diseases indirectly (Fig. 14).

Deforestation: The status of the world's forests is threatened by: conversion for crop production or pastures; road or dam building; timber extraction; and the encroachment of urban areas. Historically these activities have been associated with changes in infectious diseases in the local population. The diseases most frequently affected are

Figure 14. Natural disasters affect human and influence diseases.

Figure 15. The flood waters cause human suffering and contribute diseases.

those that exist naturally in wild ecosystems and circulate among animals, especially those with vertebrate reservoirs and invertebrate vectors. In general such changes result from factors affecting the populations of animal reservoirs, vectors, and pathogens, or from factors associated with human exposure.

Ecological influences

In addition to land-use changes, there is a host of indirect links between infectious disease and environmental conditions that are mediated through changes in ecosystems resulting from human activities (Nelson et al., 2006). Zoonoses and vector-transmitted anthroponoses, dependent on the ecology of non-human animals, will be especially sensitive to the effects of these ecological changes. An estimated 75% of emerging infectious diseases of humans have evolved from exposure to zoonotic pathogens, therefore any changes in the ecological conditions influencing wildlife diseases have the potential to impact directly on human health (Felix et al., 2006).

Control of Foodborne Diseases

Emerging foodborne problems will not be solved by individual countries acting in isolation, no matter how high their levels of expertise and food control. Emerging foodborne problems are a global issue, and a unified and joint approach by all countries and the relevant international organizations is a prerequisite for the identification and control of all emerging foodborne problems that threaten human health and international trade.

The following task and Fig. 16 mentioned information and tools required to control foodborne pathogens:

❖ Risk assessment to inform risk management
  • Prior to (major) changes in production
  • Extrapolating trends and development in production
  • Use of new substances
  • Effects of political decisions
  • Consequences of consumer preferences
❖ Simulation and scenario testing
❖ Minimizing the probability of infection (HACCP, GHP, GMP)
❖ Monitoring foodborne disease trends (inter)nationally
  • Surveillance programmes
    ○ Emergence of known pathogens
  • Outbreak investigations
    ○ Emergence of "new" pathogens
    ○ Traceback to sources of infection
    ○ Uncover routes of transmission
❖ Develop/validate/improve analytical methods
❖ Report results (inter)nationally, share expertise

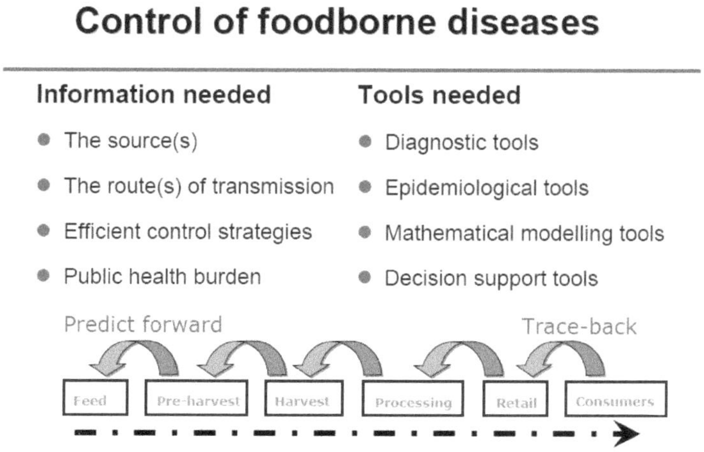

Control of foodborne diseases

Information needed	Tools needed
● The source(s)	● Diagnostic tools
● The route(s) of transmission	● Epidemiological tools
● Efficient control strategies	● Mathematical modelling tools
● Public health burden	● Decision support tools

Predict forward   Trace-back

Feed → Pre-harvest → Harvest → Processing → Retail → Consumers

Figure 16. Tools for controlling Foodborne Diseases.

Conclusion

The battle against emerging diseases is a continual process; winning does not mean stamping out every last disease, but rather getting ahead of the next one. It is likely that emerging foodborne problems will become even more significant in the coming years. Emerging foodborne problems will not be solved by individual countries acting in isolation, no matter how high their levels of expertise and food control. These problems are a global issue that requires a unified and joint approach by all countries. In order to achieve global food safety control, the appropriate international organizations, with the assistance of their members, need to elaborate a plan of action to encourage and assist countries in developing acceptable and efficient food control systems, while simultaneously indicating the minimum or basic parameters or requirements for such a purpose. Such systems should apply the three elements of risk analysis—assessment, management and communication. An initiative designed to improve the safety of food supply should focus on the hazards and foods that present the greatest risks to public health and should emphasize development and the implementation of preventive control of those risks.

References

Al-Goblan, A.S. and S. Jahan. 2010. Surveillance for foodborne illness outbreaks in Qassim, Saudi Arabia, 2006. Foodborne Pathogens and Disease, 7(12): 1559–1562. doi:10.1089/fpd.2010.0638.

Allos, B.M., M.R. Moore, P.M. Griffin and R.V. Tauxe. 2004. Surveillance for sporadic foodborne disease in the 21st century: the FoodNet perspective. Clinical Infectious Diseases, 38(s3): S115–S120. doi:10.1086/381577.

Anderson, A.L., L.A. Verrill and N.R. Sahyoun. 2011. Food safety perceptions and practices of older adults. Public Health Reports, 126(2): 220–227.

Ammon. A. and R.V. Tauxe. 2007. Investigation of multi-national foodborne outbreaks in Europe: some challenges remain, 135(6): 887–889. doi:10.1017/S0950268807008898.

Angulo, F.J. and E. Scallan. 2007. Activities, Achievements, and Lessons Learned during the First 10 Years of the Foodborne Diseases Active Surveillance Network: 1996–2005. Clinical Infectious Diseases, 44(5): 718–725. doi:10.1086/511648.

Arora, D.R. and B. Arora. 2008. Text Book of Microbiology. 3rd edn. Push Road, India, CBS Publishers & Distributors, pp. 689–664.

Baert, L., M. Uyttendaele, A. Stals, E. VAN Coillie, K. Dierick, J. Debevere and N. Botteldoorn. 2009. Reported foodborne outbreaks due to noroviruses in Belgium: the link between food and patient investigations in an international context. Epidemiology and Infection, 137(3): 316–325. doi:10.1017/S0950268808001830.

Barbara, M. Lund and Sarah J. O'Brien. 2011. The occurrence and prevention of foodborne disease in vulnerable people. Foodborne Pathog. Dis., 8(9): 961–973. doi: 10.1089/fpd.2011.0860.

Bas, M., A. Safak Ersun and G. Kıvanç. 2006. The evaluation of food hygiene knowledge, attitudes, and practices of food handlers' in food businesses in Turkey. Food Control, 17(4): 317–322. doi:16/j.foodcont.2004.11.006.

Blancou, J., B.B. Chomel, A. Beletto and F.X. Meslin. 2005. Emerging and Re-emerging bacterial zoonoses: factors of emergence, surveillance and control. Vet. Res., 36: 507–522.

Caroline, S.D. and N. Robert. 2005. Report on Global and Local: Food Safety Around the World. Center for Science in the Public Interest (CSPI) June 2005, P2–95.

Calvin, L., W. Foster, L. Solorzano, J.D. Mooney, L. Flores and V. Barrios. 2002. Response to a food safety problem in produce: a case study of a cyclosporiasis outbreak. In: B. Krissoff, M. Bohman and J. Caswell (eds.). Global Food Trade and Consumer Demand for Quality. New York: Kluwer Academic/Plenum.

Centers for Disease Control and Prevention. Provisional surveillance summary of the West Nile virus epidemic—United States, January–November 2002. MMWR Morb Mortal Wkly Rep., 51: 1129–1133.

Chugh, B.B. 2008. Emerging and re-emerging bacterial diseases in India. J. Biosci., 33: 549–555.

Dalton, H.R., R. Bendall, S. Ijaz and M. Banks. 2008. Hepatitis E: an emerging infection in developed countries. The Lancet Infectious Diseases, 8(11): 698–709. doi:16/S1473-3099(08)70255-X.

Doorduyn, Y., W.E. Van Den Brandhof, Y.T.H.P. Van Duynhoven, W.J.B. Wannet and W. Van Pelt. 2006. Risk Factors for *Salmonella enteritidis* and *Typhimurium* (DT104 and Non-DT104) infections in the Netherlands: Predominant Roles for raw eggs in enteritidis and sandboxes in *Typhimurium* infections. Epidemiology and Infection, 134(3): 617–626. doi:10.1017/S0950268805005406.

Dropulic, L.K. and H.M. Lederman. 2009. Overview of infections in the immunocompromised host. *In*: R.T. Hayden, K.C. Carroll, Y.-W. Tang and D.M. Wolk (eds.). Diagnostic Microbiology of the Immunocompromised Host. Washington, DC: ASM Press, pp. 3–43.

Eberhart-Phillips, J., R.E. Besser, M.P. Tormey, D. Feikin, M.R. Araneta, J. Wells et al. 1996. An outbreak of cholera from food served on an international aircraft. Epidemiol. Infect. 116: 9–13.

Eisenberg, J.N., J. Trostle, R.J. Sorensen and K.F. Shields. 2012. Toward a systems approach to enteric pathogen transmission: from individual independence to community interdependence. Annual Review of Public Health, 33: 239–257.

Engering, A., L. Hogerwerf and J. Slingenbergh. 2013. Pathogen–host–environment interplay and disease emergence. Emerging Microbes and Infections, 2: e1–e7.

Esmaeil Zowghi. 2008. Emerging and re-emerging zoonoses. Iranian Journal of Clinical Infectious Disease, 3(2): 109–115.

Felix, P. Amerasinghe, Kaw Bing Chua, Peter Daszak, Alex D. Hyatt, David Molyneux, Madeleine Thomson, Laurent Yameogo, Mwelecele-Malecela-Lazaro, Pedro Vasconcelos and Yasmin Rubio-Palis. 2006. Human health: ecosystem regulation of infectious diseases. pp. 391–415. *In*: Ecosystems and Human Well-being: Current State and Trends.

Gormley, F.J., C.L. Little, N. Rawal, I.A. Gillespie, S. Lebaigue and G.K. Adak. 2011. A 17-year review of foodborne outbreaks: describing the continuing decline in England and Wales (1992–2008). Epidemiology and Infection, 139(5): 688–699. doi:10.1017/S0950268810001858.

Greig, J.D. and M.B. Lee. 2009. Enteric outbreaks in long-term care facilities and recommendations for prevention: a review. Epidemiology and Infection, 137(2): 145–155. doi:10.1017/S0950268808000757.

Health Canada. Bioterrorism and Emergency Preparedness, June 2003. Available at <http://www.hc-sc. gc.ca/english/protection/biotech/bioterrorism.htm>. accessed on Nov. 15, 2014.

HPA. 2009. Guidelines for the Investigation of Zoonotic Disease, version 1. 23 April 2009.

IPCC. 2009. The Fifth Assessment Report (AR5) of the IPCC at its 31st Session of the IPCC in Bali (26–29 October 2009). available at http://www.ipcc.ch/report/ar5/index.shtml.

Jones, K.E., N.G. Patel, M.A. Levy et al. 2008. Global trends in emerging infectious diseases. Nature, 451: 990–993.

Käferstein, F.K. 1997. Food safety: a commonly underestimated public health issue. Introduction. World Health Statistics Quarterly, 50(1-2): 3–4.

Koopmans, M. and E. Duizer. 2004. Foodborne viruses: an emerging problem. International Journal of Food Microbiology, 90(1): 23–41. doi:16/S0168-1605(03)00169-7.

Kovats, R.S. et al. 2001. Early effects of climate change: do they include changes in vector-borne disease? Philosophical Transactions of the Royal Society of London Series B Sciences 356(1411): 1057–1068.

Kuchenmüller, T., S. Hird, C. Stein, P. Kramarz, A. Nanda and A.H. Havelaar. 2009. Estimating global burden of foodborne diseases—a collaborative effort. Eurosurveillance, 14(18): 191–95.

Li, J., K. Smith, D. Kaehler, K. Everstine, J. Rounds and C. Hedberg. 2010. Evaluation of a statewide foodborne illness complaint surveillance system in minnesota, 2000 through 2006. Journal of Food Protection, 73(11): 2059–2064.

Lindqvist, R., Y. Andersson, J. Lindbäck, M. Wegscheider, Y. Eriksson, L. Tideström, A. Lagerqvist-Widh et al. 2001. A one-year study of foodborne illnesses in the municipality of Uppsala, Sweden. Emerging Infectious Diseases, 7(3 Suppl.): 588–592.

Linscott, A.J. 2011. Foodborne illnesses. Clinical Microbiology Newsletter, 33(6): 41–45. doi:10.1016/j. clinmicnews.2011.02.004.

Lynch, M.F., R.V. Tauxe and C.W. Hedberg. 2009. The growing burden of foodborne outbreaks due to contaminated fresh produce: risks and opportunities. Epidemiology and Infection, 137(Special Issue 03): 307–315. doi:10.1017/S0950268808001969.

Maria De Giusti, Dario De Medici, Daniela Tufi, Marzuillo Carolina and Antonio Bocci. 2007. Epidemiology of emerging foodborne pathogens. Italian Journal of Public Health, (4)1: 24–31.

Motarjemi, Y. and A. Adams. 2006. Emerging Foodborne Pathogens. Boca Raton: CRC Press.

Motarjemi, Y. and F.K. Käferstein. 1997. Global estimation of foodborne diseases. World Health Statistics Quarterly, 50(1-2): 5–11.

Marcus, R., J.K. Varma, C. Medus, E.J. Boothe, B.J. Anderson, T. Crume, K.E. Fullerton et al. 2007. Re-assessment of risk factors for sporadic Salmonella serotype enteritidis infections: a case-control study in five FoodNet sites, 2002–2003. Epidemiology and Infection, 135(01): 84–92. doi:10.1017/S0950268806006558.

Martinez, A., A. Dominguez, N. Torner, L. Ruiz, N. Camps, I. Barrabeig, C. Arias et al. 2008. Epidemiology of foodborne norovirus outbreaks in Catalonia, Spain. BMC Infectious Diseases, 8(1): 47. doi:10.1186/1471-2334-8-47.

McCulloch, J.E. (ed.). 2000. Infection Control: Science, Management and Practice. London: Whurr Publishers.

McCabe-Sellers, B.J. and S.E. Beattie. 2004. Food safety: emerging trends in foodborne illness surveillance and prevention. Journal of the American Dietetic Association, 104(11): 1708–1717. doi:10.1016/j.jada.2004.08.028.

Mor-Mur, M. and J. Yuste. 2009. Emerging bacterial pathogens in meat and poultry: an overview. Food and Bioprocess Technology, 3(1): 24–35. doi:10.1007/s11947-009-0189-8.

Morens, D.M. and A.S. Fauci. 2013. Emerging infectious diseases: threats to human health and global stability. PLoS Pathog., 9(7): e1003467. doi:10.1371/journal.ppat.1003467.

Morens, D.M., J.K. Taubenberger and A.S. Fauci. 2009. The persistent legacy of the 1918 influenza virus. N. Engl. J. Med., 361: 225–229. doi: 10.1056/nejmp0904819.

Moore, M., P. Gould and B.S. Keary. 2003. Global urbanization and impact on health. Int. J. Hyg. Environ. Health. Aug., 206(4-5): 269–278.

Nelson, R. 2003. Antibiotic development pipeline runs dry. New drugs to fight resistant organisms are not being developed, experts say. Lancet, 362(9397): 1726–1727.

Nelson, G.C., E. Bennett, A.A. Berhe, K. Cassman, R. DeFries, T. Dietz, A. Dobermann, A. Dobson, A. Janetos, M. Levy, D. Marco, N. Nakicenovic, B. O'Neill, R. Norgaard, G. Petschel-Held, D. Ojima, P. Pingali, R. Watson and M. Zurek. 2006. Anthropogenic drivers of ecosystem change: an overview. Ecology and Society, 11(2): 29. [online] URL: http://www.ecologyandsociety.org/vol11/iss2/art29/.

Newell, D.G., M. Koopmans, L. Verhoef, E. Duizer, A. Aidara-Kane, H. Sprong, M. Opsteegh et al. 2010. Foodborne diseases—The challenges of 20 years ago still persist while new ones continue to emerge. International Journal of Food Microbiology, 139(Supplement 1): S3–S15. doi:16/j.ijfoodmicro.2010.01.021.

New Zealand Food Safety Authority. 2010. The economic cost of foodborne disease in New Zealand. Retrieved July 19, 2011, from http://www.foodsafety.govt.nz/elibrary/industry/economic-cost-foodbornedisease/foodborne-disease.pdf.

Oksenhendler, E., L. Gerard, C. Fieschi, M. Malphettes, G. Mouillot, R. Jaussaud, J.-F. Viallard, M. Gardembas, L. Galicier, N. Schleinitz, F. Suarez, P. Soulas-Sprauel, E. Hachulla, A. Jaccard, A. Gardeur, I. Theodorou, C. Rabian and P. Debre. 2008. Infections in 252 patients with common variable immunodeficiency. Clin. Infect. Dis., 46: 1547–1554.

Parish, T., D.A. Smith, S. Kendall, N. Casali, G.J. Bancroft and N.G. Stoker. 2003. Deletion of two-component regulatory systems increases the virulence of *Mycobacterium tuberculosis*. Infection and Immunity, 71: 1134–1140.

Parrish, C.R., E.C. Holmes, D.M. Morens, E.-C. Park, D.S. Burke et al. 2008. Cross-species virus transmission and the emergence of new epidemic diseases. Microbiol. Mol. Biol. Rev., 72: 457–470. doi: 10.1128/mmbr.00004-08.

Porter, R. 1997. Public medicine. pp. 412–414. *In*: The Greatest Benefit to Mankind: A Medical History of Humanity, New York: W.W. Norton.

Pozio, E. 2008. Epidemiology and control prospects of foodborne parasitic zoonoses in the European Union. Parassitologia, 50(1-2): 17–24.

Prunier, A.L., R. Schuch, R.E. Fernandez and A.T. Maurelli. 2007. Genetic structure of the *nadA* and *nadB* antivirulence loci in *Shigella* spp. Journal of Bacteriology, 189: 6482–6486.

Robinson, R.K. 2007. Emerging foodborne pathogens. International Journal of Dairy Technology, 60(4): 305–306. doi:10.1111/j.1471-0307.2007.00331.x.

Roberts, Cynthia A. "Food Safety." Encyclopedia of Food and Culture. 2003. Encyclopedia.com. 23 Feb.
2015. http://www.encyclopedia.com.
Rogers, D.J. and S.E. Randolph. 2000. The global spread of malaria in a future, warmer world. Science
289(5485): 1763–1766.
Rose, J.B. et al. 2001. Climate variability and change in the United States: potential impacts on water- and
foodborne diseases caused by microbiologic agents. Environmental Health Perspectives, 109 Suppl.
2: 211–221.
Rothman, D.J., S. Marcus and S.A. Kiceluk. 1995b. On the antiseptic principle in the practice of surgery.
pp. 247–252. *In*: Medicine and Western Civilization. New Brunswick: Rutgers University Press.
Rothman, D.J., S. Marcus and S.A. Kiceluk. 1995c. The etiology of tuberculosis. pp. 319–329. *In*: Medicine
and Western Civilization. New Brunswick: Rutgers University Press.
Rothman, D.J., S. Marcus and S.A. Kiceluk. 1995a. On the extension of the germ theory to the etiology
of certain common diseases. pp. 253–257. *In*: Medicine and Western Civilization. New Brunswick:
Rutgers University Press.
Saulat Jahan. 2012. Epidemiology of Foodborne Illness, Scientific, Health and Social Aspects of the Food
Industry, Dr. Benjamin Valdez (ed.), ISBN: 978-953-307-916-5, InTech, Available from: http://
www.intechopen.com/books/scientific-health-and-social-aspects-of-the-food-industry/epidemiology-
offoodborne-illness.
Scallan, E. 2007. Activities, achievements, and lessons learned during the first 10 Years of the foodborne
diseases active surveillance network: 1996–2005. Clinical Infectious Diseases, 44(5): 718–725.
doi:10.1086/511648.
Schmidt, H. and M. Hensel. 2004. Pathogenicity islands in bacterial pathogenesis. Clinical Microbiology
Reviews, 17: 14–56.
Smith, J.L. and P.M. Fratamico. 2005. Emerging foodborne pathogens. *In*: P.M. Fratamico, A.K. Bhunia
and J.L. Smith (eds.). Foodborne Pathogens: Microbiology and Molecular Biology. Norwich: Caister
Academic.
Sparling, P.H., C. Crowe, P.M. Griffin, D.L. Swerdlow and J.M. Rangel. 2005. Epidemiology of *Escherichia
coli* O157:H7 Outbreaks, United States, 1982–2002. Public Health Resources. Retrieved from http://
digitalcommons.unl.edu/publichealthresources/73.
Stafford, R.J., P.J. Schluter, A.J. Wilson, M.D. Kirk, G. Hall and L. Unicomb. 2008. Population-attributable
risk estimates for risk factors associated with *Campylobacter* Infection, Australia. Emerging Infectious
Diseases, 14(6): 895–901. doi:10.3201/eid1406.071008.
Taylor, L.H., S.M. Latham and M.E.J. Woolhouse. 2001. Risk factors for human disease emergence.
Philosophical Transactions of the Royal Society of London. Series B 356: 983–989.
Tatem, A.J., D.J. Rogers and S.I. Hay. 2006. Global Transport Networks and Infectious Disease Spread.
Adv. Parasitol., 62: 293–343. doi: 10.1016/S0065-308X(05)62009-X.
The Food Safety and Inspection Service [FSIS]. 2008. Disposition/Food Safety: Overview of Food
Microbiology. Retrieved July 23, 2011, from www.fsis.usda.gov/PDF/PHVt-Food_Microbiology.pdf.
Tirado, M.C. 2004. WHO Regional Office for Europe "Food safety strategies in Europe: Promoting a new
approach to for food control in the region", Second FAO/WHO Global forum of food safety regulators,
Bangkok, Thailand, 12–14 October 2004, CRD 84.
Unicomb, L.E. 2009. Food Safety: Pathogen Transmission Routes, Hygiene Practices and Prevention,
27(5): 599–601.
Villemur, R. and E. Deziel. 2005. Phase variation and antigenic variation. pp. 277–322. *In*: P. Mullany (ed.).
The Dynamic Bacterial Genome. New York: Cambridge University Press.
Wolfe, N.D., C. Panosian Dunavan and J. Diamond. 2007. Origins of major human infectious diseases.
Nature, 447: 279–283.
Woolhouse, M.E. and C. Dye. 2001. Population biology of emerging and re-emerging pathogens—preface.
Philos. Trans. R. Soc. Lond. B. Biol. Sci., 356: 981–982.
Wren, B. 2006. How bacterial pathogens evolve. pp. 3–22. *In*: Y. Motarjemi and A. Adams (eds.). Emerging
foodborne pathogens. Boca Raton: CRC Press.
World Health Organization. 1959. Zoonoses: Second report of the joint WHO/FAO Expert Committee.
World Health Organization, Food Safety Department. 2002. Food Safety Issues: Terrorist Threats to Good,
Guidance for Establishing and Strengthening Prevention and Response Systems, p. 5.
WHO Food Safety Department. 2002. Terrorist Threats to Food: Guidance for Establishing and Strengthening
Prevention and Response Systems, p. 11.
WHO. 2010. Zoonoses and Veterinary Public Health. http://www.who.int/zoonoses/en/.

World Health Organization [WHO]. 2011a. Food Safety. Retrieved June 26, 2011, from http://www.who. int/foodsafety/foodborne_disease/ferg1/en/index.html.

World Health Organization [WHO]. 2011b. Initiative to estimate the Global Burden of Foodborne Diseases. Retrieved June 26, 2011, from http://www.who.int/foodsafety/foodborne_disease/ferg/en/index1.html.

World Health Organization [WHO]. 2011c. Initiative to estimate the Global Burden of Foodborne Diseases: Information and publications. Retrieved June 26, 2011, from http://www.who.int/foodsafety/ foodborne_disease/ferg/en/index7.html.

World Tourism Organization (UNWTO). 2014. Tourism highlights 2012 edition. Calle Capitán Haya, 42 28020 Madrid, Spain. pp. 6–23. available at http://mkt.unwto.org/sites/all/files/docpdf/ unwtohighlights12enhr.pdf.

6

Emergence of Drug-Resistant Pathogens

Dinesh Babu, Kushwaha Kalpana* and Vijay K. Juneja*

Introduction

The use of antibiotic drugs to treat bacterial infections is a common practice in modern agriculture and the health industry. Antibiotics are the most frequently prescribed drugs in modern medicine. An antibiotic such as penicillin and ciprofloxacin is a type of drug that is either naturally produced by an organism to gain competitive advantage of growth in the environment or it can be designed to kill or stop the growth of other bacteria. If a bacterium is not affected by the exposure to an antibiotic drug, then it can be deemed to be resistant to that particular chemical. Thus, antibiotic resistance is a survival mechanism of bacteria and other microbes that enable the organisms to overcome or reduce and eliminate the inhibitory effects of certain chemicals or antibiotic drugs. Such organisms are referred to as "antibiotic-resistant organisms" as they continue to multiply in the presence of the given dose of antibiotics.

It is evident that the increasing use of antibiotics causes a selective pressure on the infectious microorganisms that eventually show antibiotic resistance (also termed as antimicrobial resistance or drug resistance). The ability to develop resistance to antibiotics among certain bacteria may be naturally occurring or utilize several

Department of Toxicology, School of Pharmacy, College of Health and Pharmaceutical Sciences, University of Louisiana at Monroe, Monroe, LA 71209-0497, USA.
Eastern Regional Research Center, USDA-Agricultural Research Service, 600 E. Mermaid Lane, Wyndmoor, PA 19038, USA.
* Corresponding author: babu@ulm.edu; dinesh02@hotmail.com

mechanisms to become remarkably resilient and successful in evading the effects of antibiotic drugs. As evidenced by the extensive literature in this regard, bacteria gain their ability to resist antimicrobial drugs by every possible way of overcoming the drug from affecting their survival. For example, the resistance mechanisms in bacteria may involve enzymatic degradation or inactivation of the antibiotic, modification of proteins targeted by the drug, altered membrane permeability to reduce accumulation and/or increase efflux of the antibiotic drugs and modification of metabolic pathways to overcome effect of the antibiotic drugs (Dever and Dermody, 1991; Schmieder and Edwards, 2012). Traditionally, it is known that certain bacteria may be inherently resistant to an antibiotic or the resistance can also be acquired either by mutating their DNA or by acquisition of DNA coding for resistance from another organism. Acquired resistance to antimicrobials results from genetic changes in the microbial cell through mutation or acquisition of genetic material that confer resistance genes (Russell, 1991). Such genetic material could be the encoded plasmids or transposons containing sequences for multiple drug resistance genes (integrons) (Roe and Pillai, 2003; Russell, 1991, 1996; Russell and Chopra, 1996). Some examples for this type of resistance include acquisition of genes for β-lactamase (an enzyme capable of breaking down and inactivating β-lactam antibiotics such as penicillins and cephalosporins, and mutation of a DNA gyrase subunit (the target of fluoroquinolones). Acquired resistance is the most common type and has been well studied for antibiotic drugs. Modification of the genetic material to acquire antibiotic resistance may involve vertical (from parent to progeny during multiplication) and horizontal modes of gene transfer (from same or different species of bacteria and bacteriophages in the environment). Inherent resistance may come from traits such as rigid cell wall structure that act as a barrier for the entry of a drug into the cell; or the bacteria may lack a transport system or target protein for an antibiotic.

The emergence of antibiotic resistance remains a growing problem worldwide and it is immensely important to understand the factors that influence this behavior in pathogenic bacteria. Resistance to antimicrobial drugs by infectious agents is widely accepted as a global health and economic concern, which is compounded by the multi-antibiotic resistance by increasing numbers of pathogenic bacteria. Use and misuse of antibiotics in the health and the animal industry are the most discussed topics when it comes to antibiotic resistance among the infectious agents. In this chapter, we will focus on usage of antibiotics in agricultural farming and its impact on emerging resistance to antibiotic drugs among food pathogens as we elaborate on the implications of antimicrobial resistance among food pathogens.

Antibiotic Use in Animal Agriculture

In general, use of drugs in animal agriculture is done for several reasons of controlling, treating, and preventing diseases in livestock and also to promote their nutrition efficiency (McEwen and Fedorka-Cray, 2002). Among the several reasons of antibiotics usage in agriculture, the purpose of increasing the feed efficiency (weight gain per unit quantity of food consumed for given length of time) and improving growth of farm animals is believed to be the major one. For example, the antibiotic Chlortetracycline, is normally given to treat enteric diseases in pigs, but at lower concentrations of 10 to 50 g/ton of feed, it is approved as a feed antibiotic for growing pigs (44 to 110 pounds by weight) (Animal Health Institute, 2012). Such usages of the same antibiotic drugs for therapeutic and non-therapeutic purposes have become the center of debate on the emergence of antibiotic resistance among the infectious and non-infectious bacteria.

The use of animal antibiotics in the United States started in early 1900s mainly to address the shortage of meat and increased consumer demand for meat products. At the time, growing protein demands required rearing larger quantities of animals and raising them over a short period of time. This practice was quickly faced with spread of bovine diseases among the densely raised animals and threatened the livestock production. Following the works of Selman Waksman in 1943, the discovery of a miracle drug called "streptomycin" was made. Streptomycin was found to be effective in treating many of the animal diseases including bovine tuberculosis and mastitis that were uncontrollable and widespread at the time. These outcomes were welcomed by the health and farming industry and use of antimicrobial drugs gained huge popularity as the antibiotic drug usage played a critical role in protecting the public health by saving millions of human and animal lives. During 1951, the US Food and Drug Administration (FDA) approved the sub-therapeutic and therapeutic use of antibiotics for farm animals. However, the continued use of antibiotics in livestock production steered a surprising outcome, which showed that the animals given antibiotics grew faster apart from being healthier as they showed increased feed efficiency (required less feed per unit gain in weight) that resulted into the lower costs of production. This was quickly noted by the meat industry as the heavier animals brought increasing profits. At the same time, the meat industry was struggling to meet the growing animal protein demands by the consumers, and it began to use antibiotics for the animals even when they were not sick (Boyd, 2001). By 2010, after nearly 60 years after first FDA approval of antibiotics for farm animals, the use of antibiotics in animal feed for industrial production of livestock increased significantly. Currently, antibiotics are routinely added in majority of the feeds and water given for the livestock, poultry, and fish raised in the industrial farms. Antibiotics are used to promote animal growth, feed efficiency, and reduce stress for the animals that are intensively raised in confined spaces and unsanitary conditions. Furthermore, the bacteria from the concentrated animal feeding operations (CAFOs) of intensive farming are also known to spread into the environment through several channels including air, water and flies (Gibbs et al., 2006; Rule et al., 2008; Graham et al., 2009). Research reports indicate that the "non-therapeutic" use of antibiotics for farm animals have significantly contributed to the transmission of antibiotic resistant bacteria to humans (WHO, 2001).

Use of feed antibiotics has been a typical farming practice in much of the North America and Europe for several years. This practice of modern agriculture has proven to be critical for the health and wellbeing of food animals and in making the food safer. As summarized by Giguere et al. (2006), the use of antibiotics at sub-therapeutic levels to farm animals and poultry is known to promote their growth by various mechanisms such as, stimulating increased biosynthesis of vitamins by gut bacteria, reducing competition for nutrition between host and gut bacteria by minimizing the intestinal load of bacteria, increasing energy efficiency, inhibiting potentially harmful bacteria, stimulating the immune system, increasing intestinal enzyme activity, improving nutrient absorption by the host and by modulating rumen microbial metabolism to name a few. Thus, along with the general wellbeing of the animals, the antibiotic use for food animals brings the associated economic benefits for animal producers, and safer food for consumers. In traditional system of free-range animal farming, the application of antibiotic drugs may not be essential as the animals are not subjected to the stress conditions of intensive rearing in confined spaces as typically done in the industrialized animal farming practices. The need for therapeutic use of antibiotic drugs is inevitable in the contemporary intensive animal farming due to increased likelihood of spread of bacterial infections in such practices. The relation between the non-therapeutic use of antibiotics in farm animals and the rise of antibiotic resistance among bacteria infecting humans is widely debated. The routine, non-therapeutic uses of these drugs have increased the risk of growing antibiotic resistance (WHO, 2001; Barza and Gorbach, 2002; Marshall and Levy, 2011) as evidenced by studies showing transmission of antibiotic-resistant bacteria such as tetracycline-resistant *Escherichia coli* strains from chickens to humans (Levy et al., 1976a) and the livestock-associated methicillin-resistant *Staphylococcus aureus* (MRSA) to the people working in animal production sites (Smith and Pearson, 2011). However, sufficient scientific evidences also suggest that the current mode of antibiotic usage in production of poultry and livestock has added to the growing problem of antimicrobial resistance among the infectious agents and may pose a serious risk to the health of both animals and human beings. Although the quantity of total antimicrobial agents used in the United States is unknown, substantial amounts of antimicrobials are used especially for the production of livestock and poultry animals. An estimated 80% (over 29 million pounds) of the antibiotics sold in the United States are given to farm animals (FDA, 2009). The misuse of antibiotics for nonbacterial infections is not limited to agriculture alone but may also be followed in clinical applications to treat colds and other viral infections (Levy, 2002). Together, the use and misuse of antibiotics is intense and widespread globally.

Emergence of Antibiotic Resistance among Food Pathogens

Since from the early years of antibiotics usage in agriculture and health industry, a major public health concern has been the rise of antibiotic resistance among the pathogenic bacteria that can directly infect humans and/or spread through contaminated foods. Globally, the evidences of resistant pathogenic bacteria entering from the food animals (subjected to regular antibiotic treatments) into the human food chain are growing (Anderson et al., 2003; Schroeder et al., 2004; Van Looveren et al., 2001). The estimated costs attributed to the antibiotic-resistant infections in the United States exceed US$20 billion each year (Roberts et al., 2009). According to the Centers for Disease Control and Prevention, *Salmonella* and *Campylobacter* are known to cause 1.4 and 2 million cases foodborne illnesses respectively each year in the United States.

Antibiotic resistance from the improper use of antibiotics in farm animals, animals stocked at high densities or raised with intensive livestock farming practices, rapid animal growth, and unsanitary/stress conditions, are among the several practices known to facilitate the transmission of pathogenic bacteria. These factors may also contribute to the increased animal susceptibility to infectious diseases and the spread of microorganisms resistant to antimicrobials. In general, administering the antibiotics in modern animal farming is done as the therapeutic, prophylaxis and growth promoting applications of antibiotics for cattle, pigs, poultry, sheep and goats along with salmon and trout farming practices (WHO, 1997; Gustafson and Bowen, 1997; McEwen and Fedorka-Cray, 2002). The prophylactic and metaphylactic uses of antibiotics have also caused prevalence of antibiotic-resistant bacteria such as *E. coli* and *Campylobacter* in humans (Levy et al., 1976a,b; Smith et al., 1999).

Pathogenic bacteria enter the food from several avenues and one of them is through processing of slaughtered animals for food. A huge number of bacteria are present as a normal gut flora in the intestine of food animals and when such animals are slaughtered and processed for food, they usually contaminate the meat and other animal products or enter into the environment. This gut microflora also contains several pathogenic bacteria and when the animals are routinely given with antibiotics, certain pathogens may survive due to their resistance mechanism and continue to multiply even when they enter food and the environment. When people consume such contaminated food products or directly exposed to such bacteria through alternative routes, they may be severely ill, require extended time to recover or may even die from infection that may not be controlled by antibiotic treatments. This was further evident when 55 of the foodborne disease outbreaks that happened during 1973 to 2011 showed involvement of antibiotic-resistant pathogens coming from dairy, ground beef and poultry products in more than half of the outbreaks (DeWaal and Grooters, 2013). Further, 48 of the 55 outbreaks involved *Salmonella* spp., and 56% of the outbreaks involved pathogens showing resistance to five or more antibiotics. Among these 55 outbreaks, the drug-resistance patterns indicated that nearly 70% of the outbreaks were resistant to at least one antibiotic, streptomycin, and 85% of the outbreaks showed resistance to tetracycline and 50% to cephalosporin (DeWaal and Grooters, 2013). Note that the World Health Organization and US FDA consider these antibiotics as 'critically important' or 'highly important' for human medicine, and any resistance to these antibiotics could pose serious risk to human health.

It is known that the antibiotics used in food producing animals and in health industry for treating human infections belong mostly to the same classes and therapeutic animal antimicrobials include antibiotics listed as "critically important" for human medicine (WHO, 2007; WHO, 2012). For example, in poultry, the common antibiotics used are bacitracin, chlortetracycline, erythromycin, and penicillin. In case of severe infections, the fluoroquinolones are used for treating adult human beings and food animals whereas; cephalosporins are used for treating children against the infections caused by *Salmonella* spp. This pattern of using same class of antibiotics in animals and humans may further increase the risk of emergence and spread of resistant bacteria causing infections in both human and animals (WHO, 2007, 2014).

Food-producing animals serving as the reservoirs of certain bacteria that are known to transmit antimicrobial resistance to human through contaminated food products have thus become the main cause of public health concern. Some of those commonly transmitted bacteria include *Salmonella* spp., *Campylobacter* spp., *Escherichia coli*, *Enterococcus* spp., *Listeria monocytogenes, Staphylococcus aureus* (including MRSA) and *Clostridium difficile*, and most of these organisms are known to survive in food processing conditions and environments common with humans. Further, certain non-pathogenic commensal bacteria such as *Enterococcus* and *Escherichia coli* in the human and animal gut may also contaminate food and/or transmit antimicrobial resistance to pathogenic strains. Any increase in the resistance among these commensal and foodborne pathogenic bacteria can be a health concern for humans and animals as well. Antimicrobial resistance pattern in some of the selected foodborne pathogens is discussed below.

Antimicrobial resistance phenotypes have been recognized among the majority of the foodborne pathogens. For example, during 1982, Shiga-toxin (Stx)-producing *E. coli* (STEC), caused two major outbreaks of hemorrhagic colitis related with consumption of undercooked ground beef and (Wells et al., 1983) and this pathogen is recognized as emerging human pathogens with more than 200 STEC serotypes that are isolated from food animals, meats and other sources. According to a 2011 CDC report, STEC, O157:H7 is known to cause over 2,000 (4%) hospitalizations in the United States each year (Scallen et al., 2011). However, the leading causes of hospitalization were non-typhoidal *Salmonella* spp. (35%), norovirus (26%), *Campylobacter* spp. (15%), and *Toxoplasma gondii* (8%) (Scallen et al., 2011). The resistant strains of *Campylobacter*, *Salmonella* Typhimurium definitive phage type 104 (DT104), and *Salmonella* Newport have also been reported to transmit from animals to humans through foods of animal origin (Smith et al., 1999; Ribot et al., 2002). Increasing incidences of antimicrobial resistance is reported among several serotypes of *Salmonella*. As reported by the Centers for Disease Control and Prevention, the increasing resistance is particularly seen for cephalosporins that are the drug of choice to treat severe *Salmonella* infections and most of the resistance is seen in *Salmonella* Heidelberg and *Salmonella* Newport strains.

Salmonella Typhimurium is the most recognized serotype implicated in human and animal illnesses, and the DT104 strain is of much concern due to its multidrug resistance. This strain is known to be resistant to multiple antimicrobial agents including ampicillin, chloramphenicol, streptomycin, sulfonamides and tetracycline (Bolton et al., 1999; Besser et al., 2000; Ribot et al., 2002). This penta-resistance response is known as ACSSuT-resistance pattern of *Salmonella* Typhimurium DT104. However, it is not limited to these five antibiotics as some DT104 isolates have acquired resistance to other antibiotics as well such as trimethoprim and aminoglycosides and to quinolones (Low et al., 1997; Molbak et al., 1999; Threlfall et al., 1996). The multi-drug resistance (MDR) pattern of *Salmonella* Typhimurium DT 104 is largely attributed to the arrangement of more than one drug-resistant genes on a MDR-locus. The occurrence of this locus in other strains of *Salmonella* such as *Salmonella enterica* Agona (Boyd et al., 2001), *Salmonella* Paratyphi B, and *Salmonella* Albany (Doublet et al.,

2003; Meunier et al., 2002) indicates the transmission of MDR-gene locus between serotypes. Another important multi-drug resistant *Salmonella* serotype is Newport-MDR-AmpC which is known to occur in both humans and animals (CDC, 2003). This pathogen is known to be resistant to amoxicillin/clavulanic acid, cephalothin, cefoxitin, and ceftiofur, ceftriaxone, gentamicin, kanamycin, and trimethoprim/sulfamethoxazole along with the penta-resistance as seen in DR104 phenotype. This pathogen is also transmitted from farm animals to humans as evidenced by the molecular profiles of DNA fingerprints and the indistinguishable antimicrobial susceptibility profiles (Gupta et al., 2003). These serotypes show extended spectrum beta lactamase (ESBL) resistance pattern that is associated with the resistance to narrow, expanded and broad-spectrum cephalosporins and aztreonam. The β-lactamases are also transferable (can spread) as they are plasmid associated. Major concern to public health comes from the fluoroquinolone and ceftriaxone-resistant *Salmonella* as these antibiotics along with the third generation cephalosporins such as ceftriaxone are most commonly used for treating invasive *Salmonella* infections (Angulo et al., 2000; Fey et al., 2000).

Campylobacter is another major foodborne pathogen that has become a major public health concern (Isenbarger et al., 2002; Nachamkin et al., 2002). Bacteria belonging to this genus are common inhabitants of intestinal tract of domestic and wild animals including poultry and they are usually transmitted in contaminated food, water and animal faeces. Increasingly, *Campylobacter* isolates have developed resistance to fluoroquinolones and other antimicrobials such as macrolides, aminoglycosides, and betalactams, erythromycin and ciprofloxacin. Majority of the resistance is seen in *Campylobacter jejuni* and *C. coli* (Fitzgerald, 2008). The antibiotic resistance in this pathogen is known to occur via point mutations in genes encoding DNA gyrase, 23S rRNA (Ge et al., 2003), activation of resident MDR-efflux pumps (Lin et al., 2002) or by acquiring genes that alter antibiotic or its target (Wernet et al., 2001; LeBlanc et al., 1988). Apart from mutation in DNA gyrase, a multidrug efflux pump known as CmeABC is known to confer antimicrobial resistance to the fluoroquinolones and macrolides via the efflux mechanism in *Campylobacter jejuni* (Lin et al., 2002; Pumbwe and Piddock, 2002). This efflux pump plays a key role in both intrinsic and acquired resistance of *Campylobacter* and also known to work in works in synergy with GyrA mutations in causing fluoroquinolone resistance (Luo et al., 2003). Modification of the ribosome target-binding site by 23S rRNA mutation is known to confer macrolide resistance (Pfister et al., 2004; Poehlsgaard and Douthwaite, 2005) and the resistance to tetracyclines is conferred by the ribosomal protection protein (RPP) encoding *tet(O)* gene in *Campylobacter* (Connell et al., 2003). Several studies have linked the therapeutic and non-therapeutic use of fluoroquinolones and other antimicrobials, with the emergence and spread of resistance among *Campylobacter* strains. Further, the use of therapeutic fluoroquinolone in poultry flocks has shown to select for ciprofloxacin-resistant campylobacters in poultry (Griggs et al., 2005; Humphrey et al., 2005).

Escherichia coli O157:H7 is commonly associated with cattle and usually spreads via cattle manure and contaminated beef. The Shiga toxin-producing *E. coli* O157:H7 strains (STEC) are recognized as one of the major etiologic agents in hemorrhagic colitis (HC) and hemolytic-uremic syndrome (HUS) in humans (Thielman and Guerrant, 1999). Higher prevalence of resistance to several antibiotics is also reported among *E. coli* O157 isolates. However, in comparison with other food pathogens, the antimicrobial resistance in *E. coli* O157:H7 is lower and limited to fewer antimicrobials such as tetracycline, streptomycin, sulfamethoxazole, and trimethoprim (NARMS, 1999). The resistance among these bacteria to tetracycline is most common, followed by streptomycin resistance. Multiple antimicrobial resistances in STEC and non-STEC isolates can occur from the spread of genetic elements such as plasmids, transposons, and integrons conferring resistance to several of these antimicrobials. For example, acquisition of conjugative R plasmids is attributed to the occurrence of antimicrobial resistance in *E. coli* O157:H7 (Zhao et al., 2001). Thus, increased surveillance of emerging antimicrobial resistance in *E. coli* O157:H7 and other foodborne pathogens is of primary importance to ensure safety of foods and public health.

    Listeria monocytogenes is mainly known as a foodborne pathogen that causes life-threatening infections in high-risk populations. The bacterium is responsible for causing outbreaks involving raw and ready-to-eat meats, dairy foods, non-pasteurized milk and soft or semisoft cheeses, vegetables and fish (Pesavento, 2010; Sofos and Geornaras, 2010). It is known to cause listeriosis in pregnant women and the patient recovery depends greatly on early treatments with effective antibiotics (Goulet and Marchetti, 1996; Hof, 2003). Listeriosis is caused by consumption of food contaminated with *Listeria monocytogenes* and it has a high morbidity and mortality rates from 25% to 30%. Listeriosis treatment is usually done by administering high doses of a beta-lactam antibiotic (ampicillin or amoxicillin) alone or in combination with gentamicin; sulfamethoxazole (Hof, 2004). Majority of the food contamination with *L. monocytogenes* is reported to occur from the post-processing contamination and the bacterium is known to be naturally resistant to certain quinolones, fosfomycin, and expanded-spectrum cephalosporins (Troxler, 2000). It is believed that the isolates of *Listeria monocytogenes* exhibit general susceptibility to majority of the antibiotics (Jones and MacGowan, 1999). However, environmental and foodborne strains of *L. monocytogenes* are increasingly showing resistance to antibiotics such as penicillin, ampicillin, tetracycline, streptomycin, clindamycin, and even oxacillin and vancomycin causing a major public health concern (Chen et al., 2010; Conter et al., 2009; Wlash et al., 2001). Strains of *L. monocytogenes* are known to acquire resistance genes or transfer them to enterococci, staphylococci, and streptococci bacteria (Charpentier and Courvalin, 1999; Lungu et al., 2011).

Conclusion

As discussed above, the links between the foods of animal origin and the prevailing resistance patterns for antibiotics among pathogenic bacteria implicated in foodborne outbreaks are asserting the negative effects of improper use of feed antibiotics. The need continues for the antimicrobial stewardship and higher surveillance on a global basis of antimicrobial resistant phenotypes among foodborne pathogens of animal and human origin, with specific emphasis on their susceptibility to drugs used for treatments. The successful management of drug resistance should consider the usage pattern of antibiotics for specific foodborne pathogens infecting both humans and food animals. Thus, efforts to mitigate the increasing antimicrobial resistance among food pathogens should involve approaches to detect and reduce the spread of bacteria resistant for antimicrobials, reduce selection pressure for resistance development, understand the role of commensal bacteria in transmitting the resistance and educate the people involved in animal production about the effective measures to address this problem. As the survival and distribution of resistant bacteria may vary with climatic conditions and location of animal production facilities, along with the practices of using different types and quantities of antimicrobials, it is important to educate the animal producers about the health implications of such practices. However, the animal industry on the other hand is not in total agreement with this association between the spread of resistance and the antimicrobial use in livestock. This is due to the lack of sufficient evidences and the complexity involved in transmission of antibiotic resistant bacteria that can be implicated in infectious diseases coming from farm animals, via food to humans. The farming community argues that the decision about using an antibiotic is made only after consultation between the farmers and the veterinarians and an optimum and safe administration plan for use of antibiotic drugs to the animals is made. During the production of food animals, the antibiotics are used only after the approval of FDA that does a thorough review process involving safety testing. Nevertheless, it can be agreed that the problem of established and emerging antimicrobial resistance among the infectious agents is real. It should be noted that, due to the complexity associated with the antibiotic resistance spread which is contributed from several possible sources in the environment, this problem could be well addressed with common understanding between the regulatory agencies, health and agriculture industry, the scientific community and the public as a whole. The use of antibiotics should be followed judiciously by selecting optimum drug, dose and duration levels of treatment and by eliminating their inappropriate and excessive use (Weese, 2006). It is evident that the role of agricultural antimicrobial use and the transmission and emergence of antimicrobial resistance through foodborne pathogens continues to be a growing public health concern.

References

Angulo, F.J., K.R. Johnson, R.V. Tauxe and M.L. Cohen. 2000. Origins and consequences of antimicrobial resistant nontyphoidal *Salmonella*: implications for the use of fluoroquinolones in food animals. Microb. Drug Resist., 6: 77–83.

Animal Health Institute. 2012. Additives and Their Uses. Bloomington, MN: The Animal Health Institute.

Barza, M. and S.L. Gorbach (eds.). 2002. The need to improve antimicrobial use in agriculture: ecological and human health consequences. Clin. Infect. Dis., 34(Suppl. 3): S71–144.

Besser, T.E., M. Goldoft, L.C. Pritchett, R. Khakhria, D.D. Hancock, D.H. Rice, J.M. Gay, W. Johnson and C.C. Gay. 2000. Multiresistant *Salmonella* Typhimurium DT104 infections of humans and domestic animals in the Pacific Northwest of the United States. Epidemiol. Infect., 124: 193–200.

Bolton, L.F., L.C. Kelley, M.D. Lee, P.J. Fedorka-Cray and J.J. Maurer. 1999. Detection of multi-drug resistant *Salmonella enterica* serotype Typhimurium DT104 based on a gene which confers cross-resistance to florfenicol and chloramphenicol. J. Clin. Microbiol., 37: 1348–1351.

Boyd, D., G.A. Peters, A. Cloeckaert, K.S. Boumedine, E. Chaslus-Dancla, H. Imberechts and M.R. Mulvey. 2001. Complete nucleotide sequence of a 43-kilobase genomic island associated with the multidrug resistance region of *Salmonella enterica* serovar Typhimurium DT104 and its identification in phage type DT120 and serovars. Agona. J. Bacteriol., 183: 5725–5732.

Boyd, W. 2001. Making meat: science, technology, and American meat production. Technol. and Cult., 42(4): 631–664.

CDC. 2003. Preliminary FoodNet data on the incidence of foodborne illnesses—selected sites, United States, 2002. Centers for Disease Control and Prevention. Morb. Mortal. Wkly. Rpt., 52: 340–343.

Charpentier, E. and P. Courvalin. 1999. Antibiotic resistance in *Listeria* spp. Antimicrob. Agents Chemother., 43: 2103–2108.

Chen, Y.B., R. Pyla, T.J. Kim, J.L. Silva and Y.S. Jung. 2010. Antibiotic resistance in *Listeria* species isolated from catfish fillets and processing environment. Letters Appl. Microbiol., 50: 626–632.

Connell, S.R., C.A. Trieber, C.P. Dinos, E. Einfeldt, D.E. Taylor and K.H. Nierhaus. 2003. Mechanism of Tet(O)-mediated tetracycline resistance. EMBO Journal, 22(4): 945–953.

Conter, M., D. Paludi, E. Zanardi, S. Ghidini, A. Vergara and A. Ianieri. 2009. Characterization of antimicrobial resistance of foodborne *Listeria monocytogenes*. Int. J. Food Microbiol., 128: 497–500.

Dever, L.A. and T.S. Dermody. 1991. Mechanisms of bacterial resistance to antibiotics. Arch. Intern. Med., 151(5): 886–895.

DeWaal, S.C. and V.S. Grooters. 2013. Antibiotic Resistance in Foodborne Pathogens. Washington, DC: Center for Science in the Public Interest., 1–22.

Doublet, B., R. Lailler, D. Meunier, A. Brisabois, D. Boyd, M.R. Mulvey, E. Chaslus-Dancla and A. Cloeckaert. 2003. Variant Salmonella genomic island 1 antibiotic resistance gene cluster in *Salmonella enterica* serovar Albany. Emerg. Infect. Dis., 9: 585–91.

Fey, P.D., T.J. Safranek, M.E. Rupp, E.F. Dunne, E. Ribot, P.C. Iwen, P.A. Bradford, F.J. Angulo and S.H. Hinrichs. 2000. Ceftriaxone-resistant *Salmonella* infection acquired by a child from cattle. N. Engl. J. Med., 342: 1242–1249.

Fitzgerald, F., J. Whichard and I. Nachamkin. 2008. Diagnosis and antimicrobial susceptibility of *Campylobacter* species. pp. 227–243. *In*: I. Nachamkin, C.M. Szymanski and M.J. Blaser (eds.). *Campylobacter*. American Society for Microbiology, Washington, DC, USA.

Food and Drug Administration. 2009. Summary Report on Antimicrobials Sold or Distributed for Use in Food-Producing Animals. Rockville, MD: Center for Veterinary Medicine; 9 December 2010.

Ge, B., D.G. White, P.F. McDermott, W. Girard, S. Zhao, S. Hubert and J. Meng. 2003. Antimicrobial resistant *Campylobacter* species from retail raw meats. Appl. Environ. Microbiol., 69: 3005–3007.

Giguère, S. 2006. Antimicrobial drug action and interaction: an introduction. *In*: S. Giguère, J.F. Prescott, J.D. Baggot, R.D. Walker and P.M. Dowling (eds.). Antimicrobial Therapy in Veterinary Medicine 4th edn. Blackwell Publishing, Ames Iowa, USA.

Goulet, V. and P. Marchetti. 1996. Listeriosis in 225 non-pregnant patients in 1992: clinical aspects and outcome in relation to predisposing conditions. Scand. J. Infect. Dis., 28: 367–374.

Griggs, D.J., M.M. Johnson, J.A. Frost, T. Humphrey, F. Jørgensen and L.J.V. Piddock. 2005. Incidence and mechanism of ciprofloxacin resistance in *Campylobacter* spp. isolated from commercial poultry flocks in the United Kingdom before, during, and after fluoroquinolone treatment. Antimicrobial Agents and Chemotherapy, 49(2): 699–707.

Gupta, A., J. Fontana, C. Crowe, B. Bolstorff, A. Stout, S. Van Duyne, M.P. Hoekstra, J.M. Whichard, T.J. Barrett and F.J. Angulo. 2003. Emergence of multidrug-resistant *Salmonella enterica* serotype Newport infections resistant to expanded-spectrum cephalosporins in the United States. J. Infect. Dis., 188: 1707–1716.

Gustafson, R.H. and R.E. Bowen. 1997. Antibiotic use in animal agriculture. J. Appl. Microbiol., 83: 531–41.

Hof, H. 2004. An update on the medical management of listeriosis. Expert Opin. Pharmacother., 5: 1727–1735.

Hof, H. 2003. Listeriosis: therapeutic options. FEMS Immunol. Med. Microbiol., 35: 203–205.

Humphrey, T.J., F. Jørgensen, J.A. Frost, H. Wadda, G. Domingue, N.C. Elviss, D.J. Griggs and L.J.V. Piddock. 2005. Prevalence and subtypes of ciprofloxacin-resistant *Campylobacter* spp. in commercial poultry flocks before, during, and after treatment with fluoroquinolones. Antimicrobial Agents and Chemotherapy, 49(2): 690–698.

Isenbarger, D.W., C.W. Hoge, A. Srijan, C. Pitarangsi, N. Vithayasai, L. Bodhidatta, K.W. Hickey and P.D. Cam. 2002. Comparative antibiotic resistance of diarrheal pathogens from Vietnam and Thailand, 1996–1999. Emerg. Infect. Dis., 8(2): 175–180.

Joint FAO/WHO/OIE Expert Meeting on Critically Important Antimicrobials. Rome, Italy, Food and Agriculture Organization of the United Nations/World Organization for Animal Health/World Health Organization/2007.

Jones, E.M. and A.P. MacGowan. 1995. Antimicrobial chemotherapy of human infection due to *Listeria monocytogenes*. Euro. J. Clin. Microbiol. Infect. Dis., 14: 165–175.

LeBlanc, D.J., L.N. Lee, B.M. Titmas, C.J. Smith and F.C. Tenover. 1988. Nucleotide sequence analysis of tetracycline resistance gene tetO from *Streptococcus mutans* DL5. J. Bacteriol., 170: 3618–3626.

Levy, S.B. 2002. The antibiotic paradox: how the misuse of antibiotics destroys their curative powers, 2nd edn. Perseus Publishing, Cambridge, MA.

Levy, S.B., G.B. FitzGerald and A.B. Macone. 1976a. Spread of antibiotic-resistant plasmids from chicken to chicken and from chicken to man. Nature, 260: 40–42.

Levy, S.B., G.B. FitzGerald and A.B. Macone. 1976b. Changes in intestinal flora of farm personnel after introduction of a tetracycline-supplemented feed on a farm. New Engl. J. Med., 295: 583–588.

Lin, J., L.O. Michel and Q.J. Zhang. 2002. CmeABC functions as a multidrug efflux system in *Campylobacter jejuni*. Antimicrob. Agents Chemother., 46(7): 2124–2131.

Lin, J., L.O. Michel and Q. Zhang. 2002. CmeABC functions as a multidrug efflux system in *Campylobacter jejuni*. Antimicrob. Agents Chemother., 46: 2124–31.

Low, J.C., M. Angus, G. Hopkins, D. Munro and S.C. Rankin. 1997. Antimicrobial resistance of *Salmonella enterica* typhimurium DT104 isolates and investigation of strains with transferable apramycin resistance. Epidemiol. Infect., 118: 97–103.

Lungu, B., C.A. O'Bryan, A. Muthaiyan, S.R. Milillo, M.G. Johnson, P.G. Crandall and S.C. Ricke. 2011. *Listeria monocytogenes*: Antibiotic resistance in food production. Foodborne Pathog. Dis., 8: 569–578.

Luo, N., O. Sahin, J. Lin, L.O. Michel and Q. Zhang. 2003. *In vivo* selection of *Campylobacter* isolates with high levels of fluoroquinolone resistance associated with gyrA mutations and the function of the CmeABC efflux pump. Antimicrob. Agents and Chemother., 47(1): 390–394.

Marshall, B.M. and S.B. Levy. 2011. Food animals and antimicrobials: impacts on human health. Clin. Microbiol. Rev., 24: 718–33.

McEwen, S.A. and P.J. Fedorka-Cray. 2002. Antimicrobial use and resistance in animals. Clin. Infect. Dis., 34(Suppl.) 3: S93–S106.

Meunier, D., D. Boyd, M.R. Mulvey, S. Baucheron, C. Mammina, A. Nastasi, E. Chaslus-Dancla and A. Cloeckaert. 2002. *Salmonella enterica* serotype Typhimurium DT 104 antibiotic resistance genomic island I in serotype paratyphi B. Emerg. Infect. Dis., 8: 430–433.

Molbak, K., D.L. Baggesen, F.M. Aarestrup, J.M. Ebbesen, J. Engberg, K. Frydendahl, P. Gerner-Smidt, A.M. Petersen and H.C. Wegener. 1999. An outbreak of multidrug-resistant, quinolone-resistant *Salmonella enterica* serotype typhimurium DT104. N. Engl. J. Med., 341: 1420–1425.

Nachamkin, I., H. Ung and M. Li. 2002. Increasing fluoroquinolone resistance in *Campylobacter jejuni*, Pennsylvania, USA, 1982–2001. Emerg. Infect. Dis., 8(2): 1501–1503.

NARMS. National Antimicrobial Resistance Monitoring Program: enteric pathogens. Food and Drug Administration, Centers for Disease Control and Prevention. Rockville, MD, USDA, pp. 1–115.

Pesavento, G., B. Ducci, D. Nieri, N. Comodo and A. Lo Nostro. 2010. Prevalence and Antibiotic Susceptibility of *Listeria* spp. isolated from raw meat and retail foods. Food Control, 21(5): 708–713.

Pfister, P., S. Jenni, J. Poehlsgaard, A. Thomas, S. Douthwaite, N. Ban and E.C. Bottger. 2004. The structural basis of macrolide-ribosome binding assessed using mutagenesis of 23 S rRNA positions 2058 and 2059. J. Mol. Biol., 342(5): 1569–1581.

Poehlsgaard, J. and S. Douthwaite. 2005. The bacterial ribosome as a target for antibiotics. Nat. Rev. Microbiol., 3(11): 870–881.

Pumbwe, L. and L.J.V. Piddock. 2002. Identification and molecular characterization of CmeB, a *Campylobacter jejuni* multidrug efflux pump. FEMS Microbiol. Let., 206(2): 185–189.

Ribot, E.M., R.K. Wierzba, F.J. Angulo and T.J. Barrett. 2002. *Salmonella enterica* serotype Typhimurium DT104 isolated from humans, United States, 1985, 1990, and 1995. Emerg. Infect. Dis., 8: 387–391.

Ribot, E.M., R.K. Wierzba, F.J. Angulo and T.J. Barrett. 2002. *Salmonella enterica* serotype Typhimurium DT104 isolated from humans, United States, 1985, 1990, and 1995. Emerg. Infect. Dis., 8(4): 387–391.

Roe, M.T. and S.D. Pillai. 2003. Monitoring and identifying antibiotic resistance mechanisms in bacteria. Poult. Sci., 82: 622–626.

Russell, A.D. and I. Chopra. 1996. Understanding Antibacterial Action and Resistance, 2nd ed. London: Ellis Horwood.

Russell, A.D. 1991. Mechanisms of bacterial resistance to non-antibiotics: food additives and food and pharmaceutical preservatives. J. Appl. Bacteriol., 71: 191–201.

Scallan, E., R.M. Hoekstra, F.J. Angulo, R.V. Tauxe, M.A. Widdowson, S.L. Roy, J.L. Jones and P.M. Griffin. 2011. Foodborne illness acquired in the United States-major pathogens. Emerg. Infect. Dis., 1–21.

Schmieder, R. and R. Edwards. 2012. Insights into antibiotic resistance through metagenomic approaches. Future Microbiol., 7(1): 73–89.

Smith, T.C. and N. Pearson. 2011. The emergence of *Staphylococcus aureus* ST398. Vector Borne Zoonotic Dis., 11: 327–339.

Smith, K.E., J.M. Besser, C.W. Hedberg, F.T. Leano, J.B. Bender, J.H. Wicklund, B.P. Johnson, K.A. Moore and M.T. Osterholm. 1999. Quinolone-resistant *Campylobacter jejuni* infections in Minnesota, 1992–1998. New England Journal of Medicine, 340.

Sofos, J.N. and I. Geornaras. 2010. Overview of current meat hygiene and safety risks and summary of recent studies on biofilms, and control of *Escherichia coli* O157:H7 in Nonintact, and *Listeria monocytogenes* in ready-to-eat, meat products. Meat Sci., 86(1): 2–14.

Thielman, N.M. and R.L. Guerrant. 1999. *Escherichia coli*. pp. 188–200. *In*: V.L. Yu, T.C. Merigan, Jr. and S.L. Barriere (eds.). Antimicrobial Therapy and Vaccines. The Williams and Wilkins Company, Baltimore, MD.

Threlfall, E.J., J.A. Frost, L.R. Ward and B. Rowe. 1996. Increasing spectrum of resistance in multi-resistant *Salmonella* Typhimurium. Lancet, 347: 1053–1054.

Troxler, R., A. von Graevenitz, G. Funke, B. Wiedemann and I. Stock. 2000. Natural antibiotic susceptibility of *Listeria* species: *L. grayi*, *L. innocua*, *L. ivanovii*, *L. monocytogenes*, *L. seeligeri* and *L. welshimeri* strains. Clin. Microbiol. Infect., 6: 525–535.

Walsh, D., G. Duffy, J.J. Sheridan, I.S. Blair and D.A. McDowell. 2001. Antibiotic resistance among *Listeria*, including *Listeria monocytogenes*, in retail foods. J. Appl. Microbiol., 90: 517–522.

Weese, J.S. 2006. Prudent use of antimicrobials. *In*: S. Giguère, J.F. Prescott, J.D. Baggot, R.D. Walker and P.M. Dowling (eds.). Antimicrobial Therapy in Veterinary Medicine 4th edn., Blackwell Publishing, Ames Iowa, USA.

Wells, J.G., B.R. Davis, I.K. Wachsmuth, L.W. Riley, R.S. Remis and R. Sokolow. 1983. Laboratory investigation of hemorrhagic colitis outbreaks associated with a rare *Escherichia coli* serotype. J. Clin. Microbiol., 185: 12–20.

Werner, G., B. Hildebrandt and W. Witte. 2001. Aminoglycoside-streptothricin resistance gene cluster aadE-sat4-aphA-3 disseminated among multi-resistant isolates of *Enterococcus faecium*. Antimicrob. Agents Chemother., 45: 3267–3269.

WHO. 2014. Antimicrobial resistance: global report on surveillance 2014 (http://www.who.int/drugresistance/documents/surveillancereport/en/) (Accessed January 30, 2015).

WHO. 2012. World Health Organization Advisory Group on Integrated Surveillance of Antimicrobial Resistance (AGISAR). 2011. Critically important antimicrobials for human medicine. 3rd revision (http://www.who.int/foodsafety/publications/antimicrobials-third/en/) (Accessed January 20, 2015).

World Health Organization. 2001. WHO Global Strategy for Containment of Antimicrobial Resistance, Switzerland.

Methods and Technology for Rapid and Accurate Detection of Foodborne Pathogens

Francisco B. Elegado,[1,]* *Teresita J. Ramirez,*[1]
Susumu Kawasaki,[2] *Evangelyn C. Alocilja*[3] *and Sabina Yeasmin*[4]

Introduction

Foodborne pathogens are very diverse in their nature and keep causing major public health problems worldwide. Many high-risk pathogens that cause diseases in humans are transmitted through various food items. Therefore, the microbiological safety of food has become an important concern for consumers, the industry, and the regulatory agencies. Conventional pathogen detection methods largely rely on microbiological and biochemical analyses, which are highly accurate but overly time-consuming, cost-ineffective and non-amenable to integration for on-site diagnosis. Besides, successful execution of pathogen identification and detection by conventional methods require extensive training and experience. Alternative rapid but accurate methods for pathogen detection have therefore been sought to overcome these limitations. Improved rapidity can be applied at each step of the analysis, i.e., the sampling process, sample

[1] National Institute of Molecular Biology and Biotechnology, University of the Philippines Los Banos, College, Laguna 4031, Philippines.
[2] National Food Research Institute, 2-1-12 Kannondai, Tsukuba Japan.
[3] Department of Biosystems and Agricultural Engineering, Michigan State University, East Lansing, Michigan, U.S.A.
[4] Department of Genetic Engineering and Biotechnology, University of Dhaka, Dhaka-1000, Bangladesh.
*Corresponding author: fbelegado@hotmail.com

treatment and detection/enumeration procedure. Although labour-saving and automated methods speed up the processes of sampling and sample treatment, thus improving the laboratory's output, the influence on the total analysis time is usually negligible due to the incubation time required for traditional culture-based methods. A real shortening of the analytical time can only be obtained if alternatives to the traditional incubation methods are developed.

Advances in immunological methods such as enzyme-linked immunosorbent assay (ELISA) have paved the way towards development of easier and quicker pathogen detection methods, relying on the recognition specificity of antibodies (Abs). Immunological methods however suffer from cross-reactivity of polyclonal Abs, high production cost of monoclonal Abs, need for sample pre-processing and pre-enrichment due to low processing sample volume and lower limit of detection. Polymerase chain reaction (PCR) is yet another method that leverages the nucleic acid complementarity-based specificity of pathogen detection. Recently, more sophisticated traditional analytical methods such as liquid/gas chromatography coupled with mass spectrophotometry have been used for more accurate analysis of pathogen. Although these methods have enjoyed tremendous popularity, their feasibility towards point-of-care onsite pathogen monitoring tools is hard to realize. Development of alternative tools for fast, accurate and sensitive detection of pathogens has therefore been an area of continued interest to researchers across the globe. Figure 1 outlines the steps involved in analysis of a food sample by various popular detection methods and time involved to reach a conclusive pathogen identity.

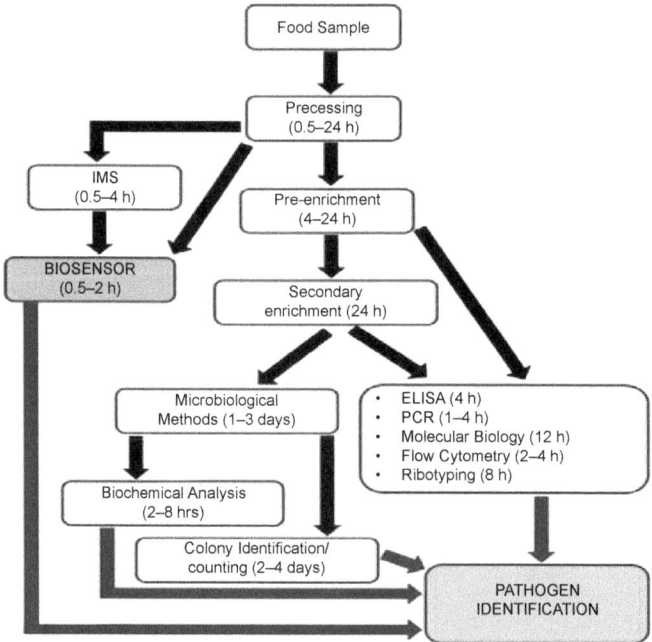

Figure 1. The steps in analysis of a food sample by various popular detection methods (adapted from Amit Singh et al., 2013).

Considering the broad range of analytical procedures available particular requirements were defined which an optimum method should meet. High sensitivity, which is defined as the lowest amount of microorganisms detectable, should be of primary importance. Likewise, high accuracy is essential. The analytical result should meet the true value and be reproducible (i.e., high precision). As explained above, rapidity is another important factor. Under practical conditions or economic considerations, the use of simple, inexpensive, universally applicable and less laborious methods are favored. Ideally, rapid methods should enable a quick estimation/detection of pathogen of interest that food manufacturers are able to take corrective actions immediately in the course of the manufacturing process. However, the majority of methods characterized as "rapid" do not meet this demand. Nevertheless, they offer a more or less pronounced advantage in analytical time compared to their conventional equivalent by eliminating laborious and/or subjective elements through mechanization and automation.

A variety of morphological, physiological and biochemical tests are used for identification of microorganisms in conventional methods. Now several commercially available kits have been developed to simplify and automate the identification of individual organisms, the result of which is comparable to that of conventional identification systems. Recently, various kinds of rapid detection, identification, and monitoring methods have been developed for foodborne pathogens, including nucleic-acid-based methods, immunological methods, and biosensor-based methods, etc. This chapter discusses the principles, characteristics, and applications of recent rapid detection methods for foodborne pathogens.

Rapid Detection Method

An effective detection method for foodborne pathogens, methods should meet the following requirements (Mandal et al., 2010):

1. Detection method must be rapid
2. The method should detect the desired specific pathogens
3. Method must be sensitive to detect small numbers of pathogens
4. The detection method should produce a quantitative analysis to help determine the severity of the hazard
5. The method should be multiplex (i.e., capable of detecting more than one contaminant simultaneously).

Rapid detection methods are generally classified as either modified and automated conventional methods, biosensors, immunological methods, or nucleic acid-based assays.

The following seven major categories of methods for detecting foodborne pathogens are available till today:

1. Classical culture-based methods
2. Microscopic-based enumeration methods
3. Methods based on growth and metabolic activity
4. Immunological methods

5. Bacteriophage-based detection methods
6. Nucleic acid-based assay methods
7. Biosensor-based methods.

Classical culture-based methods

Conventional methods for the enumeration of bacteria in food are colony count methods. In the colony count method the total number of bacteria in a product is determined by inoculating dilutions of suspensions of the sample onto the surface of a solid growth medium by the spread-plate method or by mixing the test portion with the liquefied agar medium in petri dishes (pour plate method). Enumeration is performed after incubation for fixed periods at temperatures varying from 7 to 55°C in an aerobic, micro-aerobic or anaerobic atmosphere depending upon the target organisms. During incubation each individual cell will multiply into a colony that is visible to the naked eye. Classical culture methods have a quantification limit of ca. 4 cfu/ml for liquid foods or ca. 40 cfu/g for solid foods which corresponds to ca. 4 colonies/plate if 1 ml of the primary suspension is used for plating. Based on ISO 7218, presence of 1–3 colonies/plate only indicates detection of the target organism and numbers obtained as such should only be reported as estimated numbers.

Chromogenic and fluorogenic isolation media

The recognition of presumptive colonies of target organisms has been facilitated by introducing chromogenic and fluorogenic media. These are microbiological growth media that contain enzyme substrates linked to a chromogen (colour reaction), fluorogen (fluorescent reaction) or a combination of both. The target population is characterized by enzyme systems that metabolize the substrate (sugar or amino acid) to release the chromogen/fluorogen. This results in a colour change in the medium and/or fluorescence under long wave UV light. The incorporation of such fluorogenic or chromogenic enzyme substrates into a selective medium can eliminate the need for subculture and further biochemical tests to establish the identity of certain microorganisms.

Modified cultural methods

A variety of rapid methods have been elaborated, which predominantly aim to reduce the workload and facilitate the work flow by reducing the manipulations and/or the necessity for a full lab infrastructure and not necessarily shorten the time to detection. Some of these modified cultural methods are based upon the colony count method, e.g., 3M™ Petrifilm™ and Compact Dry, Hydrophobic Grid Membrane Filter (HGMF), whereas others make use of the principle of the MPN method, e.g., TEMPO®, SimPlate®, Colilert®, Soleris™ and ISO-GRID. In addition, several modifications in sample preparation, plating techniques, counting and identification systems have made these conventional methods faster and easier.

Microscopic-based enumeration methods

Although the flow cytometry may intrinsically detect individual cells, because of the small volumes samples, it may still not be sensitive enough to detect bacterial concentrations less than 10^3–10^4 bacteria ml^{-1} because of the low inoculation volume. In these situations, an enrichment prior to flow cytometry analysis may be envisaged to increase the bacterial load of the sample to a level at which it may be detected. The principle of flow cytometry mentioned above in the part of enumeration methods. Alternatively, instead of using a flow cytometer, stained cells may be visualized and detected by means of an epifluorescent microscope. The use of epifluorescent microscopy in the frame of fluorescent *in situ* hybridization (FISH) is described in the next section on molecular-based detection methods.

Flow cytometry

Flow cytometry enables both qualitative and quantitative analysis of microbial cells in liquids. The sample is injected in a thin, rapidly moving carrier fluid which passes through a light beam. The previously fluorescently-labelled cells are detected one by one with a photoelectric unit. By using non-specific and specific fluorochromes, different wavelengths and measuring at different angles, it is feasible to discriminate between bacteria in mixed populations. The practical use of flow cytometry is still limited to a few examples. However, since the possible applications are numerous, it should be considered as a promising technology in the future. Most micro-organisms are optically too similar to resolve from each other, or from debris, and therefore labeling with fluorescent dyes can be used to probe the viability and metabolic state of micro-organisms. Flow cytometry is often used in combination with live-dead staining (SYTO-9 in combination with propidium iodide) or in combination with viability staining (e.g., ChemChrome (AES Chemunex)) indicating metabolic activity. For routine analyses of milk water, beverage or dairy industries flow cytometry is used.

Direct Epifluorescent Filter Technique (DEFT)

The DEFT is a microscopic cell counting method. A pre-treated sample (detergents and proteolytic enzymes) is filtered over a polycarbonate membrane. The microbial cells are concentrated and collected on the membrane, where they are stained with fluorescent dyes. The microscopic analysis is performed in this surface. Incident light illumination (epifluorescence) is used to examine the filter surface. This detection can be automated by linking the microscope to an image analyzing system. The actual staining and counting takes less than 0.5–1 h, but sample pre-treatment steps lengthen the total detection time. The detection limit is 10^4–10^5 cells/ml. Just as flow cytometry, this technique has been described for the determination of total count in liquid food.

Fourier Transform Infrared Spectroscopy (FT-IR)

Fourier transform infrared spectroscopy is used to generate bacterial spectral scans based on the molecular composition of a sample and mainly consists of the infrared source, the sample and the detector. It is a nondestructive rapid method and sample identification depends on the available spectral library. When IR is absorbed or transmitted through the sample to the detector, it generates a scan or fingerprint profile. A library of spectral scans can be generated for different bacterial species and strains, which can be used for future comparison. This method requires transfer of cells (biomass) from the growth media to an IR reflecting substrate for spectral collection. FT-IR has been used for classification or identification of several foodborne pathogens: *Yersinia, Staphylococcus, Salmonella, Listeria, Klebsiella, Escherichia, Enterobacter, Citrobacter*, etc. (Gupta et al., 2005; Mossoba et al., 2005; Sivakesava et al., 2004). FT-IR photoacoustic spectroscopy was used for the identification of spores of several Bacillus species with 100% accuracy (Thompson et al., 2003).

Solid Phase Cytometry (SPC)

SPC is a novel technique that allows rapid detection of bacteria at single cell level, without the need for growth phase (Haese and Nelis, 2002). The short time detection inherent in this approach is of considerable advantage over conventional plating techniques especially for slow growing bacteria. SPC combines aspects of flow cytometry and epifluorescene microscopy. The microbes are isolated from their matrix from membrane filter fluorescently labeled with argon laser excitable dye and automatically counted by laser scanning device. During a 3-min scanning process the entire membrane filter surface is scanned yielding a theoretical detection limit of one cell per membrane filler. During scanning two photomultiplier tubes with wavelength 500 to 530 nm (green) and 540–585 nm (amber) detects the fluorescent light emitted by labeled cells. The signals are processed with softwares which differentiate between viable signals (target cells) and background noises (electronic noise and fluorescent panicles). Scanned results are displayed as primary and secondary maps. Actual nature of each fluorescent spot can be further examined by epifluorescent microscope.

Methods based on growth and metabolic activity

Impedance

This method is based on the principle that bacteria actively growing in a culture medium produce positively or negatively charged end-products (early stages of breakdown of nutrients) that cause an impedance variation of the medium. This variation, which is proportional to the change in the number of bacteria in the culture, makes it possible to measure bacterial growth. The time at which growth is first detected, referred to as detection time (DT), is inversely proportional to the log number of bacteria in the sample, which means that bacterial counts can be predicted from DT. A calibration process is required initially to establish a mathematical relation experimentally between DT and the log number of target bacteria. BacTrac (Sylab) is an example of an impedance-based method. It can be used for qualitative and quantitative applications to detect the bio burden load, the quantity of selected groups of microorganisms or the presence/absence of selected pathogens.

ATP bioluminescence

This technique measures light emission produced due to the presence of ATP, which is involved in an enzyme–substrate reaction between luciferin and luciferase (bioluminescence). The quantity of light produced (measured as Relative Light Units RLUs) is proportional to the concentration of ATP and, thus, to the number of micro-organisms in the original sample. ATP bioluminescence can be used for enumeration of total count but it is only applicable if high numbers of bacteria are present (> 10,000 cfu/g). Calibration curves should be established (per type of food or surface) to correlate ATP measurements to microbial counts. Presence of ATP is not restricted to bacterial cells, but is a basic compound of any biological material. Thus ATP bioluminescence is generally used as a rapid indicator of the bio-load present. As such this technique is mostly used to estimate the total surface cleanliness, including the presence of organic debris and microbial contamination, providing results within less than 5 min.

Immunoassays

All immunoassays are based on the highly specific binding reaction between antibodies and antigens. The selection of an appropriate antibody (monoclonal or polyclonal) is the determinant factor for the method's performance. Usually, any positive result for pathogens obtained with immunoassays is considered as presumptive and requires further confirmation. Detection limit is approximately 10^4–10^5 cfu/ml, depending upon the type of antibody and its affinity for the corresponding epitope meaning that for the enrichment step, often a two-step procedure is needed. Several types of immunoassays are available in food diagnostics of which lateral flow devices (LFD), enzyme linked immunosorbent assays (ELISAs) and enzyme linked fluorescent assays (ELFAs) are widely used. Immunomagnetic separation (IMS) assays, although a sample preparation tool instead of a detection method, have been developed as an aid in reducing the time for the enrichment step prior to detection.

Lateral Flow Devices (LFD)

An LFD generally comprises a porous membrane, typically nitrocellulose, with an immobilized capture protein for the target analyte, forming a visible line in a viewing window, due to nanoparticles of gold or colored latex particles, after contact with the specified analyte. For the test to be valid a control line should form in a second viewing window. In most devices, an antibody is commonly used as capture protein, which specifically binds and captures a particular antigen if present in the sample. In most cases a sandwich assay is used. The test is fast (reading in terms of minutes) simple in use and in interpretation.

Enzyme-Linked Immunosorbent Assay (ELISA) and Enzyme-Linked Fluorescent Assays (ELFA)

ELISA is a biochemical technique that couples an immunoassay with an enzyme assay. In most of the alternative methods a sandwich ELISA is used. The sandwich ELISA comprises different steps. Specific antibodies are affixed to the surface of the walls of a 96-well microtiter plate. The sample, with an unknown amount of target antigen, is added and allowed to bind to the affixed antibodies. Unbound antigen is removed by a washing step. In a second phase antibodies targeting the antigen are added again to the wells. This step is followed by the addition of an enzyme-labeled secondary antibody. This secondary antibody is allowed to bind to the previously added antibody. Washing steps are included to remove non-bound secondary antibodies. In the final step a substrate is added that the enzyme can convert to a detectable signal. The ELISA detection itself only takes 2–3 h. Nowadays many ELISA tests are available as robotized automated systems to reduce the hands-on time but also to improve the reproducibility and standardization of each step of the assay.

Immunomagnetic separation and concentration

Super paramagnetic particles can be coated with antibodies, allowing specific capture and isolation of intact cells directly from a complex sample suspension without the need for column immobilization or centrifugation. This method is now generically referred to as immunomagnetic separation and concentration (IMS). In the ISO 16654 detection methods for *E. coli* O157:H7 from foods, IMS is incorporated, in order to pick up selectively the O157 serotype. In *Salmonella* spp. detection methods IMS is often integrated to replace the selective enrichment step, and thus approximately gaining 24 h. Monosized super paramagnetic polymer particles known as "Dynabeads" are available commercially from Invitrogen-Dynal. Pathatrix, an automated system, is a patented re-circulating immunomagnetic separation technology. This technique enables to scaleup the application of IMS from the usual 1 ml to ca. 400 ml volumes. By re-circulating the sample over a capture phase, comprising immobilized antibody coated magnetic beads, the sensitivity of the capture is increased and thus accomplishes potential significant reduction of time to detection (Lee and Deininger, 2004). A high volume wash also enables the efficient removal of the sample matrices, non-specific micro-organisms and PCR inhibitors. This technique is approved by AOAC INTERNATIONAL to perform as described.

Bacteriophage-based detection methods

Bacteriophages are viruses infecting bacteria. Phages are extremely host-specific. Most bacteria can be infected by particular phages and it is common that a given phage can recognize and infect only one or a few strains or species of bacteria. The specificity of these phages is partly mediated by tail-associated proteins that distinctively recognize surface molecules of susceptible bacteria. Bacteriophages or proteins of bacteriophages have been included in various ways in detection methods for pathogens.

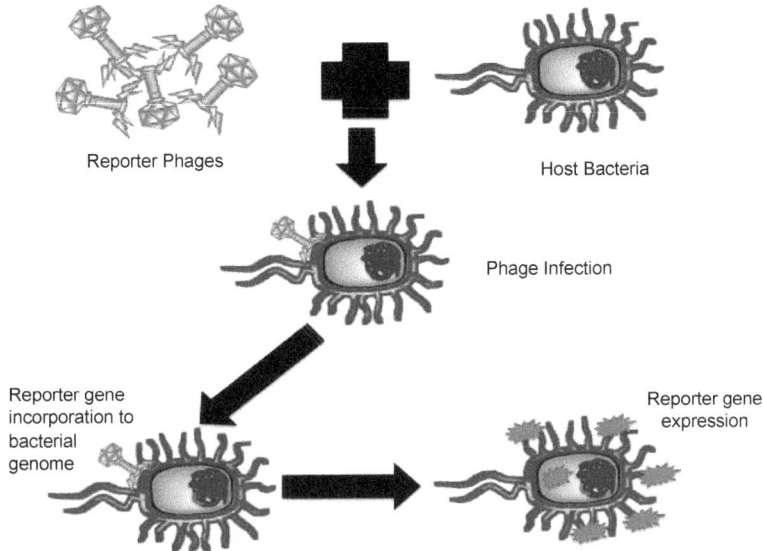

Figure 2. A schematic explaining the underlying principle of reporter phage-based detection of a pathogen of interest (adapted from Amit Singh et al., 2013).

The specific bacteriophage tail-associated proteins can be attached to paramagnetic beads to capture bacteria in suspension. The bacteria–bead complex can be integrated in fast detection protocols. Paramagnetic beads coated with phage proteins were shown to perform much better than commercially available antibody-based beads, both with respect to sensitivity and percent recovery. The *Listeria* Capture kit (Hyglos) can be integrated as part of a rapid detection method in a similar way as IMS. Another example of the integration of bacteriophage recombinant protein technology in detection methods is the new line of the VIDAS® system, called VIDAS® UP for the detection of *E. coli* O157 after 6 h. The VIDAS® UP is an automated qualitative test for the detection of *E. coli* O157 in food, feed, environmental samples and soil. Results of the test, as such are obtained within hours although also a prior enrichment takes 6–24 h depending upon the type of micro-organism.

Nucleic acid-based methods

The nucleic acid-based food pathogen detection assays is highly specific, as they detect specific nucleic acid sequences in the target organism by hybridizing them to a short synthetic oligonucleotide complementary to the specific nucleic acid sequence. The selection of a specific DNA sequence, to serve as a probe or primer, along with the conditions for hybridization, is the determinant factor for the specificity of these molecular methods.

Table 1. Nucleic acid-based techniques employed for pathogen detection (adapted from Zhao et al., 2014).

Techniques	Detected pathogens	Limit of detection	Assay time	References
Multiplex PCR	*Escherichia coli* O157:H7, *Salmonella*, and *Shigella*	8×10^{-1} CFU/g (or CFU/ml) in apple cider, cantaloupe, lettuce, tomato, and watermelon; 8×10^{1} CFU/g in alfalfa sprouts	30 h	Li and Mustapha, 2004
	Salmonella spp., *Listeria monocytogenes*, and *Escherichia coli* O157:H7	10^{3} CFU/ml for each pathogen by pure culture, 1 cell per 25 g of inoculated pork sample	30 h	Kim et al., 2012
	Escherichia coli O157:H7, *Salmonella*, and *Shigella*	10^{5} CFU/g for *Escherichia coli* O157:H7, 10^{3} CFU/g for *Salmonella*, and 10^{4} CFU/g for *Shigella*	3 h	Kawasaki et al., 2005
	Salmonella spp., *Listeria monocytogenes*, and *Escherichia coli* O157:H7	5 CFU/25 g of inoculated sample after 20 h of Enrichment	24 h	Verstraete et al., 2012
	O26, O103, O111, O145, sorbitol fermenting (SF) O157 and non-sorbitol fermenting (NSF) O157	Minced beef and sprouted seeds enrichment broths were inoculated with 5×10^{4} CFU/ml STEC O157, and raw-milk cheese enrichment broths with 5×10^{5} CFU/ml STEC O157		
Quantitative PCR	*Listeria monocytogenes*	Detect as few as 100 CFU/g and quantify as few as 1,000 CFU/g	3 h	Rodriguez-Lazaro et al., 2004
	Salmonella	10^{3} to 10^{4} CFU/ml of inoculums in broth without enrichment, < 10 CFU/ml of inoculum in broth after 18 h enrichment		Nam et al., 2005
	Shigella	0.12 to 0.74 CFU per reaction	24 h	Lin et al., 2010
	Listeria monocytogenes and *Staphylococcus aureus*	7 CFU/g in coleslaw for *L. monocytogenes* and 2 CFU/g in raw minced meat for *S. aureus*		Martinon and Wilkinson, 2011

LAMP	*Shigella* and enteroinvasive *Escherichia coli*	8 CFU per reaction	2 h	Song et al., 2005
	Streptococcus pneumonia	10 or more copies of purified *S. pneumoniae* DNA	1 h	Seki et al., 2005
	Salmonellae	3.4 to 34 viable *Salmonella* cells in pure culture and 6.1×10^3 to 6.1×10^4 CFU/g in spiked produce samples	3 h	Chen, 2008
	Vibrio parahaemolyticus	5.3×10^2 CFU/ml	1 h	Wu et al., 2007
NASBA	*Chlamydia pneumoniae* *Chlamydophila pneumoniae*	10 molecules of *in vitro* wild-type *C. pneumoniae* RNA and 0.1 inclusion-forming unit (IFU) of *C. pneumoniae*		Loens et al., 2006
	Aspergillus fumigates	10^4 copies/ml of RNA and 100 cells		Ye et al., 2011
	Mycobacterium tuberculosis	1×10^2 CFU/ml	< 5 h	Gizeli and Lowe, 1996

DNA hybridization

The identification of bacteria by DNA probe hybridization is based on the presence or absence of particular genes. A gene probe is composed of nucleic acid molecules—most often double stranded DNA. It consists of either an entire gene or a fragment of a gene with a known function. Alternatively, short pieces of single stranded DNA can be synthesized, based on the nucleotide sequence of the known gene (Laizrd et al., 1991). Double stranded DNA probes must be denatured before hybridization reaction, whereas, oligonucleotide and RNA probes, which are single stranded need not be denatured. Gene probes can be labeled with radioactive substances by two methods. Nick translation and random priming technique. Oligonucleotide probes are usually labeled at 5 with 32p, using bacteriophage T4 polynucleotide kinase and gamma AT 32p. Although, radioactive probes seem to have the greatest sensibility in the hybridization process, they are potential hazards and disposal of radioactive wastes can be expensive. Currently, labeling of probes with non-radioactive substances such as alkaline phosphatase have been used without effecting the kinetics or specificity of the hybridization (Pitcher et al., 1989).

Target nucleic acids are denatured by high temperature (above 95°C) or high pH (above 12) and then the labeled gene probe is added. If the target nucleic acid in the sample contains the same nucleotide sequence as that of the gene probe, the probe will form hydrogen bond with the target. The unreacted, labeled probe is removed by washing the solid support and presence of probe target complexes is signaled by the bound label and detected by autoradiography (Laizrd et al., 1991).

Fluorescent In Situ Hybridization (FISH)

Fluorescence *in situ* hybridization (FISH) with ribosomal RNA (rRNA) targeted oligonucleotide probes is the most commonly applied technique among the 'non-PCR-based' molecular techniques. The choice to target RNA instead of DNA results in a more sensitive technique (higher copy numbers available) and the link to viability. In the elaboration of FISH, microbial cells are treated with appropriate chemical fixatives and then hybridized under stringent conditions on a glass slide or in a solution with oligonucleotide probes. Generally, these probes are 15–25 nucleotides in length and are labelled covalently at the 5' end with a fluorescent dye. After stringent washing, to remove unbound probe, specifically stained cells are detected via epifluorescence microscopy. The limit of detection is approximately 10^4 cfu/ml. After prior enrichment (usually overnight) to attain these levels of detection, results are available in 3 h while the hands-on time required per analysis is only a few minutes. FISH is commercially exploited by for instance, Vermicon, which has detection kits for different pathogenic and non-pathogenic micro-organisms.

Conventional, real-time and multiplex Polymerase Chain Reaction (PCR)

The PCR technique is a three-step cyclic *in vitro* procedure based on the ability of the DNA polymerase to copy a strand of DNA (Fig. 3). The region to be amplified is specified by the choice of primers. Primers are short oligonucleotides, usually

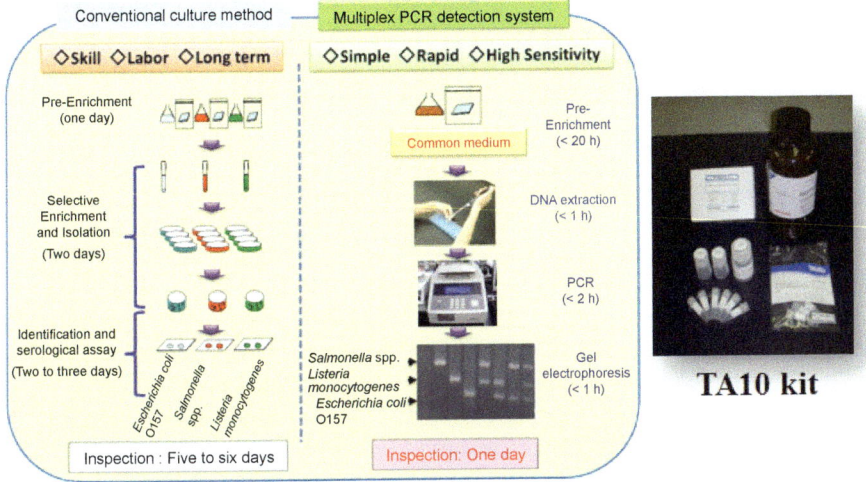

Figure 3. Conventional and Multiplex PCR detection method for pathogens.

20–30 nucleotides in length, whose sequence matches the end of the region of interest. Amplification takes place over a number of cycles. During each cycle the double-stranded DNA template is denatured by heating to produce single-strands. The reaction mixture is then cooled allowing the primers to bind to the single-strands. This provides an active site for thermo-stable DNA polymerase which synthesizes the complementary strand, producing again double-stranded DNA. In subsequent cycles, primers will bind to both the original DNA and the newly synthesized DNA resulting in an exponential increase in the numbers of copies. The presence of even 1 copy of the template within the reaction mixture can be detected within a couple of hours as about a million-fold of copies are created. The results of PCR are traditionally (conventional PCR) detected by agarose gel electrophoresis and staining. This enables the amplified DNA to be visualized as bands differing in size. Specificity of the bands may be further identified by sequencing. PCR as such is taking only ca. 30–90 min. PCR-methods for detection of pathogens in foods indeed recommended a 6–8 h up to 24 h prior enrichment step before execution of PCR as only a small volume (1 ml) is processed to extract DNA. The inclusion of an internal control is recommended to highlight inhibition of the PCR reaction. Real-time PCR is a rapid tool for screening of samples, still in case of positive PCR results it should be tempted to confirm the positive result by means of the culture-based method.

Quantitative PCR

Quantitative PCR (qPCR), also called real-time PCR, is an approach capable of continuously monitoring the PCR product formation throughout the reaction; it offers rapid, simultaneous amplification and sequence-specific-based detection of target genes and is increasingly being applied in food microbiology (Fusco et al., 2011). Using this method allows quantifying one specific microorganism in food and studying its behavior as a consequence of the influence of the environment (i.e., food composition, temperature, pH, oxygen, etc.) by studying expression of suitable target genes. Moreover, the real-time monitoring of the process means no need for post-amplification treatment of the samples, such as gel electrophoresis, reducing the time of analysis.

Loop-mediated Isothermal Amplification (LAMP)

Most recently, a novel nucleic acid amplification method known as LAMP has been demonstrated as a rapid, low-cost, easy operating, highly sensitive, and specific detection method applied in several fields (Iseki et al., 2007). This method relies on an autocycling strand displacement DNA synthesis performed by the Bst DNA polymerase large fragment, which is different from PCR in that 4–6 primers are used to target 6–8 specific regions of the target gene. The amplification is performed under isothermal conditions between 59°C and 65°C, and the amplicons are mixtures of many different sizes of stem loop DNAs with several inverted repeats of the target sequence and cauliflower-like structures with multiple loops. The reaction can be accelerated with additional one or two loop primers. LAMP reactions usually result in about 10^3-fold or higher levels of amplification product with stem-loop DNAs in 60 min than conventional PCR. LAMP products can be observed with the naked eye by employing SYBR Green I dye instead of conventional gel electrophoresis analysis; the color of the solution changes to green in the presence of LAMP amplicons, whereas it remains orange for mixtures with no amplification.

Nucleic Acid Sequence-Based Amplification (NASBA)

NASBA is an isothermal amplification reaction for the detection of RNA or DNA, which was developed after PCR had begun gaining widespread attention (Fykse et al., 2007). The reaction typically consists of three enzymes, including T7 RNA polymerase, RNase H, and avian myeloblastosis virus (AMV) reverse transcriptase (RT), all of which act together to amplify sequences from an original single-stranded RNA template. The reaction also includes buffering agents and two specific primers and takes place at approximately 41°C (Gomez et al., 2010). NASBA is specific for target RNA or DNA sequences and has been gaining popularity owing to its wide range of applications for pathogen detection in clinical, environmental, and food samples.

Microarrays

Microarrays or gene chips provide a miniaturized system for the simultaneous analysis of hybridization of fluorescent-labelled single-strand nucleotide chains to an array of oligonucleotide probes immobilized on a support such as glass or a synthetic membrane (Fig. 4). PCR amplification is often used prior to hybridization to increase sensitivity of detection. DNA microarrays may be very useful for detecting multiple bacteria simultaneously on a single glass slide. The complexity of the food matrix is a major drawback of microarrays to be used as a detection method. Therefore microarrays could be described as a tool for identification, genotyping and pathotyping (detection of appropriate virulence factors) or characterization of bacterial isolates. Since DNA arrays allow simultaneous measurements of thousands of interactions between mRNA-derived target molecules and genome-derived probes, they are rapidly producing enormous amounts of raw data never before encountered by biologists. The analysis of data on this scale is a major current challenge and therefore needs permanent attention to the issue of sample preparation and genetic material purification for successful and reliable analysis.

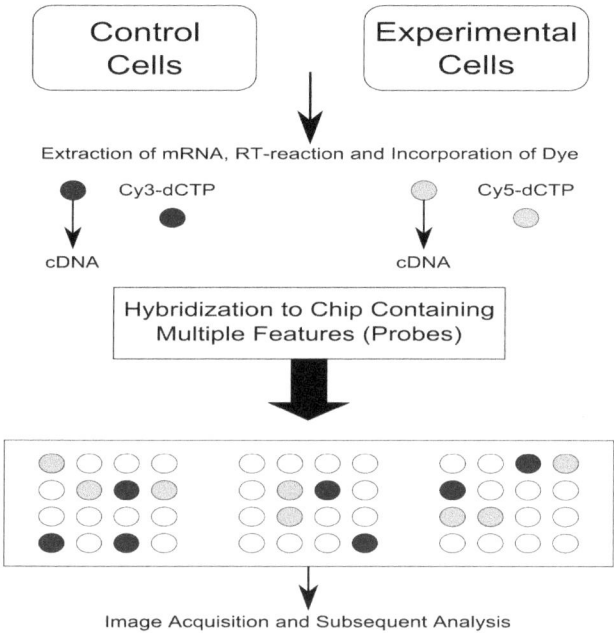

Figure 4. Microarray (Adapted from Fratamico et al., 2005).

Biosensors

Biosensors are defined as analytical devices that combine bio-specific recognition systems with physical or electrochemical signaling. Biosensing methods for pathogen detection are centered on four basic physiological or genetic properties of microorganisms: metabolic patterns of substrate utilization, phenotypic expression analysis of signature molecules by antibodies, nucleic acid analysis and the analysis of the interaction of pathogens with eukaryotic cells. Biosensors for the detection of pathogens in the food industry consist of immobilized biologically active material, like enzymes, antibodies, antigens or nucleic acids, in close proximity to a receiving transducer unit. Target recognition results in the generation of an electrical, optical or thermal signal that is proportional to the concentration of target molecules (Fig. 5). Examples of physical signals, which can report the presence of molecules, are fluorescence signals from dyes, electric fields from molecular charges, or mass changes or refractive index changes from the adsorption of molecules onto sensor surfaces.

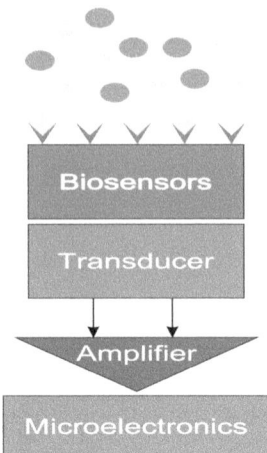

Figure 5. Principle of Biosensors.

Biosensors have the potential to shorten the time between sampling and results, but due to problems with long-term stability, reusability and sterilizability, biosensors have so far been mostly used for detecting chemical substances. Nevertheless, their future potential is enormous, since they can offer a very sensitive and accurate 'on-line' control system for food manufacturing processes.

Bioluminescence sensors

Recent advances in bio-analytical sensors have led to the utilization of the ability of certain enzymes to emit photons as a byproduct of their reaction. This phenomenon is known as bioluminescence and may be used to detect the presence and biological condition of the cells. Among the emerging technologies for rapid microbiological analysis, this technique giving results in a short time. Two distinct areas of bioluminescence are of use in food industry.

ATP bioluminescence

All living cells contain the molecule ATP. This molecule may be analyzed simply using an enzyme and coenzyme complex (Luciferase-Luciferin) found in the tail of fire fly (*Photinus pyralis*). The total light output of the sample is directly proportional to the amount of ATP present and can be quantified by luminometers. At least 10^4 cells are required to produce a signal. This system lacks specificity, but because of rapid response time for obtaining results, this system is very suitable for on-line monitoring of HACCP programs. This technique has a detection limit of 1 pg ATP which is equivalent to 1000 bacterial cells. ATP is present in both non-microbial and microbial cells. To determine microbial ATP selective extraction is used. First, non-microbial ATP is extracted with non-ionic detergents and then destroyed with high levels of potato ATPase for 5 minutes. Subsequently, microbial ATP is extracted using either trichloro-acetic acid (5%) or an organic solvent (ethanol, acetone or chloroform).

Bacterial bioluminescence

The gene responsible for bacterial bioluminescence (lux gene) has been identified and cloned. The DNA carrying this gene can be introduced into host-specific phages. These phages do not possess the intracellular biochemistry necessary to express this gene, hence they remain dark. However, on transfer of lux gene to the host bacterium during infection results in light emission that can be easily detected by luminometers (Fig. 6). Bacterial bioluminescence is capable of detecting 100 cells/h with a specificity defined by the host-specific phage, e.g., Bacteriophage p22 is specific for *Salmonella typhimurium*.

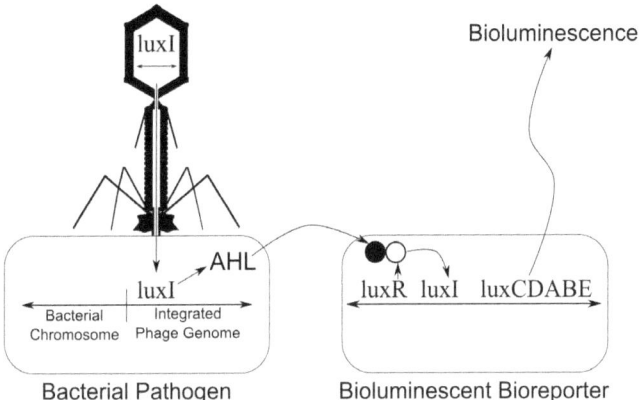

Figure 6. Bacterial Bioluminescence.

Fiber optic biosensors

A fiber optic biosensor is one of the first commercially available optical biosensors, marketed by Research International (Monroe, WA) for the detection of foodborne pathogens. The basic principle of the fiber optic sensor is that when light propagates through the core of the optical fiber, i.e., waveguide, it generates an evanescent field outside the surface of the waveguide. The waveguides are generally made up of polystyrene fibers or glass slides. When fluorescent labeled analytes such as pathogens or toxins bound to the surface of the waveguide, are excited by the evanescent wave generated by a laser (635 nm) and emit fluorescent signal (Bhunia, 2007; Taitt et al., 2005), the signal travels back through the waveguide in high order mode to be detected by a fluorescence detector in real time.

Surface Plasmon Resonance (SPR) sensors

SPR is a phenomenon that occurs during optical illumination of a metal surface and it can be used for biomolecular interaction analysis. Receptors or antibodies immobilized on the surface of a thin film of a precious metal (gold) deposited on the reflecting surface of an optically transparent waveguide are used to capture the target analyte. The sensing surface is located above or below a high index-resonant layer and a low index coupling layer. When a visible or near-infrared radiation (IR) is passed through the waveguide in such a way, it causes an internal total reflection on the surface of the waveguide. At a certain wavelength in the red or near-IR region, the light interacts with a plasma or cloud of electrons on the high-index metal surface and the resonance effect causes a strong absorbance. The exact wavelength of this absorption depends on the angle of incidence, the metal, the amount of capture molecules immobilized on the surface and the surrounding material. The presence of ligands or antigens interacting with the receptor or antibody causes a shift in the resonance to longer wavelengths and the amount of shift can be related to the concentration of the bound molecules. SPR-based sensors are governed by two basic principles: wavelength interrogation and angle interrogation. Wavelength interrogation uses a fixed angle of incidence but measures spectral changes, while in angle interrogation, a fixed wavelength is used but the angle of reflectance is monitored. Most of the commercial SPR systems are operated based on the angle interrogation mode. SPR-based sensors allow real-time or near real-time detection of binding events between two molecules. The detection system is label free, thus eliminating the need for additional reagents, assay steps and time. The sensor can be reused for the same analyte repeatedly. It is highly sensitive and it can detect molecules in the femtomolar range (Bhunia, 2007; Rasooly and Herold, 2006).

Electrical impedance biosensors

Impedance microbiology detects microbes either directly due to production of ions from metabolic end products or indirectly from liberation of CO_2. Microbial metabolism usually results in an increase in both conductance and capacitance, causing a decrease in impedance. A bridge circuit usually measures impedance. This method is well suited for detection of bacteria in clinical samples and to monitor quality and detect specific food pathogens. In this method, a population of microbes is provided with nutrients (non-electrolyte) like lactose and microbes may utilize that nutrient and convert it to lactic acid (ionic form) thus changing the impedance. This impedance is measured over a period of 20 h after inoculation in specific media. Since this does not involve serial dilution, this technique is simple to perform and faster than agar plate count. This system is capable of analyzing hundreds of sample at the same time since the instrument (Bactometer) is computer driven and automated to enable continuous monitoring. Typically most impedance analysis of food samples can be completed in 24 h. This technique is not suited for testing samples with low number of microorganisms and that the food matrix may interfere with the analysis.

Impedance-based biochip sensors

Though the concept of this detection method is old, it is now getting wider popularity. Impedance is based on the changes in conductance in a medium due to the microbial breakdown of inert substrates into electrically charged ionic compounds and acidic by-products. The principle of all impedance-based systems is that they measure the relative or absolute changes in conductance, impedance, or capacitance at regular intervals. So the threshold value for the detection of the target pathogens mainly depends on the initial inoculums and the physiological state of the cells. In media-based impedance methods, bacterial metabolism results in increased conductance and capacitance, with decreased impedance (Invitski et al., 1999). The major advantage of this system is that it allows the detection of only the viable cells, which is the major concern in food safety. The basic technical equipment required for performing impedance microbiology consists of special incubators and their culture vessels and an evaluation unit with computer, printer and appropriate software.

Piezoelectric biosensors

This system is very attractive and offers a real time output, simplicity of use and cost effectiveness. The general principle is based on coating the surface of piezoelectric sensor with a selective binding substance for example antibodies to bacteria and then placing it in a solution containing bacteria. The bacteria will bind to the antibodies and the mass of the crystal will increase while the resonance frequency of oscillation will decrease proportionally.

Cell based sensors

Cell-based assays (CBAs) continue to serve as a reliable method for detection of pathogens in food samples. The CBA systems can report perturbations in the normal physiological activities of mammalian cells as a result of exposure to an external or environmental challenge. For this, mammalian cells are used as electrical capacitors. Electrical Impedance (EI) uses the inherent electrical properties of cells to measure the parameters related to the tissue environment. The mechanical contact between cell-cell and cell-substrates is measured via conductivity or EI. The cell can be equated to a simple circuit since it is nothing more than conductive fluid encapsulated by a membrane surrounded by another conductive fluid. The conductive fluids make up the resistance elements of the circuit, while the membrane acts as a capacitor. Changes in impedance were able to detect changes in cell density, growth, or cellular behavior. These biosensors are able to provide detailed information about the growth characteristics of the tissue culture, including information on spreading, attachment and cellular morphology. Mammalian cells have been widely used for the analysis of the pathogenic potential of foodborne bacteria (Bhunia and Wampler, 2005; Gray, 2004).

Nanobiosensors

The biosensor utilizing nanotechnology is a very recent development in biosensor technology. The general principle is that, biologically synthesized nanoparticles can be used as nano-transducers in electrochemical DNA-based biosensor systems. The bio-barcode DNA assay is a promising new amplification and detection technique that makes use of short oligonucleotides as target identification strands and surrogate amplification units in both protein and nucleic acid detection (Hill and Mirkin, 2006).

Table 2. Different modes of biosensor-based foodborne pathogen detection (adapted from Zhao et al., 2014).

Mode of detection	Analyte	Limit of detection	Assay time	References
Optical biosensor	*Escherichia coli* O157:H7	5 × 105 cells/ml	45 min	Rahman et al., 2006
	Escherichia coli O157:H7, *Yersinia enterocolitica*, *Salmonella typhimurium*, and *Listeria monocytogenes*	104 CFU/ml for all bacterial species	12 h	Magliulo et al., 2007
				Hill, 1996
	Escherichia coli O157:H7	106 cells/ml		
	Salmonella typhimurium	105 CFU/ml		Seo et al., 1999
Surface plasmon resonance biosensor	*Escherichia coli* O157:H7	3 × 105 CFU/ml	5 to 7 min	Si et al., 2011
	Escherichia coli O157:H7	102 CFU/ml	2 min	Waswa et al., 2007
	Escherichia coli O157:H7	8.7 × 106 CFU/ml	35 min	Meeusen et al., 2005
	Salmonella typhimurium	1 × 106 CFU/ml		Lan et al., 2008
	Salmonella enteritidis and *Escherichia coli*	25 CFU/ml for *Escherichia coli* and 23 CFU/ml for *Salmonella*	< 1 h	Waswa et al., 2006
Piezoelectric biosensors	*Escherichia coli* O157:H7	300 spores/ml		Wu et al., 2007
	Bacillus anthracis			Campbell and Mutharasan, 2006
	Escherichia coli O157:H7, staphylococcal enterotoxin B			Ong et al., 2006
Immunosensors	*Escherichia coli* O157:H7	4.12 × 102 CFU/ml	< 45 min	Li et al., 2012
	Salmonella typhi	105 cells/ml	90 min	Singh et al., 2005
	Salmonella sp.	2.43 log CFU/ml	4 h	McEgan et al., 2009
Electrochemical biosensors	*Escherichia coli* O157:H7	102 CFU/ml	1 h 15 min	Lin et al., 2008
	Salmonella typhi	35–88 CFU/ml	6 min	Rao et al., 2005
	Bacillus cereus			Pal et al., 2008

It has been shown to have extraordinarily PCR-like sensitivity. Under controlled conditions, the assay has shown attomolar (10^{-18}) sensitivity for a variety of protein targets (Nam et al., 2003) and zeptomolar (10^{-21} M) sensitivity for a variety of target genes (Nam et al., 2004). The bio-barcode DNA assay utilizes oligonucleotide-modified gold nanoparticles (AuNPs) for signal amplification and magnetic nanoparticles (MNPs) for easy and clean separation from the sample. The technique uses the many advantageous properties of oligonucleotide-functionalized AuNPs including ease of fabrication, greater oligonucleotide binding capabilities, stability under a variety of conditions, catalytic ability, and optical properties (Lytton-Jean and Mirkin, 2005; Demers et al., 2002). The large ratio between thiolated single-strand oligonucleotide barcodes and DNA probe on AuNPs provides significant amplification. After hybridization with the target DNA, a magnetic field is used to separate the sandwich structure consisting of MNP-2pDNA/tDNA/1pDNA-AuNP-bDNA. A dithiothreitol (DTT) solution at an elevated temperature is used to release the barcode strands. The barcode strands can be identified on a microarray via scanometric detection (Stoeva et al., 2006; Taton et al., 2000), on-chip detection (Goluch et al., 2006), fluorescence (Oh et al., 2006), Raman active dye (Cao et al., 2002) if the barcodes carry with them a detectable marker. Though optical detection method for barcode DNA is commonly used and current optical sensing approaches are effective for high-density arrays, they are hard to miniaturize. Electrochemical systems can be miniaturized and integrated into micro-systems, including parts for signal processing, providing a great deal of advantage over optical detection (Templin et al., 2002). Because electrochemical reactions give an electronic signal directly, there is no need for expensive signal transduction equipment. Moreover, because immobilized probe sequences can be readily confined to a variety of electrode substrates, diagnosis can be accomplished with an inexpensive electrochemical analyzer. Indeed, portable systems for clinical testing and on-site environmental monitoring are now being developed (Wang, 2002). The basic categories of electrochemical detection of DNA include direct reduction/oxidation of DNA (Ozkan-Ariksoysal, 2008; Karadeniz et al., 2003), indirect oxidation of target DNA through the use of electrochemical mediators (Yang and Thorp, 2001), DNA-specific redox indicator detection, and DNA-mediated charge transport electrochemistry (Steel et al., 1998; Jackson and Hill, 2001). DNA-specific redox indicator detection is an analogy to fluorescence-based methods. The reporting DNA probes are labeled with redox-active molecules. Appearance of the characteristic electrochemical response of the redox reporter therefore shows the hybridization event. Detection limits in the order of $\sim 10^{10}$ molecules have been reported (Steel et al., 1998).

Future Perspectives

Traditional foodborne pathogen detection methods although sensitive enough, are often too time-consuming for practical use, taking days to a week to perform. Therefore, new methods that overcome this performance limitation are required. Recently, several methods have been explored and developed for the rapid detection of foodborne pathogens. However, most of them still require improvement in sensitivity, selectivity, or accuracy to be of any practical use.

Nucleic acid-based methods have high sensitivity and require a shorter time than conventional culture-based techniques for detection of foodborne pathogens and toxins, but most of them require trained personnel and expensive instruments, which limit their use in a practical environment. The emerging isothermal amplification methods such as LAMP and NASBA may have a good prospect for detection of pathogens and toxins in resource limit settings. The development of nucleic-acid-based methods and immunological methods helped improve the time required to yield results. The specificity and the sensitivity of immunological methods depend on the binding strength of the specific antibody to its antigen, and they work well for food matrixes without interfering factors such as other non-target cells, DNA, and proteins.

Biosensors-based methods are easy to perform without training, and yield results in real-time detection of foodborne pathogens and toxins with high sensitivity and selectivity comparable to the culture-based methods. However, they still need to be improved in food matrixes detection.

All assays available for food diagnostics require some degree of sample preparation, which is a very important factor for rapid and conventional detection methods, and also a bottleneck for the advanced rapid methods. More studies regarding the separation techniques of microorganisms from the food matrix are required, as well as for sample concentration prior to detection by immunological, nucleic acid-based, or biosensor assays. Preconcentration is the preferred choice, as it can enhance sensitivity several folds by increasing the number of target organisms per unit volume at a relatively low cost. Several available modes of preconcentration are used, including filtration, size fractionation, centrifugation, and immunomagnetic separation, or combinations of these methods.

The possibilities of combining various rapid methods, including nucleic-acid-based methods, immunological based methods, and biosensor-based methods should be further exploited. With the correct application of a number of these technologies simultaneously, broader ranging and more accurate technologies could be developed. Antibodies can be modified to capture specific cells, which may then be detected by a nucleic acid-based method. Various nucleic acid-amplified products can be quantified using immunoassays.

The trend in immunoassays and nucleic-acid-based methods should result in the quantitative detection of microorganisms and the simultaneous determination of more than one pathogen or toxin. For immunological-based methods, further study of the application of biosensor chips may result in multiplex analyte assays. Biosensors must prove that they are capable of reaching at least the same detection levels as traditional methods (between 10 and 100 CFU/ml) in order to strengthen their appeal in food microbiology applications, not to mention the cost effectiveness and time efficiency. Despite the numerous research efforts made during the past decades and in recent years for foodborne pathogen detection, current technologies still entail room for improvement. Since foodborne pathogens are mostly present in very low numbers (< 100 CFU/g) and in the presence of millions of other bacteria, they are not easily detected. Therefore, a detection method that is reliable, accurate, rapid, simple, sensitive, selective, and cost-effective would be ideal. Such methods of pathogen detection would offer a great commercial advantage in the food industry and related fields. Moreover, the trend of crossing various methods will generate novel devices or methodologies to strengthen the advantages of rapid detection methods. Moreover, the recent techniques on nanotechnology provides an interesting avenue for the development of specific, very rapid and field-operable detection systems.

In summary, there are a lot of promising applications in the field of rapid and automated detection methods for foodborne pathogens. Given the broad applicability and the great potential of such methods, there is still a great chance for further developments in the near future.

References

Abbas, A., M.J. Linman and Q.A. Cheng. 2011. New trends in instrumental design for surface plasmon resonance-based biosensors. Biosens. Bioelectron., 26: 1815–1824.

Amagliani, G., E. Omiccioli, A. Campo, I.J. Bruce, G. Brandi and M. Magnani. 2006. Development of a magnetic capture hybridization-PCR assay for *Listeria monocytogenes* direct detection in milk samples. J. Applied Microbiol., 100: 375–383.

Amit Singh, Somayyeh Poshtiban and Stephane Evoy. 2013. Recent advances in bacteriophage based biosensors for foodborne pathogen detection. Sensors, 13: 1763–1786.

Arora, P., A. Sindhu, N. Dilbaghi and A. Chaudhury. 2011. Biosensors as innovative tools for the detection of food borne pathogens. Biosens. Bioelectron., 28: 1–12.

Asiello, P.J. and A.J. Baeumner. 2011. Miniaturized isothermal nucleic acid amplification, a review. Lab. Chip., 11: 1420–1430.

Bhunia, A.K. and J.L. Wampler. 2005. Animal and cell culture models for foodborne bacterial pathogens. *In*: P. Fratamico, A.K. Bhunia and J.L. Smith (eds.). Foodborne Pathogens: Microbiology and Molecular Biology. Caister Academic Press, Norfolk, UK.

Bhunia, A.K. 2007. Biosensors and bio-based methods for the separation and detection of foodborne pathogens. Adv. Food Nutr. Res., 54: 1–44.

Bolton, F.J., E. Fritz, S. Poynton and T. Jensen. 2000. Rapid enzyme-linked immunoassay for detection of Salmonella in food and feed products: performance testing program. J. AOAC Int., 83: 299–303.

Campbell, G.A. and R. Mutharasan. 2006. Piezoelectric-excited millimeter-sized cantilever (PEMC) sensors detect *Bacillus anthracis* at 300 spores/ml. Biosens. Bioelectron., 21: 1684–1692.

Cao, Y.C., R. Jin and C.A. Mirkin. 2002. Nanoparticles with Raman spectroscopic fingerprints for DNA and RNA detection. Science, 297(5586): 1536–1540.

Cao, W., M. Su and S. Zhang. 2010. Rapid and sensitive DNA target detection using enzyme amplified electrochemical detection based on microchip. Electrophoresis, 31: 659–665.

Chapman, P.A. and R. Ashton. 2003. An evaluation of rapid methods for detecting *Escherichia coli* O157 on beef carcasses. Int. J. Food Microbiol., 87: 279–285.

Chen, H.M. and C.W. Lin. 2007. Hydrogel-coated streptavidin piezoelectric biosensors and applications to selective detection of Strep-Tag displaying cells. Biotechnol. Prog., 23: 741–748.

Chen, H.T., J. Zhang, D.H. Sun, L.N. Ma, X.T. Liu, X.P. Cai and Y.S. Liu. 2008. Development of reverse transcription loop mediated isothermal amplification for rapid detection of H9 avian influenza virus. J. Virol. Methods, 151: 200–203.

Chen, J., X.Y. Ma, Y.W. Yuan and W. Zhang. 2011. Sensitive and rapid detection of *Alicyclobacillus acidoterrestris* using loop-mediated isothermal amplification. J. Sci. Food Agric., 91: 1070–1074.

Chen, S.Y., F. Wang, J.C. Beaulieu, R.E. Stein and B.L. Ge. 2011. Rapid detection of viable salmonellae in produce by coupling propidium monoazide with loop-mediated isothermal amplification. Appl. Environ. Microbiol., 77: 4008–4016.

Chen, Z.G. 2008. Conductometric immunosensors for the detection of staphylococcal enterotoxin B based bioelectrocalytic reaction on micro-comb electrodes. Bioproc. Biosyst. Eng., 31: 345–350.

Chuang, T.L., S.C. Wei, S.Y. Lee and C.W. Lin. 2012. A polycarbonate based surface plasmon resonance sensing cartridge for high sensitivity HBV loop-mediated isothermal amplification. Biosens. Bioelectron., 32: 89–95.

Churruca, E., C. Girbau, I. Martinez, E. Mateo, R. Alonso and A. Fernandez-Astorga. 2007. Detection of *Campylobacter jejuni* and *Campylobacter coli* in chicken meat samples by real-time nucleic acid sequence-based amplification with molecular beacons. Int. J. Food Microbiol., 117: 85–90.

Compton, J. 1991. Nucleic acid sequence-based amplification. Nature, 350: 91–92.

DeCory, T.R., R.A. Durst, S.J. Zimmerman, L.A. Garringer, G. Paluca, H.H. DeCory and R.A. Montagna. 2005. Development of an immunomagnetic bead-immunoliposome fluorescence assay for rapid detection of *Escherichia coli* O157:H7 in aqueous samples and comparison of the assay with a standard microbiological method. Appl. Environ. Microbiol., 71: 1856–1864.

Demers, L.M., M. Ostblom, H. Zhang, N.H. Jang, B. Liedberg and C.A. Mirkin. 2002. Thermal desorption behavior and binding properties of DNA bases and nucleosides on gold. J. Am. Chem. Soc., 124(38): 11248–11249.

Derzelle, S., A. Grine, J. Madic, C.P. de Garam, N. Vingadassalon, F. Dilasser et al. 2011. A quantitative PCR assay for the detection and quantification of Shiga toxin-producing *Escherichia coli* (STEC) in minced beef and dairy products. Int. J. Food Microbiol., 151: 44–51.

Dey, D. and T. Goswami. 2011. Optical biosensors: a revolution towards quantum nanoscale electronics device fabrication. J. Biomed. Biotechnol., 2011: 348218.

Dwivedi, H.P. and L.A. Jaykus. 2011. Detection of pathogens in foods: the current state-of-the-art and future directions. Crit. Rev. Microbiol., 37: 40–63.

Foley, S.L. and K. Grant. 2007. Molecular techniques of detection and discrimination of foodborne pathogens and their toxins. pp. 485–510. *In*: S. Simjee (ed.). Infectious Disease: Foodborne Diseases. Humana Press Inc., Totowa, NJ.

Fratamico, P.M., A.K. Bhunia and J.L. Smith. 2005. Foodborne Pathogens: Microbiology and Molecular Biology. Caister Academic Press.

Fu, Z., S. Rogelj and T.L. Kieft. 2005. Rapid detection of *Escherichia coli* O157: H7 by immunomagnetic separation and real-time PCR. Int. J. Food Microbiol., 99: 47–57.

Fusco, V., G.M. Quero, M. Morea, G. Blaiotta and A. Visconti. 2011. Rapid and reliable identification of *Staphylococcus aureus* harbouring the enterotoxin gene cluster (egc) and quantitative detection in raw milk by real time PCR. Int. J. Food Microbiol., 144: 528–537.

Fykse, E.M., G. Skogan, W. Davies, J.S. Olsen and J.M. Blatny. 2007. Detection of *Vibrio cholerae* by real-time nucleic acid sequence-based amplification. Appl. Environ. Microbiol., 73: 1457–1466.

Gill, P. and A. Ghaemi. 2008. Nucleic acid isothermal amplification technologies: a review. Nucleosides Nucleotides Nucleic Acids, 27: 224–243.

Gill, P., R. Ramezani, M.V.P. Amiri, A. Ghaemi, T. Hashempour, N. Eshraghi et al. 2006. Enzyme-linked immunosorbent assay of nucleic acid sequence-based amplification for molecular detection of M-tuberculosis. Biochem. Biophys. Res. Commun., 347: 1151–1157.

Gizeli, E. and C.R. Lowe. 1996. Immunosensors. Curr. Opin. Biotechnol., 7: 66–71.

Goluch, E.D., J.M. Nam, D.G. Georganopoulou, T.N. Chies, K.A. Shaikh, K.S. Ryu, A.E. Barron, C.A. Mirkin and C. Liu. 2006. A bio-barcode assay for on-chip attomolar-sensitivity protein detection Lab Chip, 6(10): 1293–1299.

Gomez, P., M. Pagnon, M. Egea-Cortines, F. Artes and J. Weiss. 2010. A fast molecular nondestructive protocol for evaluating aerobic bacterial load on fresh-cut lettuce. Food Sci. Technol. Int., 16: 409–415.

Gray, K.M. 2004. Cytotoxicity and cell-based sensors for detection of *Listeria monocytogenes* and *Bacillus* species. Ph.D. Thesis, Purdue University, West Lafayette.

Grothaus, G.D., M. Bandla, T. Currier, R. Giroux, G.R. Jenkins, M. Lipp et al. 2006. Immunoassay as an analytical tool in agricultural biotechnology. J. AOAC Int., 89: 913–928.

Gupta, M.J., J.M. Irudayaraj, C. Debroy, Z. Schmilovitch and A. Mizrach. 2005. Differentiation of food pathogens using FTIR and artificial neural networks. Trans. ASAE, 48: 1889–1892.

Haese, E.D. and H.J. Nelis. 2002. Rapid detection of single cell bacteria as a novel approach in food microbiology. J. AOAC. Int., 85: 979–983.

Hahm, B.K. and A.K. Bhunia. 2006. Effect of environmental stresses on antibody-based detection of *Escherichia coli* O157:H7, *Salmonella enterica* serotype Enteritidis and *Listeria monocytogenes*. J. Appl. Microbiol., 100: 1017–1027.

Hahn, M.A., P.C. Keng and T.D. Krauss. 2008. Flow cytometric analysis to detect pathogens in bacterial cell mixtures using semiconductor quantum dots. Anal. Chem., 80: 864–872.

Hill, W.E. 1996. The polymerase chain reaction: applications for the detection of foodborne pathogens. Crit. Rev. Food Sci. Nutr., 36: 123–173.

Hill, H.D. and C.A. Mirkin. 2006. The bio-barcode assay for the detection of protein and nucleic acid targets using DTT-induced ligand exchange. Nat. Protoc., 1(1): 324–326.

Hudson, J.A., R.J. Lake, P. Savill, P. Scholes and R.E. Mc-Connick. 2001. Rapid detection of *L. monocytogenes* in harn sample using immunomagnetic separation followed by PCR. J. Applied Microbiol., 90: 614–621.

Ikeda, S., K. Takabe, M. Inagaki, N. Funakoshi and K. Suzuki. 2007. Detection of gene point mutation in paraffin sections using *in situ* loop-mediated isothermal amplification. Pathol. Int., 57: 594–599.

Invitski, D., L.A. Harnid, P. Atanasov and E. Wilkins. 1999. Biosensors for detection of pathogenic bacteria. Biosensors Bioelectronics, 14: 599–624.

Iseki, H., A. Alhassan, N. Ohta, O.M.M. Thekisoe, N. Yokoyama, N. Inoue et al. 2007. Development of a multiplex loopmediated isothermal amplification (mLAMP) method for the simultaneous detection of bovine Babesia parasites. J. Microbiol. Methods, 71: 281–287.

Jackson, N.M. and M.G. Hill. 2001. Electrochemistry at DNA-modified surfaces: new probes for charge transport through the double helix. Curr. Opin. Chem. Biol., 5(2): 209–215.

Jasson, V., L. Jacxsens, P. Luning, A. Rajkovic and M. Uyttendaele. 2010. Alternative microbial methods: an overview and selection criteria. Food Microbiol., 27: 710–730.

Jordan, D., T. Vancov, A. Chowdhury, L.M. Andersen, K. Jury, A.E. Stevenson and S.G. Morris. 2004. The relationship between concentration of a dual marker strain of *Salmonella typhimurium* in bovine faeces and its probability of detection by immunomagnetic separation and culture. J. Applied Microbiol., 97: 1054–1062.

Jung, B.Y., S.C. Jung and C.H. Kweon. 2005. Development of a rapid immunochromatographic strip for detection of *Escherichia coli* O157. J. Food Prot., 68: 2140–2143.

Karadeniz, H., B. Gulmez and F. Sahinci. 2003. Disposable electrochemical biosensor for the detection of the interaction between DNA and lycorine based on guanine and adenine signals. J. Pharm. Biomed. Anal., 33(2): 295–302.

Kawasaki, S., P.M. Fratamico, N. Horikoshi, Y. Okada, K. Takeshita, T. Sameshima and S. Kawamoto. 2009. Evaluation of a multiplex PCR system for simultaneous detection of *Salmonella* spp., *Listeria monocytogenes*, and *Escherichia coli* O157:H7 in foods and in food subjected to freezing. Foodborne Pathog. Dis., 6: 81–89.

Kawasaki, S., N. Horikoshi, Y. Okada, K. Takeshita, T. Sameshima and S. Kawamoto. 2005. Multiplex PCR for simultaneous detection of *Salmonella* spp., *Listeria monocytogenes*, and *Escherichia coli* O157:H7 in meat samples. J. Food Prot., 68: 551–556.

Kim, H.J., H.J. Lee, K.H. Lee and J.C. Cho. 2012. Simultaneous detection of pathogenic Vibrio species using multiplex realtime PCR. Food Control, 23: 491–498.

Laizrd, P.W., A. Zijderveld, K. Linders, M.A. Rudnicki and A. Berns. 1991. Simplified mammalian DNA isolation procedure. Nucleic Acid Res., 19: 4293–4293.

Lan, Y.B., S.Z. Wang, Y.G. Yin, W.C. Hoffmann and X.Z. Zheng. 2008. Using a surface plasmon resonance biosensor for rapid detection of *Salmonella typhimurium* in chicken carcass. J. Bionic. Eng., 5: 239–246.

Laura, A., D. Gilda, B. Claudio, G. Cristina and G. Gianfranco. 2011. A lateral flow immunoassay for measuring ochratoxin A: development of a single system for maize, wheat and durum wheat. Food Control, 22: 1965–1970.

Lazcka, O., F.J. Del Campo and F.X. Munoz. 2007. Pathogen detection: A perspective of traditional methods and biosensors. Biosens. Bioelectron., 22: 1205–1217.

Lee, J.Y. and R.A. Deininger. 2004. Detection of *E. coli* in beach water within 1 hour using immunomagnetic separation and ATP bioluminescence. Luminescence, 19: 31–36.

Li, Y., P. Cheng, J.H. Gong, L.C. Fang, J. Deng, W.B. Liang and J.S. Zheng. 2012. Amperometric immunosensor for the detection of *Escherichia coli* O157:H7 in food specimens. Anal. Biochem., 421: 227–233.

Li, Y. and A. Mustapha. 2004. Simultaneous detection of *Escherichia coli* O157:H7, S*almonella*, and *Shigella* in apple cider and produce by a multiplex PCR. J. Food Prot., 67: 27–33.

Lin, W.S., C.M. Cheng and K.T. Van. 2010. A quantitative PCR assay for rapid detection of *Shigella* species in fresh produce. J. Food Prot., 73: 221–233.

Lin, Y.H., S.H. Chen, Y.C. Chuang, Y.C. Lu, T.Y. Shen, C.A. Chang and C.S. Lin. 2008. Disposable amperometric immuno sensing strips fabricated by Au nano particles-modified screen printed carbon electrodes for the detection of foodborne pathogen *Escherichia coli* O157:H7. Biosens. Bioelectron., 23: 1832–1837.

Liu, Y., C.K. Chuang and W.J. Chen. 2009. *In situ* reverse transcription loop-mediated isothermal amplification (*in situ* RT-LAMP) for detection of Japanese encephalitis viral RNA in host cells. J. Clin. Virol., 46: 49–54.

Loens, K., T. Beck, H. Goossens, D. Ursi, M. Overdijk, P. Sillekens and M. Ieven. 2006. Development of conventional and real-time nucleic acid sequence-based amplification assays for detection of *Chlamydophila pneumoniae* in respiratory specimens. J. Clin. Microbiol., 44: 1241–1244.

Lytton-Jean, A.K.R. and C.A. Mirkin. 2005. A thermodynamic investigation into the binding properties of DNA functionalized gold nanoparticle probes and molecular fluorophore probes. J. Am. Chem. Soc., 127(37): 12754–12755.

Magliulo, M., P. Simoni, M. Guardigli, E. Michelini, M. Luciani, R. Lelli and A. Roda. 2007. A rapid multiplexed chemiluminescent immunoassay for the detection of *Escherichia coli* O157:H7, *Yersinia enterocolitica*, *Salmonella typhimurium*, and *Listeria monocytogenes* pathogen bacteria. J. Agric. Food Chem., 55: 4933–4939.

Mandal, P.K., A.K. Biswas, K. Choi and U. Pal. 2011. Methods for rapid detection of foodborne pathogens: an overview. Am. J. Food Technol., 6: 87–102.

Martinon, A. and M.G. Wilkinson. 2011. Selection of optimal primer sets for use in a duplex SYBR green-based, real-time polymerase chain reaction protocol for the detection of *Listeria monocytogenes* and *Staphylococcus aureus* in foods. J. Food Saf., 31: 297–312.

Mercanoglu, B. and M.W. Griffiths. 2005. Combination of immunomagnetic separation with real-time PCR for rapid detection of *Salmonella* in milk, ground beef and alfalfa sprouts. J. Food Prot., 68: 557–561.

McEgan, R., T.J. Fu and K. Warriner. 2009. Concentration and detection of *Salmonella* in mung bean sprout spent irrigation water by use of tangential flow filtration coupled with an amperometric flow through enzyme-linked immunosorbent assay. J. Food Prot., 72: 591–600.

Meeusen, C.A., E.C. Alocilja and W.N. Osburn. 2005. Detection of *E. coli* O157:H7 using a miniaturized surface plasmon resonance biosensor. Trans. ASAE, 48: 2409–2416.

Mossoba, M.M., S.F. Al-Khaldi, J. Kirkwood, F.S. Fry, J. Sedman and A.A. Ismail. 2005. Printing microarrays of bacteria for identification by infrared micro spectroscopy. Vibrational Spectroscopy, 38: 229–235.

Mukhopadhyay, A. and U.K. Mukhopadhyay. 2007. Novel multiplex PCR approaches for the simultaneous detection of human pathogens: *Escherichia coli* O157:H7 and *Listeria monocytogenes*. J. Microbiol. Methods, 68: 193–200.

Muldoon, M.T., G. Teaney, J. Li, D.V. Onisk and J.W. Stave. 2007. Bacteriophage-based enrichment coupled to immuno chromatographic strip-based detection for the determination of *Salmonella* in meat and poultry. J. Food Prot., 70: 2235–2242.

Nam, H.M., V. Srinivasan, B.E. Gillespie, S.E. Murinda and S.P. Oliver. 2005. Application of SYBR green real-time PCR assay for specific detection of *Salmonella* spp. in dairy farm environmental samples. Int. J. Food Microbiol., 102: 161–171.

Nam, J.M., S.I. Stoeva and C.A. Mirkin. 2004. Bio-bar-code-based DNA detection with PCR-like sensitivity. J. Am. Chem. Soc., 126(19): 5932–5933.

Nam, J.M., C.S. Thaxton and C.A. Mirkin. 2003. Nanoparticle-based bio-bar codes for the ultrasensitive detection of proteins. Science, 301(5641): 1884–1886.

Naravaneni, R. and K. Jamil. 2005. Rapid detection of foodborne pathogens by using molecular techniques. J. Med. Microbiol., 54: 51–54.

Nemoto, J., M. Ikedo, T. Kojima, T. Momoda, H. Konuma and Y. Hara-Kudo. 2011. Development and evaluation of a loop-mediated isothermal amplification assay for rapid and sensitive detection of *Vibrio parahaemolyticus*. J. Food Prot., 74: 1462–1467.

Notomi, T., H. Okayama, H. Masubuchi, T. Yonekawa, K. Watanabe, N. Amino and T. Hase. 2000. Loop-mediated isothermal amplification of DNA. Nucleic Acids Res., 28: E63.

Oh, B.K., J.M. Nam, S.W. Lee and C.A. Mirkin. 2006. A fluorophore-based bio-barcode amplification assay for proteins. Small, 2(1): 103–8.

Ong, K.G., K.F. Zeng, X.P. Yang, K. Shankar, C.M. Ruan and C.A. Grimes. 2006. Quantification of multiple bioagents with wireless, remote-query magnetoelastic microsensors. IEEE Sensors J., 6: 514–523.

Ozkan-Ariksoysal, D., B. Tezcanli, B. Kosova and M. Ozsoz. 2008. Design of electrochemical biosensor systems for the detection of specific DNA sequences in PCR-amplified nucleic acids related to the catechol-O-methyltransferase Val108/158Met polymorphism based on intrinsic guanine signal. Anal. Chem., 80(3): 588–596.

Pal, S., W. Ying, E.C. Alocilja and F.P. Downes. 2008. Sensitivity and specificity performance of a direct-charge transfer biosensor for detecting *Bacillus cereus* in selected food matrices. Biosys. Eng., 99: 461–468.

Park, Y.S., S.R. Lee and Y.G. Kim. 2006. Detection of *Escherichia coli* O157:H7, *Salmonella* spp., *Staphylococcus aureus* and *Listeria monocytogenes* in kimchi by multiplex polymerase chain reaction (mPCR). J. Microbiol., 44: 92–97.

Pedrero, M., S. Campuzano and J.M. Pingarron. 2009. Electroanalytical sensors and devices for multiplexed detection of foodborne pathogen microorganisms. Sensors, 9: 5503–5520.

Pitcher, D.G., N.A. Saunders and R.J. Owen. 1989. Rapid extraction of bacterial genomic DNA with guanidium thiocyanate. Lett. Applied Microbiol., 8: 151–156.

Planche, T., A. Aghaizu, R. Holliman, P. Riley, J. Poloniecki, A. Breathnach and S. Krishna. 2008. Diagnosis of *Clostridium difficile* infection by toxin detection kits: a systematic review. Lancet Infect. Dis., 8: 777–784.

Posthuma-Trumpie, G.A., J. Korf and Amerongen A. van. 2009. Lateral flow (immuno) assay: its strengths, weaknesses, opportunities and threats. A literature survey. Anal. Bioanal. Chem., 393: 569–582.

Qiao, Y.M., Y.C. Guo, X.E. Zhang, Y.F. Zhou, Z.P. Zhang, H.P. Wei et al. 2007. Loop-mediated isothermal amplification for rapid detection of *Bacillus anthracis* spores. Biotechnol. Lett., 29: 1939–1946.

Rahman, S., R.J. Lipert and M.D. Porter. 2006. Rapid screening of pathogenic bacteria using solid phase concentration and diffuse reflectance spectroscopy. Anal. Chim. Acta, 569: 83–90.

Rasooly, A. and K.E. Herold. 2006. Biosensors for the analysis of food and waterborne pathogens and their toxins. J. AOAC Int., 89: 873–883.

Rao, V.K., G.P. Rai, G.S. Agarwal and S. Suresh. 2005. Amperometric immunosensor for detection of antibodies of *Salmonella typhi* in patient serum. Anal. Chim. Acta, 531: 173–177.

Rodriguez-Lazaro, D., A. Jofre, T. Aymerich, M. Hugas and M. Pla. 2004. Rapid quantitative detection of *Listeria monocytogenes* in meat products by real-time PCR. Appl. Environ. Microbiol., 70: 6299–6301.

Salmain, M., M. Ghasemi, S. Boujday, J. Spadavecchia, C. Techer, F. Val et al. 2011. Piezoelectric immunosensor for direct and rapid detection of staphylococcal enterotoxin A (SEA) at the ng level. Biosens. Bioelectron., 29: 140–144.

Sankaran, S., S. Panigrahi and S. Mallik. 2011. Olfactory receptor based piezoelectric biosensors for detection of alcohols related to food safety applications. Sensors Actuat. B Chem., 155: 8–18.

Schlosser, G., P. Kacer, M. Kuzma, Z. Szilagyi, A. Sorrentino, C. Manzo et al. 2007. Coupling immunomagnetic separation on magnetic beads with matrix-assisted laser desorption ionization-time of flight mass spectrometry for detection of staphylococcal enterotoxin B. Appl. Environ. Microbiol., 73: 6945–6952.

Seki, M., Y. Yamashita, H. Torigoe, H. Tsuda, S. Sato and M. Maeno. 2005. Loop-mediated isothermal amplification method targeting the lytA gene for detection of *Streptococcus pneumoniae*. J. Clin. Microbiol., 43: 1581–1586.

Seo, K.H., R.E. Brackett, N.F. Hartman and D.P. Campbell. 1999. Development of a rapid response biosensor for detection of *Salmonella typhimurium*. J. Food Prot., 62: 431–437.

Shah, J., S. Chemburu, E. Wilkins and I. Abdel-Hamidb. 2003. Rapid amperometric immunoassay for *Escherichia coli* based on graphite coated nylon membranes electroanalysis. Electroanalysis, 15: 23–24.

Sharma, H. and R. Mutharasan. 2013. Review of biosensors for foodborne pathogens and toxins. Sensors Actuat. B. Chem., 183: 535–549.

Si, C.Y., Z.Z. Ye, Y.X. Wang, L. Gai, J.P. Wang and Y.B. Ying. 2011. Rapid detection of *Escherichia coli* O157:H7 using surface plasmon resonance (SPR) biosensor. Spectrosc. Spectral Anal., 31: 2598–2601.

Singh, C., G.S. Agarwal, G.P. Rai, L. Singh and V.K. Rao. 2005. Specific detection of *Salmonella typhi* using renewable amperometric immunosensor. Electroanalysis, 17: 2062–2067.

Sivakesava, S., J. Irudayaraj and C. DebRoy. 2004. Differentiation of microorganisms by FTIR-ATR and NIR spectroscopy. Trans. ASAE, 47: 951–957.

Song, T.Y., C. Toma, N. Nakasone and M. Iwanaga. 2005. Sensitive and rapid detection of *Shigella* and enteroinvasive *Escherichia coli* by a loop-mediated isothermal amplification method. FEMS Microbiol. Lett., 243: 259–263.

Steel, A.B., T.M. Herne and M.J. Tarlov. 1998. Electrochemical Quantitation of DNA Immobilized on Gold. Anal. Chem., 70(22): 4670–4677.

Stoeva, S.I., J.S. Lee, C.S. Thaxton and C.A. Mirkin. 2006. Multiplexed DNA detection with biobarcoded nanoparticle probes. Angew. Chem. Int. Ed. Engl., 45(20): 3303–6.

Taitt, C.R., G.P. Anderson and E.S. Ligler. 2005. Evanescent wave fluorescence biosensors. Biosensors Bioelectronics, 20: 2470–2487.

Taton, T.A., C.A. Mirkin and R.L. Letsinger. 2000. Scanometric DNA array detection with nanoparticle probes. Science, 289(5485): 1757–60.

Templin, M.F., D. Stoll, M. Schrenk, P.C. Traub, C.F. Vöhringer and T.O. Joos. 2002. Protein microarray technology. Trends Biotechnol., 20(4): 160–166.

Thompson, S.E., N.S. Foster, T.J. Johnson, N.B. Valentine and J.E. Amonette. 2003. Identification of bacterial spores using statistical analysis of Fourier transform infrared photoacoustic spectroscopy data. Applied Spectroscopy, 57: 893–899.

Tokarskyy, O. and D.L. Marshall. 2008. Immunosensors for rapid detection of *Escherichia coli* O157: H7-perspectives for use in the meat processing industry. Food Microbiol., 25: 1–12.

Tsaloglou, M.N., M.M. Bahi, E.M. Waugh, H. Morgan and M. Mowlem. 2011. On-chip real-time nucleic acid sequence-based amplification for RNA detection and amplification. Anal. Methods, 3: 2127–2133.

Ueda, S., T. Maruyama and Y. Kuwabara. 2006. Detection of *Listeria monocytogenes* from food samples by PCR after IMS-plating. Biocontrol Sci., 11: 129–134.

Velusamy, V., K. Arshak, O. Korostynska, K. Oliwa and C. Adley. 2010. An overview of foodborne pathogen detection: in the perspective of biosensors. Biotechnol. Adv., 28: 232–254.

Verstraete, K., J. Robyn, J. Del-Favero, P. De Rijk, M.A. Joris and L. Herman. 2012. Evaluation of a multiplex-PCR detection in combination with an isolation method for STEC O26, O103, O111, O145 and sorbitol fermenting O157 in food. Food Microbiol., 29: 49–55.

Wang, J. 2002. Portable electrochemical systems. TrAC Trends in Analytical Chemistry, 21(4): 226–232.

Wang, C.H., K.Y. Lien, J.J. Wu and G.B. Lee. 2011. A magnetic bead-based assay for the rapid detection of methicillinresistant *Staphylococcus aureus* by using a microfluidic system with integrated loop-mediated isothermal amplification. Lab. Chip, 11: 1521–1531.

Wang, D.G., Y.Z. Wang, J.H. Wang, X.W. Zhang and F.G. Xiao. 2011. Rapid detection of viable *Listeria monocytogenes* in raw milk using loop-mediated isothermal amplification with the aid of ethidium monoazide. Milchwissenschaft Milk Sci. Int., 66: 426–429.

Wang, L.X., Y. Li and A. Mustapha. 2007. Rapid and simultaneous quantitation of *Escherichia coli* O157:H7, *Salmonella*, and *Shigella* in ground beef by multiplex realtime PCR and immunomagnetic separation. J. Food Prot., 70: 1366–1372.

Wang, Y.X., Z.Z. Ye, C.Y. Si and Y.B. Ying. 2011. Subtractive inhibition assay for the detection of *E. coli* O157:H7 using surface plasmon resonance. Sensors, 11: 2728–2739.

Waswa, J., J. Irudayaraj and C. DebRoy. 2007. Direct detection of *E. coli* O157:H7 in selected food systems by a surface plasmon resonance biosensor. LWT Food Sci. Technol., 40: 187–192.

Waswa, J.W., C. Debroy and J. Irudayaraj. 2006. Rapid detection of *Salmonella enteritidis* and *Escherichia coli* using surface plasmon resonance biosensor. J. Food Process Eng., 29: 373–385.

Wu, V.C.H., S.H. Chen and C.S. Lin. 2007. Real-time detection of *Escherichia coli* O157:H7 sequences using a circulating-flow system of quartz crystal microbalance. Biosens. Bioelectron., 22: 2967–2975.

Yamazaki, W., Y. Kumeda, R. Uemura and N. Misawa. 2011. Evaluation of a loop-mediated isothermal amplification assay for rapid and simple detection of *Vibrio parahaemolyticus* in naturally contaminated seafood samples. Food Microbiol., 28: 1238–1241.

Yang, H., X.Y. Ma, X.Z. Zhang, Y. Wang and W. Zhang. 2011. Development and evaluation of a loop-mediated isothermal amplification assay for the rapid detection of *Staphylococcus aureus* in food. Eur. Food Res. Technol., 232: 769–776.

Yang, L. and R. Bashir. 2008. Electrical/electrochemical impedance for rapid detection of foodborne pathogenic bacteria. Biotechnol. Adv., 26: 135–150.

Yang, I.V. and H.H. Thorp. 2001. Modification of indium tin oxide electrodes with repeat polynucleotides: electrochemical detection of trinucleotide repeat expansion. Anal. Chem., 73(21): 5316–5322.

Ye, Y.X., B. Wang, F. Huang, Y.S. Song, H. Yan, M.J. Alam et al. 2011. Application of *in situ* loop-mediated isothermal amplification method for detection of Salmonella in foods. Food Control, 22: 438–444.

Yoo, J.H., S.M. Choi, J.H. Choi, E.Y. Kwon, C. Park and W.S. Shin. 2008. Construction of internal control for the quantitative assay of *Aspergillus fumigatus* using real-time nucleic acid sequence-based amplification. Diagn. Microbiol. Infect. Dis., 60: 121–124.

Zhao, X.H., Y.M. Li, L. Wang, L.J. You, Z.B. Xu, L. Li et al. 2010. Development and application of a loop-mediated isothermal amplification method on rapid detection *Escherichia coli* O157 strains from food samples. Mol. Biol. Rep., 37: 2183–2188.

Xihong, Zhao, Chii-Wann Lin, Jun Wang and Oh Deog Hwan. 2014. Advances in rapid detection methods for Foodborne Pathogens. J. Microbiol. Biotechnol., 24(3): 297–312.

8

Impact of Climate Change on Foodborne Diseases

Dam Sao Mai* and Nguyen Khanh Hoang

Introduction

There are three sectors of the triad (hosts, microbes, and environment) which climate impacts (Fig. 1). The climate change has a dramatic effect on infectious disease and on the transmission of food and waterborne diseases that comes from a number of sources (e.g., the seasonality of foodborne and diarrheal disease), changes in disease patterns that occur as a consequence of temperature, and associations between increased incidence of food and waterborne illness and severe weather events (Hall et al., 2002; Rose et al., 2001). The predictable food and waterborne diseases of the developing countries (e.g., bacillary dysentery, cholera) are less frequent in developed countries, due to stringent public health measures such as proper sewage disposal, clean water and hygiene.

There are theoretical and unintended consequences of global climate change on food safety. Climatic factors influence the growth and survival of pathogens, as well as transmission pathways (Vladimir and Dragan, 2012). Higher ambient temperatures increase replication cycles of foodborne pathogens, and prolonged seasons may augment the opportunity for food handling mistakes—in 32% of investigated food-borne outbreaks in Europe "temperature misuse" is considered a contributing factor (Tirado and Schmidt, 2001).

Climate change associated diseases are estimated already to comprise 4.6% of all environmental risks. It has been estimated that climate change in the year 2000 contributed to about 2.4% of all diarrhea outbreaks in the world, 6% of malaria outbreaks in certain developing countries and 7% of the episodes of dengue fever

Industrial University of HCMC, Institute of Biotechnology and Food Technology, 12 Nguyen Van Bao Str. Go Vap District, HCM City, Vietnam.
* Corresponding author: damsaomai@foodtech.edu.vn

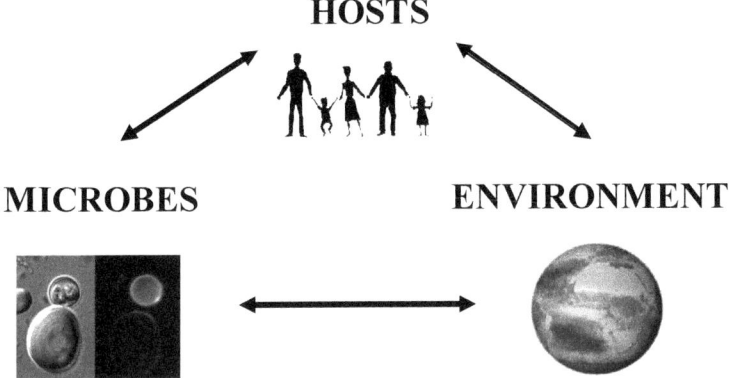

Figure 1. Cycle of infection: The cycle of infection is made up of microbes, human hosts and environmental factors. All of these factors are changing. The population is aging and people are living longer. As a result, there are many people with compromised immune systems, who are more susceptible to foodborne illnesses. The environment is also changing and new nichés are evolving where microbes can survive, grow and contaminate foods. The microorganisms are changing, too. There are more virulent pathogens, where few cells will cause serious illness. Several bacteria can grow slowly at refrigerator temperatures, while others have become more resistant to antibiotics.

in some industrial countries. In total, the estimates show that mortality due to climate change has been 0.3%, whereas the related burden of disease has been 0.4% (World Health organization, 2002).

Climate Change and Its Impact

Climate change refers to any significant change in the measures of climate lasting for an extended period of time. In other words, climate change includes major changes in temperature, precipitation, or wind patterns, or water demand, or other effects that occur over several decades or longer. According to the Fourth Assessment Report of the Intergovernmental Panel on Climate Change IPCC (IPCC, 2007) the increase in global temperatures observed since mid-20th century is predominantly due to human activities such as fossil fuel burning and land use changes. Projections for the 21st century showed that global temperature will increase from 1.8°C to 4°C. Climate change endangers human health, affecting all sectors of society, domestically, industrially, and globally. The environmental consequences of climate change has already observed that are anticipated, such as rise in sea level, increased frequency of heavy precipitation events resulting in flood, drought and extended dry periods, heat waves, more intense hurricanes and storms, and degraded air quality, will affect human health both directly and indirectly. These changes are anticipated to continue into the foreseeable future, which have implications for food production, food security and food safety. It is widely understood that the risks of global climate change occurring as a consequence of human activities are inequitably distributed. Most of the actions causing climate change originate from the developed countries, but the less developed countries are likely to bear the brunt of the public health burden (Campbell-Lendrum et al., 2007). For example, some populations or regions may face particular health challenges related to high exposure to climate hazards (Seguin, 2008). The extreme heat events in Europe in 2003 and more recently in Russia in 2010, which together caused an estimated 125,000 deaths (Robine et al., 2008; Barriopedro et al., 2011) revealed that countries not well prepared for climate-related events can be severely impacted. Climate change poses significant risks to human health and well-being, with impacts from water-, food-, vector- and rodent-borne diseases (Seguin, 2008; Costello et al., 2009). The economic costs of climate change on communities are expected to increase (Stern, 2006; NRTEE, 2011) and the economic costs of weather-related incidents worldwide have been rising rapidly for decades.

Climate Change and Foodborne Diseases

Food being a source of essential nutrients can be a source of exposure for foodborne illness. Illnesses resulting from the ingestion of spoiled or contaminated food for instance with microbes, such as *E. coli* O157:H7, *Salmonella*, *Listeria monocytogenes*, etc., or chemical residues such as pesticides, heavy metal, biotoxins, or other toxic substances. It is estimated that there are 38 million cases of foodborne illnesses in the United States each year, resulted over 180,000 hospitalizations and 2,700 deaths (Mead et al., 1999). Seafood contaminated with metals, biotoxins, toxicants, or pathogens; crops burdened with chemical pesticide residues or microbes; extreme shortages of staple foods; and malnutrition are among the possible effects of climate change on the production, quality, and availability of food (IPCC, 2007). The potential effects of climate change on foodborne illness are mostly indirect, but a large number of people are likely to be affected and consequently will cause human sufferings globally. The Intergovernmental Panel on Climate Change (IPCC) projected increase in malnutrition and consequent disorders, including those related to child growth and development, as a result of climate change. Some of these effects are already observed during extreme weather events such as the recent droughts, flooding, and hurricanes worldwide. The World Health Organization estimated that there have been over 77,000 deaths from malnutrition and 47,000 deaths from diarrhea (many from foodborne exposures) due to climate change (Campbell-lendrum et al., 2007).

Impact of Climate Change on Foodborne Diseases

The impact of climate change is likely to increase the spread of several foodborne pathogens, depending on the pathogens' survival, persistence, habitat range, and transmission in a changing environment (Gamble et al., 2008). Rising temperatures and impacts on other environmental parameters such as ocean acidification may also lead to more virulent strains of existing pathogens and changes in their distribution, or the emergence of new pathogens (Smolinski et al., 2003). Increased risks from animal-borne disease pathogens could be especially acute in human populations that are highly dependent on marine-based diets for subsistence and who live where environmental effects resulting from climate change are pronounced (for example

Table 1. Climate change factors related to foodborne pathogens.

Related factors	Aspects	Negative impact to foods	Adapted solution	Essence of agents	Risk of diseases caused by foods
Global warming		Food damage ratio is increase during production and processing	Use preservatives	Chemical	Poisoning of preservatives
	Agriculture; aquaculture; animal husbandry	Adaptation of foodborne pathogens	Use antibiotics	Chemical; Biological	Chemicals accumulation; antibiotic resistance of pathogenic bacteria
		More insect appear to infect plant and domestic animals	Use antibiotics and pesticides	Chemical	Accumulation of pesticides; toxic by-products from insects
Ice melting		Salt invasion due to the increase in sea level	Convert plant and animal structure; create new breeds	Chemical; Biological	Accumulation of drugs used for plants and domestic animals. Risk implicit when using gene-modified foods
		Immigration of pathogens	Increase food demands	Biological	Diseases caused by unsaved foods
Acidic rain		Damage the living environment for plants and domestic animals	Suitable breeds; Use pesticides, antibiotics	Chemical; Biological	Accumulation of drugs used for cultivation. Implicit in the risk when using gene-modified foods
Ozone cracking		Decrease production and ecological system	Create adaptable breeds by modifying genes	Biological	Implicitin the risk when using gene-modified foods
Forest fire		Regression of ecosystem	Recover ecosystem	Biological	Implicit in the risk when using gene-modified foods
Flood		Lack of fresh water; Deficiencies of foods; Increase of disease risk	Prevent floods; dyke		Diseases related to unsaved water and foods
Atmosphere drought		Failure of crops; Lack of water	Create new breeds	Biological	Implicit in the risk when using gene-modified foods
Desertification		Regression of land; decrease plant production	Create adaptable breeds by modifying genes	Biological	Implicit in the risk when using gene-modified foods

in certain native populations in Alaska) (Sokurenko et al., 2006). Increased acidity of water associated with climate change may alter environmental conditions leading to greater proliferation of microbes of a public health concern. This is a significant concern in molluscan shellfish, because ocean acidification may affect formation of their carbonate shells and immune responses, making them more vulnerable to microbial infection. The combined impact of potential contaminant-induced immune suppression and expanding ranges of disease-causing pathogens and biotoxins on food supply could be significant.

Climate Change and Zoonotic Diseases

Zoonotic diseases are infectious diseases whose transmission cycles involve animal hosts or vectors. Vector-borne diseases are those in which organisms, typically blood-feeding arthropods (insects, ticks, or mites) carry the pathogen from one host to another, generally with increased virulence in the vector (e.g., malaria). Zoonoses are diseases that can be transmitted from animals to humans by either contact with animals or by vectors that can carry zoonotic pathogens from animals to humans (for example, avian flu). Both domestic animals and wildlife, including marine mammals, fish, sea turtles, and seabirds may play a role in the transmission by serving as zoonotic reservoirs for human pathogens or as means of interspecies transmission of pathogens. The epidemiology of zoonotic diseases in the United States has changed significantly over the past century, and many diseases that previously caused significant illness and death, including malaria (Faust, 1949; Adler and Wills, 2003; Petri, 2004; White, 1965) dengue, yellow fever, and murine typhus, are now rarely seen in the US. This dramatic change is a result of intentional programmes to control vectors, vaccination against certain diseases, and detection and treatment with additional benefits from improvements in sanitation, development, and environmental modification. Examples of vector-borne diseases currently prevalent in the United States include Lyme disease (Bacon et al., 2008) and ehrlichiosis, bacterial diseases that are transmitted primarily by ticks. Other important zoonoses in the United States, some of which are also vector-borne, include rabies, Q fever, anthrax, pathogenic *E. coli*, tularemia, hantavirus pulmonary syndrome (Douglass et al., 2005) and plague (Blanton et al., 2008; McQuiston et al., 2006). Although zoonotic diseases are currently not a leading cause of morbidity or mortality in the United States, there are some cases for urgency on this issue. The population is directly susceptible to the zoonotic diseases that circulate in warmer climates, and therefore vulnerable as a result of global trade and travel. The ability to respond to such threats on both national and international level is currently limited. Many vector-borne diseases that have been virtually eliminated from the industrialized world are still prevalent in developing countries, which cause significant morbidity and mortality globally. For example, in 2006, there were 247 million cases of malaria and 881,000 malaria-related deaths worldwide (WHO, 2008). The WHO estimates that malaria is responsible for 2.9% of the world's total disability-adjusted life years (DALYs) (WHO, 2002).

In the long term, climate change's potential to cause social upheaval and population displacement may provide opportunities for the resurgence of certain zoonotic diseases worldwide (Carroll, 2007). Disruption of economies, transportation routes, agriculture, and environmental services could result in large-scale population movements within and between countries, which will lead to decrease in the standards of living (Costello et al., 2009). A severe degradation of rural and urban climate and sanitation conditions could bring malaria, epidemic typhus, plague, dengue and yellow fever to their former prominence. The exact projections on the impacts of climate change on zoonotic diseases are lacking, and a scientific consensus has yet to emerge. Even though the technical ability to treat or vaccinate against many zoonotic diseases has been developed in industrialized countries, the lack of these technologies in developing countries, may led to higher mortality from certain disease outbreaks and could reach as high as 20–50% (Orenstein et al., 2005). Therefore, the projections should be specific to location, altitude, ecosystem, and host or vector. Health impacts from changing distributions of zoonotic diseases are likely to unfold over the next several decades, and prevention and control activities must be developed and honed prior to significant expansion. Emerging zoonotic disease outbreaks are increasing worldwide, as well as significant emerging diseases such as SARS, Nipah virus, and HIV/AIDS, originating in animals (IMC et al., 2009). A recent report noted that the United States remains the world's largest importer of wildlife, both legal and illegal; these animals represent a potential source of zoonotic pathogen introduction into U.S. communities (Pavlin et al., 2009). Interactions of wildlife with domestic animals and people will likely increase due to changes of ecosystems and disease transmission resulting from climate change, adaptation and response strategies.

(a) Impacts on zoonotic diseases

Zoonotic diseases' ecology is complex, and weather and climate are among several factors that influence transmission cycles and human disease incidence (Boxall et al., 2009). The extrinsic incubation period of pathogens in invertebrate vectors is highly dependent on ambient temperature. Since the lifespan of vector species is relatively constant, changes in the incubation period due to precipitation and temperature significantly alter the likelihood of transmission (Strickman and Kittayapong, 2003).

Coastal and marine ecosystems will be particularly impacted by increasing temperatures, changes in precipitation patterns, sea-level rise, altered salinity, ocean acidification, and more frequent and intense extreme weather events. These changes will directly and indirectly affect ocean and coastal ecosystems by influencing community structure, biodiversity, and the growth, survival, persistence, distribution, transmission, and severity of disease-causing organisms, vectors, and reservoirs (Niemi et al., 2004). The time scale of this threat will be continuous unless mitigating measures are taken. Economic and regulatory restrictions continue to slow the development and use of new modes of action against vectors.

Table 2 gives a list of some zoonotic agents that are expected to be affected by climate change and their mode of transmission.

Table 2. Examples of some zoonotic agents that are expected to be affected by climate change and their mode of transmission.

Virus	Host	Mode of transmission to humans
Rift Valley fever virus	Multiple species of livestock and wildlife	Blood or organs of infected animals (handling of animal tissue), unpasteurized or uncooked milk of infected animals, mosquito, hematophagous flies.
Nipah virus	Bats and pigs	Directly from bats to humans through food in the consumption of date palm sap (Luby et al., 2006). Infected pigs present a serious risk to farmers and abattoir workers.
Hendra virus	Bats and horses	Secretions from infected horses.
Hantavirus	Rodents	Aerosol route from rodents. Outbreaks from activities such as clearing rodent infested areas and hunting.
Rotavirus	Humans	Faecal-oral route, spread through contaminated water and also by infected food-handlers who do not wash their hands properly.
Hepatitis E virus	Wild and domestic animals	Faecal-oral pig manure is a possible source through contamination of irrigation water and shellfish in coastal waters.
Bacterium	**Host**	**Mode of transmission**
Salmonella	Poultry and pigs	Faecal/oral
Campylobacter	Poultry	Faecal/oral
E. coli O157	Cattle and other ruminants	Faecal/oral
Anaerobic sporeforming bacteria	Birds, mammals and livestock	Ingestion of spores through environmental routes, water, soil and feeds. This has been associated with outbreaks of anthrax in livestock and wild animals, blackleg (*Clostridium chauvoei*) in cattle and botulism in wild birds after droughts. The meat and milk from cattle that have botulism should not be used for human consumption.
Yersinia	Birds and rodents with regional differences in the species of animal infected. Pigs are a major livestock reservoir	Handling pigs at slaughter is a risk to humans.
Listeria monocytogenes	Livestock	In the northern hemisphere, listeriosis has a distinct seasonal occurrence in livestock probably associated with feeding of silage.
Leptospirosis	All farm animal species	Leptospirae shed in urine to contaminate pasture, drinking water and feed.
Protozoan	**Host**	**Mode of transmission**
Toxoplasma gondii	Cats, pigs, sheep	Cat faeces are a major source of infection. Handling and consuming raw meat from infected sheep and pigs pose a zoonotic risk.

Table 2. contd....

Table 2. contd.

Protozoan	Host	Mode of transmission
Cyptosporidium and *Giardia*	Cattle, sheep	Faecal-oral transmission. (Oo)cysts are highly infectious and with high loadings, livestock faeces pose a risk to animal handlers.

Parasite	Host	Mode of transmission
Tapeworm *(Cysticercus bovis)*	Cattle	Faecal-oral.
Liver fluke *(Fasciola hepatica)*	Sheep, cattle	Eggs are excreted in faeces, and life cycle involves lymnaeid snail hosts. Human cases generally associated with the ingestion of marsh plants such as watercress.

Climate Change and Waterborne Diseases

Waterborne diseases are caused by a wide variety of pathogenic microorganisms, biotoxins, and toxic contaminants found in the water we drink. Waterborne microorganisms include protozoa that cause cryptosporidiosis, parasites that cause schistosomiasis, bacteria that cause cholera and legionellosis, viruses that cause viral gastroenteritis, amoebas that cause amoebic meningoencephalitis, and algae that cause neurotoxicity (Batterman et al., 2009). The majority of waterborne disease is gastrointestinal, though waterborne pathogens affect most human organ systems and the epidemiology is dynamic. A recent shift has been seen in waterborne disease outbreaks from gastrointestinal toward respiratory infections such as that caused by Legionella, which lives in cooling ponds and is transmitted through air conditioning systems (Yoder et al., 2008). In addition to diarrheal disease, waterborne pathogens are implicated in other illnesses with immunologic, neurologic, hematologic, metabolic, pulmonary, ocular, renal and nutritional complications (Meinhardt, 2006). The World Health Organization estimates that 4.8% of the global burden of disease and 3.7% of all mortality attributable to the environment is due to diarrheal disease (Mathers et al., 2008). Most of these diseases produce serious symptoms and greater risk of death in children and pregnant women. For most waterborne pathogens in the United States, surveillance is spotty, diagnoses are not uniform, and the impact of normal weather and climate variation on disease incidence, as well as illness and death burdens, is not firmly established. Impacts of any intensifying climate events at local, regional, national, and global levels are a growing concern. Experts estimate that there is a high incidence of mild symptoms from waterborne pathogens and a relatively small, but not negligible mortality burden (Craun and Calderon, 2006).

Globally, the impact of waterborne diarrheal disease is high and expected to climb with climate change. Improving domestic surveillance is a high priority, as this would enhance epidemiologic characterization of the drivers of epidemic disease. In particular, weather and climate-related drivers are not well understood. Waterborne disease outbreaks are highly correlated with extreme precipitation events (Curriero et al., 2001) but this correlation is based on limited research and needs further investigation and confirmation. Prevention and treatment strategies for waterborne disease are well established throughout the developed world, is not likely to impact greatly of the prevention strategies in the United States. However, climate change is very likely to increase global diarrheal disease incidence, and changes in the hydrologic cycle including increases in the frequency and intensity of extreme weather events and droughts may greatly affect due to inadequate prevention efforts. Enhanced understanding and reinvigorated global prevention efforts are very important. Ocean-related diseases are those associated with direct contact with marine waters (aerosolized in some cases) or sediments (including beach sands), ingestion of contaminated seafood, or exposure to zoonosis (Heaney et al., 2009).

Pathogenic microorganisms (bacteria, viruses, protozoa, and fungi) that may occur naturally in ocean, coastal, and Great Lakes waters, or as a result of sewage pollution and runoff, are the primary etiologic agents (Mos et al., 2006). Human exposure to these agents may result in a variety of infectious diseases including serious wound and skin infections, diarrhea, respiratory effects, and others (Stewart et al., 2008). Research has concluded that the antibiotic resistant methicillin resistant *Staphylococcus aureus* (MRSA) is persistent in both fresh and seawater and could become waterborne if released into these waters in sufficient quantities (Tolba et al., 2008). While this has yet to emerge as a significant public health concern, the potential for recreational exposure is significant, as nearly one billion people make trips to the beach annually in the United States alone. In contrast to diarrheal disease, there are few effective preventive strategies for marine-based environmental exposures beyond closing beaches to the public, and these areas need immediate additional research.

The effects of climate changes on the distribution and bioaccumulation of chemical contaminants in marine food webs are poorly understood and may be significant for vulnerable populations of humans and animals. The U.S. Climate Change Science Program (CCSP) reported a likely increase in the spread of waterborne pathogens depending on the pathogens' survival, persistence, habitat range, and transmission in a changing environment (Ebi et al., 2008). Furthermore, recent findings demonstrate that pathogens that can pose disease risks to humans occur widely in marine vertebrates and regularly contaminate shellfish and aquacultured finfish (Moore et al., 2008).

(a) Impacts on waterborne diseases

Climate directly impacts the incidence of waterborne disease through effects on water temperature and precipitation frequency and intensity. These effects are pathogen and pollutant specific, and risks for human disease are markedly affected by local conditions, including regional water and sewage treatment capacities and practices. Domestic water treatment plants may be susceptible to climate change leading to human health risks. For example, droughts may cause problems with increased concentrations of effluent pathogens and overwhelm water treatment plants; aging water treatment plants are particularly at risk (Kistermann et al., 2002). Urbanization of coastal regions may lead to additional nutrient, chemical, and pathogen loading in runoff (Dwight et al., 2004). Our understanding of weather and climate impacts on specific pathogens is incomplete.

Climate also indirectly impacts water-borne disease through changes in ocean and coastal ecosystems including changes in pH, nutrient and contaminant runoff, salinity, and water security. These indirect impacts are likely to result in degradation of fresh water available for drinking, washing food, cooking, and irrigation, particularly in developing and emerging economies where much of the population still uses untreated surface water from rivers, streams, and other open sources for these needs. Even in countries that treat water, climate-induced changes in the frequency and intensity of extreme weather events could lead to damage or flooding of water and sewage treatment facilities, increasing the risk of waterborne diseases. Severe outbreaks of cholera, in particular, have been directly associated with flooding in Africa and India (Sidley, 2008). A rise in sea level, combined with increasingly severe weather events, is likely to make flooding events commonplace worldwide. A 40 cm rise in sea level is expected to increase the average annual numbers of people affected by coastal storm surges from less than 50 million at present to nearly 250 million by 2080 (Ford et al., 2009).

Several secondary impacts are also a concern. Ecosystem degradation from climate change will likely result in pressure on agricultural productivity, crop failure, malnutrition, starvation, increasing population displacement, and resource conflict, all of which are predisposing factors for increased human susceptibility and increased risk of waterborne disease transmission due to surface water contamination with human waste and increased contact with (Shultz et al., 2009; Diaz, 2007) such waters through washing and consumption.

Climate change may also affect the distribution and concentrations of chemical contaminants in coastal and ocean waters, for example through release of chemical contaminants previously bound up in polar ice sheets or sediments, through changes in volume and composition of runoff from coastal and watershed development, or through changes in coastal and ocean goods and services. Both naturally occurring and pollution-related ocean health threats will likely be exacerbated by climate change (Sandifer et al., 2007).

Other climate-related environmental changes may impact marine food webs as well, such as pesticide runoff, leaching of arsenic, fluoride, and nitrates from fertilizers, and lead contamination of drinking and recreational waters through excess rainfall and flooding.

Climate Change and Natural Hazards

Climate change is affecting extreme weather and climate events around the world (WMO, 2013) and is expected to make weather more variable, affecting the frequency, intensity, spatial extent, duration, and timing of these events (IPCC, 2012). The IPCC Special Report on Managing the Risks of Extreme Events and Disasters to Advance Climate Change Adaptation (2012) notes that these changes can result in "unprecedented" extreme events that can have severe impacts on individuals and communities in both developed and developing countries.

(a) Air quality and its impact

The direct and indirect influence of climate on air quality is substantial and well established (McMichael et al., 2006; Lamy and Bouchet, 2008; IOM, 2011). Recent studies increase confidence that climate change will exacerbate existing health risks associated with poor air quality through heat and other meteorologically-related increases in ambient air pollutants (e.g., O_3 and particulate matter (PM)) (Frumkin et al., 2008; Bambrick et al., 2011), aeroallergens, and biological contaminants and pathogens (Greer et al., 2008; Schenck et al., 2010). Climate is an important factor in the formation of some air pollutants (e.g., ozone) that cause harm to health (IPCC, 2007), but the degree to which air pollutant levels are attributed to climate change is unclear. While uncertainty still exists regarding the potential impacts of climate change on air quality and variability in meteorology include both ozone and particulate matter (PM) (e.g., Tagaris et al., 2007). Simulations air quality data, of ten summer seasons of year 2000 and 2045 in North America suggested that O_3 concentrations are expected to increase up to 9–10 ppb by volume with climate change, when anthropogenic air pollutant emissions are kept constant (Kelly et al., 2012). The same simulations forecast showed decreased PM2.5 (< 0.2 μg m^{-3}) in North America suggesting that while climate change negatively affects air quality, the impact can be modulated through reductions in air pollutant emissions. Reducing air pollution would contribute to reductions in acute air-quality episodes, acidifying deposition and ozone deposition, and their associated impacts (e.g., increased mortality, damage to buildings and crops, etc.) (Kelly et al., 2012).

(b) Aeroallergens and its impact

Aeroallergens such as pollens from trees, grasses or weeds, molds (indoor and outdoor), and dust mites are air-borne substances that once inhaled trigger allergic responses in sensitized individuals. Increased aeroallergen formation has been associated with exacerbation of respiratory diseases (Frumkin et al., 2008), such as asthma and chronic obstructive pulmonary disease (COPD) leading to increased hospital admissions (Hess et al., 2009). Climate change is expected to impact aeroallergens by leading to an earlier onset of the pollen season in temperate zones, increasing the amount of pollen produced and the allergenicity or severity of allergic reaction (US EPA, 2008; Rosenzweig et al., 2011; Ziska et al., 2011). Earlier flower blooming resulting from temperature increases and increased carbon dioxide (CO_2) concentrations affects timing of aeroallergens distribution such as pollen through plant photosynthesis and metabolism. There is also a possibility that certain aeroallergens may become more allergenic as temperatures and CO_2 concentrations increase. Precipitation-affected aeroallergens such as mold spores also are of concern, as 5% of individuals are predicted to have some respiratory allergic airway symptoms from molds over their lifetime (CDC, 2010b). In North America, the ragweed season is becoming longer, a pattern most prevalent in northern latitudes. Ragweed is pervasive in highly populated areas of Canada and the leading cause of seasonal allergic rhinitis in north-eastern North America, responsible for approximately 75% of seasonal allergy symptoms (Ziska et al., 2011). Between 1995 and 2009, the length of the ragweed season increased by 27 days in Saskatoon and by 25 days in Winnipeg (Ziska et al., 2011).

Higher temperatures and drier conditions due to climate change could facilitate the establishment of fungal pathogens in new locations (Greer et al., 2008). For example, *Cryptococcus gattii*, a fungal pathogen typically found in tropical and sub-tropical regions, was identified on Vancouver Island in 1999 and has since spread to the British

Columbia mainland. Its prevalence may be linked to warmer and drier summers in western Canada (Kidd et al., 2007; CDC, 2012). Sensitive populations who are exposed to this fungus may become sick with cryptococcal disease (cryptococcosis) which can be serious and result in pneumonia or meningitis (CDC, 2012).

(c) Extreme weather and its impact

The extreme weather events, including extreme heat wave, hurricanes, flood, blizzards, and droughts events can lead to severe infrastructure damage and high rates of morbidity (illness) and mortality (death). Wildfires are a good example of the complexities of these extreme weather events and their potential connections to climate change. Wildfires can occur naturally and play a long-term role in the health of these ecosystems, but climate change threatens to increase the frequency, extent, and severity of fires through increased temperatures and drought. Extreme weather conditions (e.g., flooding, drought, hurricanes, etc.) can impact on the transmission of disease. For example, periods of excessive precipitation and periods of drought influence both the availability and quality of water and have been linked to the transmission of water and food borne disease. Climate change is expected to increase the frequency and intensity of these events, including floods, droughts, and heat waves. The health impacts of these events can be severe, and include direct impacts such as injury, deaths, and mental health impacts, as well as indirect, such as population displacement and outbreaks of waterborne diseases. Intensive and frequent precipitation events can lead to flooding, increasing exposure to toxic chemicals in runoff, waterborne diseases, and ecosystem changes such as loss of wetlands. More intense and frequent hurricanes could result in death and injury, infrastructure damage, and increases in stress and anxiety in vulnerable populations. Furthermore, extreme weather events can result in forced evacuation of refugees into close quarters. This frequently results in extreme stress, malnutrition, and limited access to medical care, all of which contribute to increased susceptibility and severity of disease.

(d) Seasonality/temperature and its impact

The seasonality of some infectious diseases is well documented, but its relationship to potential long-term warming effects is poorly characterized. Seasonal shifts in immunity and host susceptibility, exacerbated by increased exposure through crowds during the colder months, will also increase patterns of infectious disease spread (Altizer et al., 2006). There are different ways in which weather conditions can affect the incidence of foodborne diseases. Firstly, the prevalence of specific pathogenic organisms in animals may increase with higher temperatures. Secondly, the food cooling chain is harder to maintain in higher temperatures and prolonged warm weather increases the risk of mistakes in food handling. Thirdly, higher air temperatures may speed up the replication cycles of foodborne pathogenic organisms, which leads to a higher degree of contamination. Higher temperatures, in interaction with inadequate hygiene conditions, improper food handling, and lack of hand-washing, may lead to an increased number of epidemics resulting from consumption of unsafe food.

Foodborne and climate-sensitive pathogenic organisms causing the greatest concern in the context of climate change include the following:

Salmonellosis

The second largest foodborne disease is caused by the *Salmonella* spp. In 2007, the European Union incidence was 31.1 cases per 100,000 population (151,995 confirmed cases), with eggs being the biggest contributors to these outbreaks, followed by fresh poultry and pork. Higher ambient temperatures have been associated with 5–10% higher salmonellosis notifications for each degree increase in weekly temperature, for ambient temperatures above 5°C. Roughly one-third of the transmission of salmonellosis in England and Wales, Poland, the Netherlands, the Czech Republic, Switzerland and Spain can be attributed to temperature influences (EFSA, 2009).

Campylobacter

The risk of infections caused by *Campylobacter* is directly proportional to the increase in temperature. Recent studies show increased incidence of campylobacteriosis at 2–5% per each degree Celsius rise of temperature, based on weekly temperature data. Notwithstanding that it is mandatory to report cases of campylobacteriosis in the Republic of Macedonia, there is currently no reliable information on its distribution, although estimates indicate that its incidence exceeds 18,000 cases annually.

Other foodborne pathogenic organisms

These include *Brucella*, *Hepatitis A*, *E. coli* O157 H7 (EHEC) and bacteria causing bacterial food poisoning (e.g., *Clostridium perfringens*). As far as these pathogenic organisms are concerned, the effect of climate change remains within the area of speculation. Hepatitis A is constantly present in the Republic of Macedonia and there were 290 reported cases in 2009, 243 reported cases in 2008 and 257 reported cases in 2007. Higher temperature and humidity in the week before infection has been correlated with decreased hospitalization rates for children diagnosed with rotavirus. This is particularly interesting because survival of the virus is favored at lower temperature and humidity (D'Souza et al., 2004). Rotavirus is considered a significant cause of foodborne illness (FAO, 2008a).

Taken together, it appears that changes in both ambient temperature and humidity do have a role in foodborne disease transmission that is independent of other factors such as population behavior and susceptibility. It is likely that some of the first detectable changes of global climate change on food safety will be seen as longer summertime peaks of foodborne disease and/or increased geographic range (Watson and McMichael, 2001).

Extreme heat and its impact

Prolonged exposure to extreme heat can cause heat exhaustion, heat cramps, heat stroke, and death, as well as exacerbate preexisting chronic conditions, such as various respiratory, cerebral, and cardiovascular diseases. Extreme heat resistance food borne pathogens including *Campylobacter jejuni, Escherichia coli,* and *Salmonella typhimurium* can survive and causes human sufferings. These serious health consequences usually affect more vulnerable populations such as the elderly, children, and those with existing cardiovascular and respiratory diseases. Socioeconomic factors, such as economically disadvantaged and socially isolated individuals, are also at risk from heat-related burdens. As global temperatures rise and extreme heat events increase in frequency due to climate change we can expect to see more heat-related illnesses and mortality. Increased temperatures and increase in extreme heat events cause heat exhausting, heat stroke, and death, especially in vulnerable populations. High concentrations of buildings in urban areas cause urban heat island effect, generation and absorbing heat, making the urban center several degrees warmer than surrounding areas.

Droughts and its impact

Droughts have significant impacts on pathogens (Wheaton et al., 2008) by lowering groundwater levels and stream flows, increasing wind erosion of soils, and causing cracking of cisterns and septic tanks, creating the potential for increased sediment levels in water. They can also result in an increase in water-borne pathogens and water contamination (English et al., 2009; Ostry et al., 2010; Wittrock et al., 2011) leading to gastroenteritis (US CDC et al., 2010). Certain vector-borne diseases may spread more easily during periods of drought (Frumkin et al., 2008) and health can be impacted through wind erosion and dust storms (Wheaton et al., 2008). Droughts can decrease agricultural and crop production (Wheaton et al., 2008) leading to suboptimal nutrition due to food shortages, lack of food availability, and high costs (Horton et al., 2010), particularly for low income people and those relying on fishing or agriculture for their livelihoods (US CDC et al., 2010). Increased stress and mental health issues, particularly among farmers, have been linked to drought conditions (US CDC et al., 2010; Poulain et al., 2011; Wittrock et al., 2011). Summer continental interior drying, drought risk and areas impacted by drought are all projected to increase in some places in the world, such as Canada (Wheaton et al., 2008; Wittrock et al., 2011).

Ecological factors and its impact

Cholera is perhaps the best model for understanding the potential for climate-induced changes in the transmission of foodborne disease. *Vibrio cholerae* is the causative agent of this disease, which produces substantial morbidity and mortality, particularly in the developing world. In certain regions, cholera is endemic, displaying characteristic waves of epidemic peaks followed by periods of relative quiescence. Peaks of disease are seasonal and associated with higher water temperature. There are three major factors driving cholera endemicity (i.e., abiotic, phytoplankton, and zooplankton) that have been described by Colwell and Huq (Lipp et al., 2002) in their hierarchical model of the transmission of cholera. The interaction between these factors are extremely complex. Following the role of intrinsic parameters and microbial competition in these dynamics are illustrated:

Some investigators believe that photosynthetic phytoplankton proliferation results in elevated environmental pH. This alkaline pH gives *V. cholerae* a competitive advantage over other marine bacteria since it thrives at higher pH and promotes attachment of *V. cholerae* cells to zooplankton (particularly copepods); this protects *V. cholerae* cells from external stresses. As phytoplankton populations disintegrate, additional nutrient sources are available to stimulate the growth of the organism. Together, the interplay of these factors results in extremely high populations of *V. cholerae* in tight association with copepods.

The human factor is another important part of the equation. While cholera is predominantly a waterborne disease, foodborne transmission can occur though the use of contaminated water for food preparation or irrigation, or from consumption of molluscan shellfish. Furthermore, sewage presents an additional risk factor for disease transmission, particularly in parts of the world in which water and sewage treatment remains substandard.

Table 3. Key health concerns from climate change (Source: Adapted from Seguin, 2008).

Health Impact Categories	Potential Changes	Projected/Possible Health Effects
Temperature extremes	- More frequent, severe and longer heat waves - Overall warmer weather, with possible colder conditions in some locations	- Heat-related illnesses and deaths - Respiratory and cardiovascular disorders - Possible changed patterns of illness and death due to cold
Extreme weather events and natural hazards	- More frequent and violent thunderstorms, more severe hurricanes and other types of severe weather - Heavy rains causing mudslides and floods - Rising sea levels and coastal instability - Increased drought in some areas, affecting water supplies and agricultural production, and contributing to wildfires - Social and economic changes	- Death, injury and illness from violent storms, floods, etc. - Psychological health effects, including mental health and stress-related illnesses - Health impacts due to food or water shortages - Illnesses related to drinking water contamination - Effects of the displacement of populations and crowding in emergency shelters - Indirect health impacts from ecological changes, infrastructure damages and interruptions in health services
Air quality	- Increased air pollution: higher levels of ground-level ozone and airborne particulate matter, including smoke and particulates from wildfires - Increased production of pollens and spores by plants	- Eye, nose and throat irritation, and shortness of breath - Exacerbation of respiratory conditions - Chronic obstructive pulmonary disease and asthma. - Exacerbation of allergies - Increased risk of cardiovascular diseases (e.g., heart attacks and ischemic heart disease) - Premature death

Table 3. contd....

Table 3. contd.

Health Impact Categories	Potential Changes	Projected/Possible Health Effects
Infectious diseases transmitted by insects, ticks and rodents	- Changes in the biology and ecology of various disease-carrying insects, ticks and rodents (including geographical distribution) - Faster maturation for pathogens within insect and tick vectors - Longer disease transmission season	- Increased incidence of vector-borne infectious diseases native to Canada (e.g., eastern & western equine encephalitis, Rocky Mountain spotted fever) - Introduction of infectious diseases new to Canada - Possible emergence of new diseases, and re-emergence of those previously eradicated in Canada
Stratospheric ozone depletion	- Depletion of stratospheric ozone by some of the same gases which are responsible for climate change (e.g., chloro- and fluorocarbons) - Temperature-related changes to stratospheric ozone chemistry, delaying recovery of the ozone hole - Increased human exposure to UV radiation owing to behavioral changes resulting from a warmer climate	- More cases of sunburns, skin cancers, cataracts and eye damage - Various autoimmune disorders

Climate Changes and Bacterial Intrinsic Factors

There are a number of intrinsic factors that has potential impacts on foodborne pathogens. These include: (i) impacts on microbial evolution and stress response; (ii) pathogen emergence; (iii) changes in water availability and quality; and (iv) other considerations (Tirado et al., 2010). These will be discussed briefly.

(a) Microbial evolution and stress response

Over the course of time, many bacterial agents have developed mechanisms that allow them to survive and grow under unfavorable or "stressful" conditions. Many of these conditions are the manipulations of the intrinsic and extrinsic parameters. Stress responses are encoded genetically and in many cases, initial exposure to a sub-lethal dose of a stress environment, the bacterial cell, allow themselves to survive even harsher conditions. This is well documented for *E. coli* O157:H7, where, for instance, the organism is able to survive an acid shock as low as pH 2.0 after previous exposure to pH 5.0. In addition, as microorganisms acquire increased tolerance after pre-exposure to a sublethal stress, they frequently develop enhanced resistance to other types of stress, a phenomenon referred as cross-protection (Rodriguez-Romo and Yousef, 2005). This is relevant to global climate change in that climate-induced changes in intrinsic factors may induce stress responses that make certain bacteria more resistant.

Gene transfer between related and even unrelated bacterial species is now considered a common occurrence. This may be facilitated when multiple species are present in large and diverse communities (such as in the environment, in raw foods, or in the gut) through horizontal gene transfer or by infection with bacteriophage (Tirado et al., 2010). Gene transfer is also an important contributor to the emergence of antibiotic resistance. We know very little about the triggers or dynamics of gene transfer events, but they are likely to be impacted by the environment. In fact, this has been documented for *V. cholerae*. For example, non-toxigenic strains of this organism can acquire genes for the cholera toxin though bacteriophage mediated transfer events, which occur naturally in the environment. More specifically, both phage infection and prophage induction seem to be sensitive to environmental triggers (temperature, sunlight, pH) (Lipp et al., 2002). This means that environmental changes may have a substantial impact on pathogen evolution and/or pathogenicity (Tirado et al., 2010).

(b) Pathogen re-emergence

Climatic change can also impact the emergence or re-emergence of infectious disease agents. Emerging foodborne pathogens are defined as infectious agents, transmitted by foodborne routes, which have: (i) newly appeared in a population; (ii) were thought to be controlled but are now resurging; or (iii) have existed but are rapidly increasing in incidence, geographic range, or by some other factor (For more information see Chapter 3).

Rarely, if ever, do foodborne pathogens re-emerge without a reason. There are some general principles of pathogen emergence, which are associated with changes in the following sectors: (i) ecology and agriculture; (ii) technology and industry; (iii) globalization; (iv) human behavior and demographics; (v) epidemiological surveillance; and (vi) microbial adaptation (Tauxe, 2002). It is important to recognize that pathogen re-emergence usually occurs as a consequence of a combination of two or more specific factors. Further, re-emergence may not necessarily be predictable. For example, climate-induced changes in the movement of animal populations may facilitate the spread of a pathogen, which previously was of low prevalence or little consequence.

(c) Water availability and quality

Periods of excessive precipitation and drought can influence both the availability and the microbiological quality of water. Furthermore, new demands on existing water sources could occur if sea levels rise as predicted, adversely impacting water availability (Charron et al., 2004; Kovatts et al., 2005). An emerging environmental health threat is the decline in global freshwater resources caused mostly by increasing rates of water extraction and contamination. This has resulted in a decline in both water quality and quantity, especially in arid regions such as the Mediterranean and Northern Africa (Campbell-Lendrum et al., 2007). Needless to say, limited access to safe water has a negative effect on hygiene practices throughout the food chain.

(d) Other considerations

There are many other potential impacts of global climate change on food safety. For example, climate change could result in movement of crop production areas, resulting in very different ecosystem exposures, including microbes, pests (insects), and wilds animals (rodents, reptiles and amphibians). Intermingling or crowding of food animals caused in response to natural disaster or climate induced changes in animal husbandry practices might promote the transmission of pathogens between animals, resulting in greater pathogen load in faeces and increased prevalence of carcass contamination. The list of possibilities is nearly endless, making prioritization of risk very difficult. Nonetheless, there are a few characteristics of pathogens that may predispose them to being more sensitive to the impacts of global climate change (Tirado et al., 2010). For example, those foodborne pathogens that cause disease at very low doses (enteric viruses, parasitic protozoa, *Shigella* spp., enterohemorrhagic *E. coli* strains) and/or have notable environmental persistence (enteric viruses and parasitic protozoa) will likely to be of great concern, particularly after adverse weather events. Those pathogens with documented stress tolerance responses (temperature, pH), such as enterohemorrhagic *E. coli* and *Salmonella* are likely to compete better in the event of climate change.

Climate Change and Its Impact on Mold and Mycotoxin

The factors governing exposure of man to dietary mycotoxins form a complex interconnected system that starts with fungi interacting with crop plants. The performance of each 'partner' is affected by the condition of the other at the same time as both respond to the prevailing conditions of weather and soil. Climate change affects all components of the system. Due to this complexity only qualitative indications can be provided on how climate change might affect toxigenic fungi and mycotoxin contamination.

Mycotoxins are produced by a large variety of fungi, each of them being characterized by its own ecological requirements. Although the impact of climate change on fungal colonization has not been yet specifically and thoroughly addressed, temperature, humidity and precipitation are known to have an effect on toxigenic molds and on their interaction with the plant hosts. In general we know that fungi have temperature ranges within which they perform better and therefore increasing average temperatures could lead to changes in the range of latitudes at which certain fungi are able to compete. Since 2003, frequent hot and dry summers in Italy have resulted in increased occurrence of *A. flavus*, the most xerophilic of the *Aspergillus* genus, with consequent unexpected and serious outbreak of aflatoxin contamination, uncommon in Europe, even in the southern regions. Also in United States serious outbreaks of *A. flavus* have been reported for similar reasons. Generally, moist, humid conditions favor mold growth—moist conditions following periods of heavy precipitation or floods would be expected to favor mold growth. Generally speaking, conditions adverse to the plant (drought stress, stress induced by pest attack, poor nutrient status, etc.) encourages the fungal partner to develop more than underfavorable plant conditions with the expectation of greater production of mycotoxins.

Table 4. Regional variations in climate change health risks and impacts (Sources: Lemmen et al., 2008; Seguin, 2008; ICLR, 2012; Public Safety Canada, 2013).

Climate Risk Category	Examples of Regions at Highest Risk	Examples of Climate-related Risk Factors and Health Impacts
Extreme Temperatures	**Extreme Heat** Windsor to Quebec corridor (e.g., Windsor, Hamilton, Toronto, Kingston, Montreal), regions along Lake Erie, Lake Ontario and St. Lawrence River, Prairies (e.g., Winnipeg), Atlantic Canada (e.g., Fredericton) and British Columbia (e.g., Vancouver)	More frequent and severe heat waves and longer periods of warmer weather, possible colder conditions in some locations
	Extreme Cold Arctic, Prairies, Ontario, eastern Canada	Increase in annual heat-related mortality in Quebec projected for 2020, 2050 and 2080 to be 150, 550, and 1400 excess annual deaths respectively, based on mean temperature increase
Extreme Weather Events and Natural Hazards	**Thunderstorms, Lightning, Tornadoes, Hailstorms** Canada: Wide, low-lying areas of southern Canada, Saskatchewan, Manitoba, Nova Scotia, Ontario, Quebec, Alberta	More frequent and violent extreme weather events, land-shifts, rising sea levels, increased floods, drought and wildfires
	Freezing Rain, Winter Storms Atlantic Canada, Ontario, southern Saskatchewan, southern and northwestern Alberta, southwestern interior British Columbia	In January 2012, a freezing rain event in Montreal resulted in 50 road accidents that included 1 fatality and followed by a similar event a few weeks earlier which led to several traffic accidents, hospitalizations and road closures
	Hurricanes, Storm Surges, Sea-Level Rise Eastern Canada (particularly Atlantic Canada), Arctic, British Columbia	
	Mud-Rock and Landslides, Debris Flows, Avalanches Rocky Mountains, Alberta, British Columbia, Yukon, southern and northeastern Quebec and Labrador, Atlantic coastline, Great Lakes, St. Lawrence shorelines	
	Floods New Brunswick, southern Ontario, southern Quebec and Manitoba	Between 2003 and 2011, there were 60 extreme flood events in Canada which resulted in the evacuation of 44,255 people. A June 2012 flood event in British Columbia resulted in 1 fatality, at least 350 evacuations, treacherous travel conditions and road closures
	Drought Prairies, southern Canada	

Air Quality	**Wildfires**	
	Ontario, Quebec, Manitoba, Saskatchewan, British Columbia, Northwest Territories, Yukon	34 wildfires occurred from 2003–2011 resulting in 113,996 evacuations, and in one event, 2 fatalities; wildfires in Kelowna BC in 2003 were implicated in increased physician visits for respiratory disease up to 78% relative to previous years
	Outdoor Air Pollutants (Ozone and Particulate Matter)	
	Ontario (Great Lakes Region), particularly urban areas of southern Ontario (Toronto), southern Quebec (e.g., Montreal) British Columbia (Vancouver, Lower Fraser Valley), Alberta (Calgary, Edmonton, Fort McMurray) and Manitoba (Winnipeg)	Higher ground-level ozone levels, airborne dust, increased production of pollens and spores by plants
	Aeroallergens (Ragweed) and Fungal Pathogens	
	Southern Quebec and southern Ontario, central and southern Saskatchewan (e.g., Saskatoon) and Manitoba (e.g., Winnipeg), British Columbia	Increased average temperature in Canada could lead to an increase in ozone concentration and could result in an overall increase of 312 premature deaths. On the Island of Montreal, nearly 40,000 children suffer from ragweed-related allergic reactions
Contamination of Food and Water and Food Security	**Water Contamination**	
	Canada: Wide, marine or freshwater coastal regions or watersheds vulnerable to sea-level rise and/or exposure to toxic or pathogenic surface run-off (West Coast, East Coast, Arctic, Great Lakes), regions that are vulnerable to drought (e.g., Prairies), overland flow or flooding leading to surface or groundwater contamination (e.g., rural agricultural areas, urban centers)	Contamination of drinking and recreational water due to run-off from heavy rainfall, and coastal algal blooms in coastal regions
	Food Contamination	
	Canada: Wide, agricultural regions (e.g., Prairies, Ontario, Quebec), regions with communities that are vulnerable to power outages and heat waves (e.g., urban centers such as Toronto), are exposed to toxic marine biota (coastal regions of British Columbia and Atlantic Provinces), are reliant on outdoor cold temperatures for food storage (e.g., Arctic)	4 million people suffer from food-related illnesses each year in Canada; 7 provinces were implicated in the 2008 Canadian listeriosis outbreak that, of the 57 confirmed cases, lead to 23 deaths (75% occurring in Ontario)

Table 4. contd....

Table 4. contd.

Climate Risk Category	Examples of Regions at Highest Risk	Examples of Climate-related Risk Factors and Health Impacts
Contamination of Food and Water and Food Security	**Food Security** Arctic and agricultural regions	In 2006, 30% of Inuit children in Canada had experienced hunger at some point because the family had run out of food or money to buy food 16. In 2007–2008, 9.7% of Canadian households with children experienced food insecurity
Infectious Diseases Transmitted by Insects, Ticks and Rodents	**Lyme Disease** Southern and southeastern Quebec, southern and eastern Ontario, and southeastern Manitoba, New Brunswick and Nova Scotia, southern British Columbia	Changes in the biology and ecology of various disease-carrying insects, ticks and rodents, faster maturation for pathogens within insect vectors and longer disease transmission season
	West Nile Virus Urban and semi urban areas of southern Quebec and southern Ontario, rural populations in the Prairies, rural and semi-urban areas of British Columbia	Range expansion of the tick vector for Lyme disease is expected in the coming decade at an increased rate with climate change; this is expected to increase human Lyme disease risk, particularly in eastern Canada
	Eastern Equine Encephalitis From Ontario to Nova Scotia	
	Rodent-borne Diseases (e.g., Hantavirus) British Columbia, Alberta, Saskatchewan and Manitoba, Northwest Territories, Ontario, Quebec	
Stratospheric Ozone Depletion	Canada wide, in particular regions: at high altitudes; with highly reflective surfaces (e.g., Arctic); with limited natural or built-form shade or air particulates (i.e., smog) that may block UV radiation (e.g., rural areas), in southern Canada (i.e., lower latitude regions closer to the equator	Increased human exposure to UV radiation owing to behavioral changes resulting from a warmer climate In 2008, the estimated number of new cases of non-melanoma skin cancer among men was 40,000 and 33,000 among women. The estimated number of deaths was 160 among men and 100 among women in that year

The most widespread and studied mycotoxins are metabolites of some genera of molds such as *Aspergillus, Penicillium and Fusarium.* Valuable reviews on mycotoxin formation have been published, one of the most recent and complete being the CAST Report (Cast Report 2003). The discussion below provides an illustration of how the climate change factors might be expected to affect mycotoxin contamination by these three main genera of molds with a focus on temperature and precipitation. Following this, some attention will be paid to some of the less considered climate influenced factors (insect and other pest attack, soil, fertilizers and trace elements) that should be recognized and studied as potential and indirect triggers of fungi colonization and mycotoxin production.

Table 5 gives a list of molds and mycotoxins of worldwide importance.

Table 5. Molds and mycotoxins of worldwide importance.

Mold Species	Mycotoxins Produced
Aspergillus parasiticus	Aflatoxins B1, B2, G1, G2
Aspergillus flavus	Aflatoxins B1, B2
Fusarium sporotrichioides	T-2 toxin
Fusarium graminearum	Deoxynivalenol (or nivalenol) Zearalenone
Fusarium moniliforme (F. verticillioides)	Fumonisin B1
Penicillium verrucosum	Ochratoxin A
Aspergillus ochraceus	Ochratoxin A

Conclusion

Climate change research needs to be properly coordinated and the benefits optimized to meet the needs of policy-makers in the country. Attention needs to be focused on data that will assist with mitigation of, and adaptation to, climate change and address specific areas of vulnerability. Further, national data are required to show the advantages and acceptability of a variety of technologies related to climate change. A variety of methodologies of assessments of the potential health effects of climate variability and change should be used. Both qualitative and quantitative approaches should be used, as appropriate, depending on the data availability, level and type of knowledge.

A basic assessment should be conducted using readily available information and data, such as previous assessment, literature reviews and others available region-specific data. Limited analysis should be conducted of regional events data, such as plotting the data against weather variables over time. A more comprehensive assessment included a literature search focused on the goals of the assessment, some quantitative assessment using available data, more involvement by experts, some quantification of effects and the formal peer review of results. Some comprehensive assessment included a detailed literature review, collecting new data and/or generating new models to estimate impact, extensive analysis of quantification and sensitivity.

Health effects related to communicable diseases in the context of climate change are generally preventable, provided that the health care system is prepared and the population informed. The health care system should strengthen its functions as a leading sector that needs to have the capacity to protect the population and to work together with multidisciplinary sector. The health care system has an important role in establishing adaptation, health promotion, prevention and response measures against the health risks related to climate change and communicable diseases. The activities encompassed by the health care sector should include strengthening the capacities of health care practitioners and strengthening the laboratory diagnostic system for identification and diagnosis; obtaining knowledge; adaptation; and health promotion.

The variations in rainfall will most probably compromise the supply of fresh drinking water, thereby increasing the risk of waterborne diseases. Climate change may prolong the season of transmission of certain vector-borne diseases and will tend to change their geographical distribution, potentially allowing them to spread into regions characterized by lack of immunity among the population and/or lack of well-organized health care infrastructure.

Measures for adaptation to climate-change-related health risks are aimed at reducing the effects of climate change on human health.

References

Adler, P. and W. Wills. 2003. The History of Arthropod-Borne Human Disease in South Carolina. American Entomologist, 49: 216–228.

Altizer, S., A. Dobson, P. Hosseini, P. Hudson, M. Pascual and P. Rohani. 2006. Seasonality and the dynamics of infectious diseases. Ecol. Lett., 9: 467–484.

Bacon, R.M., K.J. Kugeler and P.S. Mead. 2008. Surveillance for Lyme disease—United States (1992–2006). MMWR Surveill Summ, 57(10): 1–9.

Bambrick, H.J., A.G. Capon, G.B. Barnett, R.M. Beaty and A.J. Burton. 2011. Climate change and health in the urban environment: adaptation opportunities in Australian cities. Asia Pacific Journal of Public Health, 23(2): 67S–79S.

Barriopedro, D., E.M. Fischer, J. Luterbacher, R.M. Trigo and R. Garćıa-Herrera. 2011. The hot summer of 2010: redrawing the temperature record map of Europe Science, 332: 220–4.

Batterman, S., J. Eisenberg, R. Hardin, M.E. Kruk, M.C. Lemos, A.M. Michalak, M. Mukherjee, E. Renne, H. Stein, C. Watkins, M. Wilson and L. Sustainable. 2009. Control of water-related infectious diseases: a review and proposal for interdisciplinary health-based systems research. Environ. Health Perspect., 117(7): 1023–32.

Blanton, J.D., D. Palmer, K.A. Christian and C.E. Rupprecht. 2008. Rabies Surveillance in the United States during 2007. J. Am Vet Med. Assoc., 233(6): 884–897.

Boxall, A.B.A., A. Hardy, S. Beulke, T. Boucard, L. Burgin, P.D. Falloon, P.M. Haygarth, T. Hutchinson, R.S. Kovats, G. Leonardi et al. 2009. Impacts of climate change on indirect human exposure to pathogens and chemicals from agriculture. Environmental Health Perspectives, 117(4): 508–514.

British Columbia Centers for Disease Control. 2012. *Cryptococcus gattii*. http://www.bccdc.ca/dis cond/a-z/_c/CryptococcalDisease/overview/default.htm.

Campbell-Lendrum, D., R. Woodruff and WHO. 2007. Climate change: quantifying the health impact at national and local levels. *In*: A. Pruss-Ustun and C. Corvalan (eds.). Environmental Burden of Disease Series. Geneva: World Health Organization. 66.

Campbell-Lendrum, D., C. Corvalan and M. Neira. 2007. Global climate change: implications for international public health policy. Bull. WHO, 85: 235–237.

Carroll, J.F. 2007. A note on the occurrence of the lone star tick, *Amblyomma americanum* in the greater Baltimore-Washington area. Proc. of the Entomological Society of Washington, 109(1): 253–256.

CAST. 2003. Mycotoxins: Risks in Plant, Animal and Human Systems, Task Force Report, ISSN 0194; N. 139.

Charron, D.F., M.K. Thomas, d. Waltner-Toews, J.J. Aramini, T. Edge, R.A. Kent, A.R. Maarouf and J. Wilson. 2004. Vulnerability of waterborne diseases to climate change in Canada: a review. J. Tox. Environ. Health, 67: 1667–1677.

Costello, A., M. Abbas, A. Allen et al. 2009. Managing the health effects of climate change. The Lancet, 373: 1693–1733.

Craun, G.F. and R.L. Calderon. 2006. Observational epidemiologic studies of endemic waterborne risks: cohort, case-control, time-series, and ecologic studies. J. Water Health, 4 Suppl. 2: 101–19.

Curriero, F.C., J.A. Patz, J.B. Rose and S. Lele. 2001. The association between extreme precipitation and waterborne disease outbreaks in the United States, 1948–1994. Am. J. Public Health, 91(8): 1194–9.

D'Souza, R.M., N.G. Becker, G. Hall and K.B. Moodie. 2004. Does Ambient Temperature Affect Foodborne Disease? Epidemiology, 15 1 8692.

Diaz, J.H. 2007. The influence of global warming on natural disasters and their public health outcomes. Am. J. Disaster Med., 2(1): 33–42.

Douglass, R.J., C.H. Calisher and K.C. Bradley. 2005. State-by-state incidences of hantavirus pulmonary syndrome in the United States, 1993–2004. Vector-borne Zoonotic Dis., 5(2): 189–92.

Dwight, R.H., D.B. Baker, J.C. Semenza and B.H. Olson. 2004. Health effects associated with recreational coastal water use: urban versus rural California. American Journal of Public Health, 94(4): 565–567.

Ebi, K.L., J. Balbus, P.L. Kinney, E. Lipp, D. Mills, M.S. O'Neill et al. 2008. Effects of global change on human health. In: J.L. Gamble (ed.). Analyses of the Effects of Global Change on Human Health and Welfare and Human Systems A Report by the US Climate Change Science Program and the Subcommittee on Global Change Research. Washington, DC: U.S. Environmental Protection Agency, 2-1–2-78.

EFSA (European Food Safety Authority). 2009. The community summary report on trends and sources of zoonoses and zoonotic agents in the European union in 2007. The EFSA Journal, 223: 1–215.

English, P.B., A.H. Sinclair, Z. Ross, H. Anderson, V. Boothe, C. Davis and E. Simms. 2009. Environmental health indicators of climate change for the United States: findings from the State Environmental Health Indicator Collaborative. Environ Health Perspect., 117(11): 1673–81. doi: 10.1289/ehp.0900708.

FAO. 2008. Viruses in Food: Scientific Advice to support risk management activities. Microbiological Risk Assessment Series No.7. In press.

Faust, E. 1949. Malaria incidence in North America. In: M.F. Boyd (ed.). Malariology; A Comprehensive Survey of all Aspects of this Group of Diseases from a Global Standpoint. Saunders: Philadelphia, p. 2v. (xxi, 1643p.).

Ford, T.E., R.R. Colwell, J.B. Rose, S.S. Morse, D.J. Rogers and T.L. Yates. 2009. Using satellite images of environmental changes to predict infectious disease outbreaks. Emerging Infectious Diseases, 15(9): 1341–1346.

Frumkin, H., J. Hess, G. Luber, J. Malilay and M. McGeehin. 2008. Climate change: the public health response. Am. J. Public Health, 98: 435–445; doi:10.2105/AJPH.2007.119362.

Gamble, J.L., K.L. Ebi, A.E. Grambsch, F.G. Sussman and T.J. Wilbanks. 2008. Analyses of the effects of global change on human health and welfare and human systems : final report, synthesis and assessment product 4.6: report by the U.S. Climate Change Science Program and the Subcommittee on Global Change Research. 2008, Washington, D.C.: U.S. Climate Change Science Program. ix, 204p.

Greer, A., V. Ng and D. Fisman. 2008. Climate change and infection diseases in North America: the road ahead. Canadian Medical Association Journal, 178: 715–722.

Hall, G.V., R.M. D'Souza and M.D. Kirk. 2002. Foodborne disease in the new millennium: out of the frying pan and into the fire? Med. J. Aust., 177: 614–618.

Heaney, C.D., E. Sams, S. Wing, S. Marshall, K. Brenner, A.P. Dufour et al. 2009. Contact with beach sand among beachgoers and risk of illness. Am. J. Epidemiol., 170(2): 164–72.

Hess, J., K.L. Heilpern, T. Davis and H. Frumkin. 2009. Climate change and emergency medicine: Impacts and opportunities; Academic Emergency Medecine, 16(8): 782–794.

Horton, G., L. Hanna and B. Kelly. 2010. Drought, drying and climate change: Emerging health issues for ageing Australians in rural areas. Australasian Journal on Ageing, 29: 2–7.

ICLR [Institute for Catastrophic Loss Reduction]. 2012. Telling the weather story. <http://iclr.org/images/Telling_the_weather_story.pdf>.

IOM [Institute of Medicine]. 2011. Climate Change, the Indoor Environment and Health; The National Academies Press, Washington DC.

IPCC (Intergovernmental Panel on Climate Change). 2007. Contribution of Working Group I to the Fourth Assessment Report of the Intergovernmental Panel on Climate Change. Cambridge University Press: Cambridge, United Kingdom; New York.

IPCC (Intergovernmental Panel on Climate Change). 2007. Summary for policymakers. In: M.L. Parry, O.F. Canziani, J.P. Palutikot, P.J. van der Linden and C.E. Hanson (eds.). Climate Change: Impacts,

Adaptation and Vulnerability. Contribution of Working Group II to the Fourth Assessment Report of the Intergovernmental Panel on Climate Change. Cambridge University Press: Cambridge, UK.

IPCC (Intergovernmental Panel on Climate Change). 2012. Summary for policymakers; in Managing the Risks of Extreme Events and Disasters to Advance Climate Change Adaptation; Cambridge University Press: Cambridge, United Kingdom. <http://www.ipcc wg2.gov/SREX/images/uploads/SREX-All_FINAL.pdf>.

Kidd, S.E., Y. Chow, S. Mak, P.J. Bach, H. Chen, A.O. Hingston, J.W. Kronstad and K.H. Bartlett. 2007. Characterization of environmental sources of the human and animal pathogen *Cryptococcus gattii* in British Columbia, Canada, and the Pacific Northwest of the United States; Applied and Environmental Microbiology, 73(5): 1433–1443.

Lamy, S. and V. Bouchet. 2008. Chapter 4: air quality, climate change and health. pp. 45–111. *In*: J. Seguin (ed.). Human Health in a Changing Climate: a Canadian Assessment of Vulnerabilities and Adaptive Capacity. Ottawa: Health Canada.

Lemmen, D.S., F.J. Warren, J. Lacroix and E. Bush. 2008. From Impacts to Adaptation: Canada in a Changing Climate 2007; Government of Canada, Ottawa, Ontario.

Lipp, E.K., A. Hau and R.R. Colwell. 2002. Effects of global climate on infectious disease: the cholera model. Clin. Microbiol. Rev., 15: 757–770.

Luby, S.P., M. Rahman, M.J. Hossain, L.S. Blum et al. 2006. Foodborne transmission of Nipah virus, Bangladesh. Emerg. Infect. Dis., 12: 1888–1894.

Mathers, C., D.M. Fat, J.T. Boerma and World Health Organization. 2008. The global burden of disease : 2004 update. Geneva, Switzerland: World Health Organization. vii, 146p.

McMichael, A.J., R.E. Woodruff and S. Hales. 2006. Climate change and human health: Present and future risks. Lancet, 367: 859–869.

Mcquiston, J.H., R.C. Holman, C.L. McCall, J.E. Childs, D.L. Swerdlow and H.A. Thompson. 2006. National surveillance and the epidemiology of human Q fever in the United States, 1978–2004. The American Journal of Tropical Medicine and Hygiene., 75(1): 36–40.

Mead, P.S., L. Slutsker, V. Dietz, L.F. McCaig, J.S. Bresee, C. Shapiro, P.M. Griffin and R.V. Tauxe. 1999. Food-related illness and death in the United States. Emerg. infect. Dis., 5(5): 607–25.

Meinhardt, P.L. 2006. Recognizing waterborne disease and the health effects of water contamination: a review of the challenges facing the medical community in the United States. J. Water Health, 4: 27–34.

Ministry of Health of the Republic of Macedonia. 2011. Adaptation Strategy in Health Sector from Climate Change of the Republic of Macedonia. Skopje.

Moore, S.K., V.L. Trainer, N.J. Mantua, M.S. Parker, E.A. Laws, L.C. Backer and L.E. Fleming. 2008. Impacts of climate variability and future climate change on harmful algal blooms and human health. Environmental Health: a Global access science source, 7(Suppl. 2).

Mos, L., B. Morsey, S.J. Jeffries, M.B. Yunker, S. Raverty, S.D. Guise et al. 2006. Chemical and biological pollution contribute to the immunological profiles of tree ranging harbor Seals. Environmental Toxicology and Chemistry, 25(12): 3110–3117.

Niemi, G.J., D. Wardrop, R. Brooks, S. Anderson, V. Brady, H. Paerl, C. Rakocinski, M. Brouwer, B. Levinson and M. McDonald. 2004. Rationale for a new generation of ecological indicators for coastal waters. Environmental Health Perspectives, 112(9): 979–986.

NRTEE (National Round Table on the Environment and the Economy). 2011. Paying the Price: The Economic Impacts of Climate Change for Canada.

Orenstein, W.A., R.G. Douglas, L.E. Rodewald and A.R. Hinman. 2005. Immunizations in the United States: success, structure, and stress. Health Aff (Millwood) May–Jun., 24(3): 599–610.

Ostry, J.D., G. Atish, H. Karl, C. Marcos, S.Q. Mahvash and B.S.R. Dennis. 2010. Capital Inflows: The Role of Controls. IMF Staff Position Note 10/04. (Washington: International Monetary Fund).

Pavlin, B.I., L.M. Schloegel and P. Daszak. 2009. Risk of importing zoonotic diseases through wildlife trade, United States. Emerg. Infect. Dis., 15: 1721–6.

Petri, W.A., J.R. 2004. America in the world: 100 years of tropical medicine and hygiene. Am. J. Trop. Med. Hyg., 71(1): 2–16.

Poulain, L., Y. Iinuma, K. Müller, W. Birmili, K. Weinhold, E. Brüggemann, T. Gnauk, A. Haus-mann, G. Löschau, A. Wiedensohler and H. Herrmann. 2011. Diurnal variations of ambient particulate wood burning emissions and their contribution to the concentration of Polycyclic Aromatic Hydrocarbons (PAHs) in Seiffen, Germany, Atmos. Chem. Phys., 11: 12697–12713, doi:10.5194/acp-11-12697-2011.

Robine, J.M., S.L. Cheung, S. Le Roy, H. Van Oyen, C. Griffiths, J.P. Michel et al. 2008. Death toll exceeded 70,000 in Europe during the summer of 2003. C. R. Biol. 331(2): 171–178.

Rodriguez-Romo, L. and A. Yousef. 2005. Cross-protective effects of bacterial stress. *In*: M. Griffiths (ed.). Understanding Pathogen Behaviour. Woodhead Publishing, Cambridge, U.K.

Rose, J.B., P.R. Epstein, E.K. Lipp, B.H. Sherman, S.M. Bernard and J.A. Patz. 2001. Climate variability and change in the United States: potential impacts on water- and foodborne diseases caused by microbiologic agent. Environ. Health Perspectives, 109: 211–221.

Rosenzweig, C., W.D. Solecki, S.A. Hammer and S. Mehrotra. 2011. Climate Change and Cities: First Assessment Report of the Urban Climate Change Research Network; Cambridge University Press, Cambridge, United Kingdom.

Sandifer, P., C. Sotka, D. Garrison and V. Fay. 2007. Interagency Oceans and Human Health Research Implementation Plan: A Prescription for the Future. Interagency Working Group on Harmful Algal Blooms, Hypoxia, and Human Health of the Joint Subcommittee on Ocean Science and Technology. Washington, DC.

Schenck, P., K.A. Ahmed, A. Bracker and R. DeBernardo. 2010. Climate change, indoor air quality and health; U.S. Environmental Protection Agency.

Seguin, J. 2008. Human Health in a Changing Climate: A Canadian Assessment of Vulnerabilities and Adaptive Capacity. Health Canada. Ottawa, Ontario.

Shultz, A., J.O. Omollo, H. Burke, M. Qassim, J.B. Ochieng et al. 2009. Cholera outbreak in Kenyan refugee camp: risk factors for illness and importance of sanitation. American Journal of Tropical Medicine & Hygiene, 80: 4640–4645.

Sidley, P. 2008. Floods in southern Africa result in cholera outbreak and displacement. BMJ, 336(7642): 471.

Smolinski, M.S., M.A. Hamburg and J. Lederberg. 2003. Washington, D.C.: National Academies Press. xxviii, 367p.

Sokurenko, E.V., R. Gomulkiewicz and D.E. Dykhuizen. 2006. Source-sink dynamics of virulence evolution. Nat. Rev. Microbiol., 4(7): 548–555.

Stern, N. 2006. Stern Review on the Economics of Climate Change. UK Government Economic Service, London. www.sternreview.org.uk.

Stewart, J.R., R.J. Gast, R.S. Fujioka, H.M. Solo-Gabriele, J.S. Meschke, L.A. Amaral-Zettler, E.D. Castillo, M.F. Polz, T.K. Collier, M.S. Strom et al. 2008. The coastal environment and human health: microbial indicators, pathogens, sentinels and reservoirs. Environ. Health., 7: S3.

Strickman, D. and P. Kittayapong. 2003. Dengue and its vectors in Thailand: Calculated transmission risk from total pupal counts of *Aedesaegypti* and association of wing-length measurements with aspects of the larval habitat. American Journal of Tropical Medicine and Hygiene, 68: 209–217.

Tagaris, E., K. Manomaiphiboon, K.-J. Liao, L.R. Leung, J.-H. Woo, S. He, P. Amar and A.G. Russell. 2007. Impacts of global climate change and emissions on regional ozone and fine particulate matter concentrations over the United States. Journal of Geophysical Research, 112: doi:10.1029/2006JD008262.

Tauxe, R.V. 2002. Emerging foodborne pathogens. Int. J. Food Microbiol., 78: 31–41.

The World Meteorological Organisation's (WMO) state of the *climate* report for 2013.

Tirado, M.C., R. Clarke, L.A.A. Jaykus, Abigail L. McQuatters-Gollop and J.M. Frank. 2010. Climate Change and Food Safety: A Review. Food Research International, 43(7): 1745–1765.

Tirado, C. and K. Schmidt. 2001. WHO surveillance programme for control of foodborne infections and intoxications: preliminary results and trends across greater Europe. J. Infect., (43)1: 804.

Tolba, O., A. Loughrey, C.E. Goldsmith, B.C. Millar, P.J. Rooney and J.E. Moore. 2008. Survival of epidemic strains of healthcare (HA-MRSA) and community-associated (CA-MRSA) methicillin-resistant *Staphylococcus aureus* (MRSA) in river-, sea- and swimming pool water. Int. J. Hyg. Environ. Health, 211(3-4): 398–402.

US CDC et al. 2010. Centers for Disease Control and Prevention. Extreme Weather Events. http://www.cdc.gov/climatechange/effects/extremeweather.htm.

US EPA (United States, Environmental Protection Agency). 2008. Health Security through Healthy Environments; Briefing Note 01.

Vladimir, K. and G. Dragan. 2012. Chapter 7. Climate Change: Implication for Food-Borne Diseases (Salmonella and Food Poisoning Among Humans in R. Macedonia). Structure and Function of Food Engineering. Edited by Ayman AmerEissa. Publisher: InTech.

Watson, R.T. and A.J. McMichael. 2001. Global climate change-the latest assessment: does global warming warrant a health warning? Global Change and Human Health, 2: 64–75.

Wheaton, J.M., S.E. Darby and D. Sear. 2008. The scope of uncertainties in river restoration. pp. 21–39. *In*: S.E. Darby and D. Sear (eds.). River Restoration: Managing the Uncertainty in Restoring Physical Habitat. John Wiley and Sons, Chichester, U.K.

White, Pc, Jr. 1965. Murine Typhus in Fulton County, Georgia. Mil. Med., 130: 386–388.

WHO. 2008. Global Malaria Programme. Geneva: World Health Organization, xx, 190p.

WHO. 2002. Geneva: World Health Organization, xx, 190p.

Wittrock, V., E. Wheaton and E. Siemens. 2011. Drought and Excessive Moisture Saskatchewan's Nemesis: Characterizations for the Swift Current Creek, North Saskatchewan River, Assiniboine River and Upper Souris River Watersheds. Prepared for Saskatchewan Watershed Authority for the Prairie Regional Adaptation Collaborative Project. Saskatchewan Research Council (SRC), Saskatoon, SK. SRC Publication No. 13022-6E11.

Yoder, J., V. Roberts, G.F. Craun, V. Hill, L.A. Hicks, N.T. Alexander et al. 2008. Surveillance for waterborne disease and out breaks associated with drinking water and water not intended for drinking—United States, 2005–2006. MMWR Surveill Summ, 57(9): 1–29.

Ziska, L., K. Knowlton, C. Rogers, D. Dalan, N. Tierney, M.A. Elder, W. Filley, J. Shropshire, L.B. Ford, C. Hedberg, P. Fleetwood, K.T. Hovanky, T. Kavanaugh, G. Fulford, R.F. Vrtis, J.A. Patz, J. Portnoy, F. Coates, L. Bielory and D. Frenz. 2011. Recent warming by latitude associated with increased length of ragweed pollen season in central North America; Proceedings of the National Academy of Sciences of the United States of America, 108(10): 4248–4251.

9a

Food Safety Status in Developing Countries

Chiraporn Ananchaipattana,[1] *Rithy Chrun*[2] and
Yasuhiro Inatsu[3],*

Introduction

Food safety refers to the conditions and practices that preserve the quality of foods to prevent contamination from biological, chemical or physical hazards at level that can cause adverse effects on human's health. The potential effects on people due to unsafe practices when preparing, cooking and serving are that it could cause food poisoning (foodborne illness) or can even lead to death. This threat to life from foodborne illness is especially more in young children, older adults, pregnant women, and people with weakened immune system. So, large proportions of income of developing countries may be lost by individuals due to reduced productivity and expenditures on medical care. With increasing globalization and cooperation among nations, problems and dangers could easily and quickly spread within nations and across nations faster than ever before. The situation is worse in some developing countries where borders are highly porous due to institutional and policy weaknesses and gaps so that some illegal foods such as meats from diseased and unsound animals or banned agricultural chemicals, etc. are transported and distributed to the markets.

Trade in food among developing countries have become important because they play an increasingly important role in export markets in the world. However, developing

---

[1] Department of Biology, Faculty of Sciences and Technology, Rajamangala University of Technology Thanyaburi, 39 Muh 1, Rangsit-Nakhonnayok Rd. Klong Hok, Thanyaburi Pathum Thani 12110, Thailand.
[2] Faculty of Agro-Industry, Royal University of Agriculture, Chamkardaung, Dangkor District, P.O. Box 2696, Phnom Penh, Cambodia.
[3] National Food Research Institute, National Food and Agricultural Organization, 2-1-12, Kannondai, Tsukuba-shi, Ibaraki-305-8642, Japan.
* Corresponding author: inatu@affrc.go.jp

countries may lack the capacity to address gaps in their food safety systems, or to educate producers and processors on safe food handling behaviors that could reduce the risks (Fig. 1). There are many arguments presented about food poisoning from developing countries. For example, Bird flu virus H5N1, which originated from East Asia has spread to many parts of the world. The same report outbreaks of *Shigella sonnei* infections in Denmark (Lewis et al., 2007) and Australia (Stafford et al., 2007) after eating baby

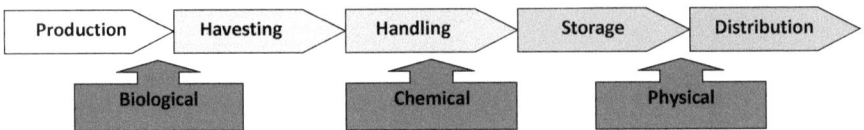

Figure 1. Produce is vulnerable to hazards at every step of the chain.

corn imported from Thailand, was linked to contamination from packing house in Thailand. Vegetable salad with raw octopus led to *Vibrio parahaemolyticus* enteritis outbreak following a wedding banquet in a rural village Kampong Speu in Cambodia (Som et al., 2012). In 2009 outbreaks with *Shigella* connected to sugar snaps from Kenya were reported from Norway and Denmark (Löfdahl et al., 2009). An outbreak of foodborne salmonellosis from breads with mayonnaise contaminated eggs in Vietnam (Thuan et al., 2014). At least 10 people in Hyogo and Chiba Prefectures in Japan have suffered food poisoning after consuming frozen "gyoza" meat and vegetable dumplings from China that contained the pesticide, organo-phosphate. Unsafe food and water is one of the major causes of illness and death in developing countries and also spread to the developed countries. Foodborne and waterborne diarrheal diseases kill an estimated 2 million people annually, mostly children in developing countries (WHO, 2014). According to the UNICEF, there are about 6.3 million people (nearly half of population) still lack access to clean water and basic sanitation in Cambodia. About 3.9 million of those without access to safe drinking water in Cambodia are poor and live in rural areas. The lack of clean drinking water and basic sanitation, children (41 per cent of the population) are especially vulnerable to water-borne diseases (UNICEF, 2014). These examples demonstrate the impact of food safety to the public health security.

Characteristics of Foodborne Diseases

Unsafe food is food that contains toxins, pesticides, chemical and physical contaminants, and especially microbiological pathogens such as bacteria, parasites, and viruses at any stage of the chain that can cause illness. However, the people who are working with food safety are concerned more about microbial foodborne diseases. The symptoms associated with foodborne diseases are nausea, vomiting, diarrhea, abdominal pain, fever, and headache. Most people have had foodborne illness that occurred the symptoms in short time after eating foods and that is caused by ingesting food containing toxins formed by bacteria which resulted from the bacterial growth in the food item, it is called "Foodborne intoxication". Sometimes it is difficult to find out the sources of poisonous foods because the symptoms of foodborne illness occurs few days later after consuming, they might forget what they ate. This type of disease is caused by the ingestion of food containing live bacteria which grow and establish themselves in the human intestinal tract and is call "foodborne infection". Usually symptoms of foodborne illnesses have a short duration and are self-limiting. However, some foodborne diseases are associated with long-term chronic illness such as *Salmonella, Shigella, Yersinia,* and *Campylobacter* spp. are linked to reactive arthritis and cause inflammation; *E. coli* O157:H7 and a few other strains, can result in hemorrhagic (very bloody) diarrhea, kidney failure (termed hemolytic-uremic syndrome), and occasionally, death.

In developing countries, diarrhea is among the leading causes of childhood morbidity and mortality. About 20% of 2.5 million deaths in children fewer than 5 years of age were caused by diarrhea as determined by active surveillance (Kosek et al., 2003). In Thailand, the mortality rate has been declining considerably due to improved and extensive coverage of health services as well as the success of the campaign on oral rehydration therapy (ORT). Acute diarrhea is still a crucial public health problem with a relatively slight change in incidence among both children and adults, particularly among children under five years of age whose incidence is higher than that in adults as the data shows in Fig. 2 (Bureau of Epidemiology, 2006).

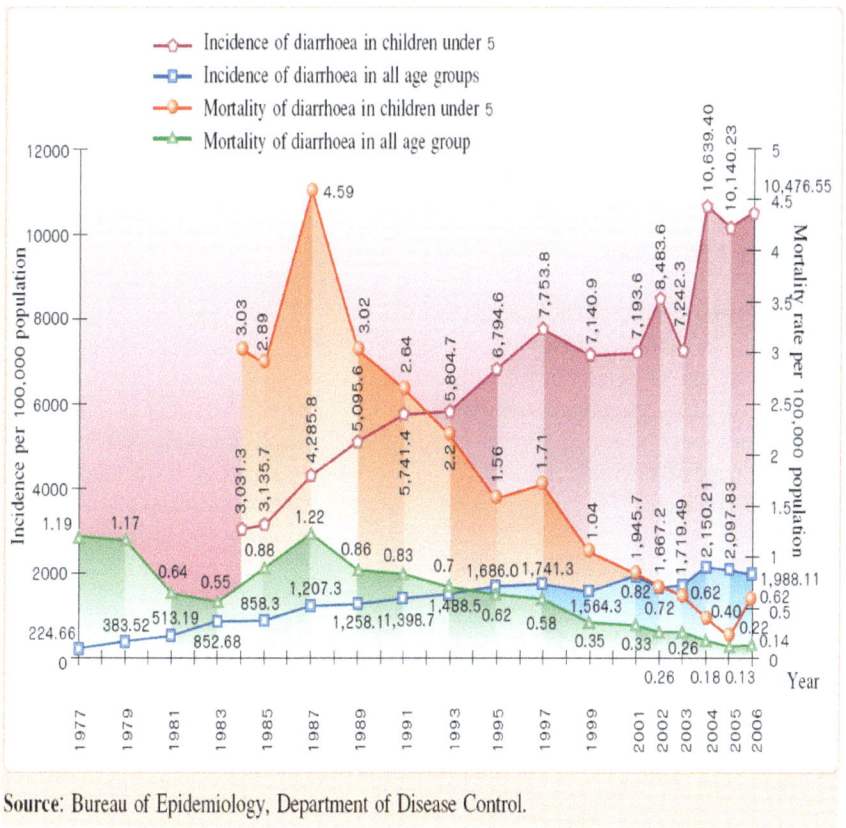

Source: Bureau of Epidemiology, Department of Disease Control.

Figure 2. Incidence and mortality rate of diarrhea in Thailand, 1977–2006.

Enteropathogenic *Escherichia coli* (EPEC), rotavirus and *Shigella* spp. are found to be common pathogens causing diarrhea among hospitalized children in Vietnam (Hung, 2006).

Common Causes of Foodborne Disease

In developing countries, especially in tropical and developing countries, enteric bacteria and parasites are more prevalent than viruses. It is a serious problem among older children and adults as well as in infants and young children. Diarrhea caused by bacteria and parasites can be common when traveling in developing countries and is often called "traveler's diarrhea". The range of causative foodborne microorganisms is very large but common bacterial, viral infection and parasites are shown in Table 1.

The prevalence of foodborne bacteria isolated from various kind of raw food in Thailand were *E. coli*, *Salmomonella* spp., *Enterococcus* spp., *Staphylococcus* spp. and *Bacillus cereus* (Vindigni et al., 2007; Ananchaipattana et al., 2012a). The data are show in Tables 2 and 3.

Table 1. Overview of causative agents in diarrhea.

Bacteria	Viruses	Parasites
E. coli	Rotavirus	*Giardia lamblia*
Salmonella spp.	Norwalk virus	*Entamoeba histolytica*
Shigella	Norwalk-like viruses	*Cyclospora cayetanensis*
Campylobacter	Enteric adenoviruses	*Cryptosporidium*
Yersinia	Caliciviruses	
Vibrio	Astroviruses	
Clostridium difficile		

Source: Hung, 2006, modified.

The increasing antimicrobial resistance of *Salmonella* spp. has become a significant public health concern in developing countries. Antibiotic resistance of *Salmonella* spp. has led to failure of treatment for salmonellosis (Butt et al., 2003). The isolation of antibiotic-resistant *Salmonella* spp. from chicken meat and human stool in Thailand has been previously reported (Boonmar et al., 1998; Kulwichit et al., 2007). Recently high percentage of *Salmonella* isolates were resistant to streptomycin 78%, tetracycline 59% and ampicillin 51%. The data is show in Fig. 3 (Ananchaipattana et al., 2014).

However, very rarely *Listeria monocygenes* conaminaton was food in raw food in Thailand (Table 3) and (Table 4) (Minami et al., 2010; Ananchaipattana et al., 2012a).

Factors Causing Existing Unsafe Foods in the Developing World

Foodborne diseases are major health problems in developed and developing countries. The problem is more noticeable in developing countries due to

Table 2. Prevalence of foodborne microorganisms isolated from fresh meat and eggs by markets in Bangkok, Thailand, 2003.

Meat samples	Bacteria strains	Contaminated samples sold in			
		Open market*		Supermarket*	
		Number	(%)	Number	(%)
Chicken	*Salmonella* spp.	23/27	85	8/23	35
	Campylobacter jejuni	4/27	15	8/23	35
	Campylobacter coli	6/27	22	8/23	35
	Acrobacter butzleri	16/27	59	3/23	13
	Enterococcus spp.	26/27	96	23/23	100
Beef	*Salmonella* spp.	26/26	100	15/24	63
	Campylobacter jejuni	1/26	4	0/24	0
	Campylobacter coli	0/26	0	0/24	0
	Acrobacter butzleri	9/26	35	5/24	21
	Enterococcus spp.	26/26	100	24/24	100
Pork	*Salmonella* spp.	25/27	93	17/23	74
	Campylobacter jejuni	0/27	0	2/23	9
	Campylobacter coli	0/27	0	2/23	9
	Acrobacter butzleri	5/27	19	3/23	13
	Enterococcus spp.	27/27	100	23/23	100
Chicken eggs	*Salmonella* spp.	5/27	19	2/23	9
	Campylobacter jejuni	0/27	0	0/23	0
	Campylobacter coli	0/27	0	0/23	0
	Acrobacter butzleri	1/27	4	0/23	0
	Enterococcus spp.	26/27	96	13/23	57

*Number of positive samples/total samples and percentage in parentheses; Vindigni et al., 2007.

Table 3. Prevalence of foodborne microorganisms isolated from various kinds of food by markets in Bangkok, Thailand, 2012.

Categories of Foods	Pathogens	Contaminated samples sold in				Fisher's Direct P-value
		Open market		Supermarket		
		Numbers	(%)	Numbers	(%)	
Meat	*Escherichia coli*	30/36	(83)	12/15	(80)	0.92
	Salmonella spp.	30/36	(83)	10/15	(67)	0.95
	Yersinia spp.	1/36	(3)	0/15	(0)	0.71
	Cronobacter sakazakii	1/36	(3)	0/15	(0)	0.71
	Listeria spp.	4/36	(11)	2/15	(13)	0.80
	Bacillus cereus	1/36	(3)	0/15	(0)	0.71
	Staphylococcus spp.	20/36	(56)	9/15	(60)	0.51
Fish or seafood	*Escherichia coli*	13/17	(76)	9/20	(45)	0.10
	Salmonella spp.	11/17	(65)	2/20	(10)	0.00
	Yersinia spp.	1/17	(6)	1/20	(5)	0.71
	Cronobacter sakazakii	1/17	(6)	1/20	(5)	0.71
	Listeria spp.	1/17	(6)	0/20	(0)	0.46
	Bacillus cereus	1/17	(6)	1/20	(5)	0.71
	Staphylococcus spp.	1/17	(41)	1/20	(5)	0.01
Vegetable	*Escherichia coli*	8/18	(44)	3/20	(15)	0.42
	Salmonella spp.	1/18	(6)	1/20	(5)	0.73
	Yersinia spp.	0/18	(0)	0/20	(0)	1.00
	Cronobacter sakazakii	3/18	(17)	1/20	(5)	0.26
	Listeria spp.	0/18	(0)	0/20	(0)	1.00
	Bacillus cereus	7/18	(39)	6/20	(30)	0.41
	Staphylococcus spp.	12/18	(67)	5/20	(25)	0.01

Ananchaipattana et al., 2012a.

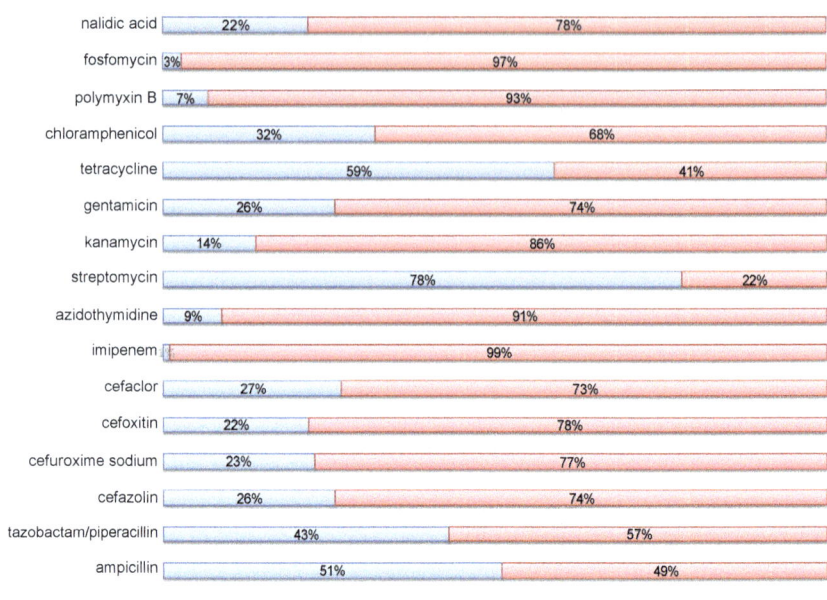

☐ Resistant ☐ Sensitive

Figure 3. Antibiotic profile of *Salmonella* spp. isolated for raw food in Thailand (Ananchaipattana et al., 2014).

Table 4. Prevalence of foodborne pathogens in open markets and supermarkets in Thailand during June 2006 to July 2007 (Minami et al., 2010).

Sample	Isolate	Supermarket*	Prevalence (%)	Open market	Prevalence (%)
Beef	*L. monocytogenes*	2/68	3	0/40	0
	E. coli O157	0/46	0	0/33	0
	Salmonella spp.	6/25	24	0/4	0
Chicken	*L. monocytogenes*	1/28	4	5/81	6
	E. coli O157	0/17	0	0/44	0
	Salmonella spp.	4/7	57	13/27	48
Pork	*L. monocytogenes*	14/44	32	1/36	3
	E. coli O157	0/24	0	0/22	0
	Salmonella spp.	2/17	12	0/13	0
Shrimp	*L. monocytogenes*	0/26	0	0/17	0
	E. coli O157	0/26	0	9/17	53
	Salmonella spp.	0/26	0	0/17	0
Oyster	*L. monocytogenes*	0/5	0	0/43	0
	E. coli O157	0/5	0	0/43	0
	Salmonella spp.	0/5	0	0/43	0

*Number of positive samples/total of samples.

Water-related factors

Water may be the vector of unsafe foods when contaminated with viruses, bacteria, parasites, or chemicals. The people living in rural areas in developing countries usually use unsafe water sources such as rivers, ponds, dams, lakes, streams, wells and other surface water sources (Karim et al., 2001; Tumwine et al., 2002; Brooks et al., 2003). The hous holds obtaining water from storage containers by dipping showed statistically significant association with diarrheal morbidity (Teklemariam et al., 2000). Faecal contamination levels in household water containers were generally high even when the source water was of good quality (Jensen et al., 2004). In Bangkok Thailand, most of the raw food were contaminated with coliform bacteria (Ananchaipattana et al., 2012b). The data is show in Table 5.

Table 5. Contamination rate of bacteria in raw food in 2011 (Ananchaipattana et al., 2012b).

Categories of foods	Total Numbers of tested samples	Gram negative bacteria							
		Coliform bacteria		Faecal coliform		*Escherichia coli*		*Pseudomonas* spp.	
		Contaminated samples		Contaminated samples		Contaminated samples		Contaminated samples	
		Number	(%)	Number	(%)	Number	(%)	Number	(%)
Meat	51	51	(100)	48	(94)	36	(71)	16	(31)
Fish or seafood	37	37	(100)	37	(100)	26	(70)	22	(60)
Vegetable	38	35	(100)	35	(92)	13	(34)	29	(76)
Fermented meat and fish	11	6	(55)	5	(45)	5	(45)	1	(9)
Total	137	129	(94)	125	(91)	80	(58)	68*	(50)

* indicates a statistically significant differences ($p<0.05$) in the contamination rate among the four categories of food.

Most rural areas in developing countries lack clean water to irrigate vegetables as well as to process foods. Therefore, the level of unsafe foods associated with contaminated water is very high. Each year, more than 20% children up to 14 years of age die as a result of inadequate access to safe water, sanitation, and hygiene (Prüss-Üstün et al., 2008).

Socio-economic factors

The risk of foodborne diseases were associated with poor environmental conditions, having no kitchen or having an unhygienic toilet, low income, lack of financial resources to invest on safer equipment, lack of education for food handlers (Etiler et al., 2004) and lack of quality control in food factories that found *B. cereus* in package tofu. The source of contamination might be the presence of *B. cereus* spore on the soybean seeds, which survived after heat treatment step of soy milk production (Ananchaipattana et al., 2012c).

Hygiene practices

One of the most important routes for transmission of infection is via the hands. There are many contamination sources such as raw food, pets, soil, contaminated surfaces and reservoir sites such as toilets, sneezing, coughing and transfer of nasal secretions to the hands. A systematic review reported handwashing could reduce diarrhea risk by 47% (Curtis and Cairncross, 2003). Microbes such as *Salmonella*, *Shigella*, *E. coli* and *Staphylococcus aureus*, and viruses such as rotavirus and norovirus can be transferred from unwashed hands to other people, surfaces and food. Microbes can also be transferred directly from hand to mouth. This is known as "faecal-oral transmission". Unhygienic kitchen, improper food storage, presence of pets or animals inside the house, presence of flies inside the house, were associated with risk of diarrhea morbidity in children (Curtis et al., 2000). A study pointed about 50% of food handlers of Dangila town in North West Ethiopia had good food handling practice when they have more monthly income, knowledge status, existence of shower facility, existence of separate dressing room and absence of insects and rodents in processing rooms (Tessema et al., 2014). Most market places in developing countries are traditional open-air markets where foodstuffs are sold by individual vendors or farmers and usually displayed unwrapped and at ambient temperature. Same categories of foods (such as vegetables, raw meats, fishes) tend to be sold by vendor in the open market. These markets naturally have multiple sources of potential contamination (rodents, insects, sewage and water).

Poor consumption behaviors

In some regions of some developing country, such as the north-eastern region of Thailand, people like to consume raw or undercooked meat, which is a cause of diarrheal diseases.

References

Ananchaipattana, C., Y. Hosotani, S. Kawasaki, S. Pongswat, M.B. Latiful, S. Isobe and Y. Inatsu. 2012a. Prevalence of foodborne pathogens in retailed foods in thailand. Journal of Foodborne Pathogen and Disease, 9(9): 835–840.

Ananchaipattana, C., Y. Hosotani, S. Kawasaki, S. Pongswat, M.B. Latiful, S. Isobe and Y. Inatsu. 2012b. Bacterial contamination in retail foods purchased in Thailand. Food Science and Technology Research, 18(5): 705–712.

Ananchaipattana, C., Y. Hosotani, S. Kawasaki, S. Pongswat, M.B. Latiful, S. Isobe and Y. Inatsu. 2012c. Bacterial contamination of soybean curd (Tofu) sold in Thailand. Food Science and Technology Research, 18(6): 843–848.

Ananchaipattana, C., Y. Hosotani, S. Kawasaki, K.A. Yamaguchi, M.B. Latiful, S. Isobe and Y. Inatsu. 2014. Serotyping, RAPD grouping and antibiotic susceptibility testing of *Salmonella enterica* isolated from retail foods in Thailand. Food Science and Technology Research, 20(4): 905–913.

Boonmar, S., A. Bangtrakulnonth, S. Pornruangwong, S. Samosomsuk, K. Kaneko and M. Ogawa. 1998. Significant increase in antibiotic resistance of Salmonella isolates from human beings and chicken meat in Thailand. Vet. Microbiol., 62: 73–80.

Brooks, J.T., R.L. Shapiro, L. Kumar, J.G. Wells, P.A. Phillips-Howard, Y.P. Shi, J.M. Vulule, R.M. Hoekstra, E. Mintz and L. Slutsker. 2003. Epidemiology of sporadic bloody diarrhea in rural western Kenya. Am. J. Trop. Med. Hyg., 68(6): 671–7.

Bureau of Epidemiology. 2006. Report on Diseases under Epidemiological Surveillance System. Bureau of Epidemiology, Department of Disease Control, MoPH.

Butt, T., R. Ahmad, M. Mahmood and S. Zaidi. 2003. Ciprofloxacin treatment failure in typhoid fever case, Pakistan. Emerg. Infect. Dis., 9: 1621–1622.

Curtis, V., S. Cairncross and R. Yonli. 2000. Review: Domestic hygiene and diarrhoea—pinpointing the problem. Trop. Med. Int. Health, 5(1): 26–30.

Curtis, V. and S. Cairncross. 2003. Effect of washing hands with soap on diarrhoea risk in the community: a systematic review. Lancet Infect. Dis., 3(5): 275–81.

Etiler, N., S. Velipasaoglu and M. Aktekin. 2004. Risk factors for overall and persistent diarrhea in infancy in Antalya, Turkey. Public Health, 118(1): 62–9.

Hung, V.B. 2006. The most common causes of and risk factors for diarrhea among children less than five years of age admitted to Dong Anh Hospital, Hanoi, Northern Vietnam. A thesis submitted to University of Oslo as a partial fulfillment for the degree Master of Philosophy in International Community Health.

Jensen, P.K., G. Jayasinghe, W. van der Hoek, S. Cairncross and A. Dalsgaard. 2004. Is there an association between bacteriological drinking water quality and childhood diarrhea in developing countries? Trop. Med. Int. Health, 9(11): 1210–15.

Karim, A.S., S. Akhter, M.A. Rahman and M.F. Nazir. 2001. Risk factors of persistent diarrhea in children below five years of age. Indian J. Gastroenterol., 20(2): 59–61.

Kosek, M., C. Bern and R.L. Guerrant. 2003. The global burden of diarrhoeal disease, as estimated from studies published between 1992 and 2000. Bull World Health Organ, 81(3): 197–204.

Lewis, H., S. Ethelberg, M. Lisby, S. Madsen, K. Olsen, P. Rasmussen, C.K. Kjelsø, M. Howitz and K. Mølbak. 2007. Outbreak of shigellosis in Denmark associated with imported baby corn, August 2007. Euro Surveill., 12(8): E070830.1. Available from: http://www.eurosurveillance.org/ew/2007/070830.asp#1.

Löfdahl, M., S. Ivarsson, J. Långmark and L. Plym-Forshell. 2009. An outbreak of Shigella dysenteriae in Sweden, May–June 2009, with sugar snaps as the suspected source. Eurosurveillance, 14(28), 16 July 2009. Available from: http://www.eurosurveillance.org/ ViewArticle.aspx?ArticleId=19268.

Minami, A., W. Chaicupa, M. Chongsa-Nguan, S. Samosornsuk, S. Monden, K. Takeshi, S. Makino and K. Kawamoto. 2010. Prevalence of foodborne pathogens in open markets and supermarkets in Thailand, Food Control, 21: 221–226.

Prüss-Üstün, A., R. Bos, F. Gore and J. Bartram. 2008. Safer water, better health: costs, benefits and sustainability of interventions to protect and promote health. Geneva: World Health Organization.

Stafford, R., M. Kirk, C. Selvey, D. Staines, H. Smith, C. Towner and M. Salter. 2007. On behalf of the Outbreak Investigation Team. An outbreak of multi-resistant *Shigella sonnei* in Australia: possible link to the outbreak of shigellosis in Denmark associated with imported baby corn from Thailand. Euro Surveill, 12(9): E070913.1. Available from: http://www. eurosurveillance.org/ew/2007/070913.asp#1.

Teklemariam, S., T. Getaneh and F. Bekele. 2000. Environmental determinants of diarrheal morbidity in under-five children, Keffa-Sheka zone, south west Ethiopia. Ethiop. Med. J., 38(1): 27–34.

Tessema, A.G., K.A. Gelaye and D.H. Chercos. 2014. Factors affecting food handling Practices among food handlers of Dangila town food and drink establishments, North West Ethiopia. BMC Public Health, 14: 571.

Thuan Huu Vo, Ninh Hoang Le, Thuy Thanh Diem Cao, J. Pekka Nuorti and Nguyen Nhu Tran Minh. 2014. An outbreak of foodborne salmonellosis linked to a bread takeaway shop in Ben Tre City, Vietnam. International Journal of Infectious Diseases. Volume 26, September 2014, Pages 128–131.

Tumwine, J.K., J. Thompson, M. Katua-Katua, M. Mujwajuzi, N. Johnstone and I. Porras. 2002. Diarrhoea and effects of different water sources, sanitation and hygiene behaviour in East Africa. Trop. Med. Int. Health, 7(9): 750–756.

Som, V., L. Som, Ph Has, D. Justin and M.C. Roces. 2012. *Vibrio parahaemolyticus* enteritis outbreak following a wedding banquet in a rural village—Kampong Speu, Cambodia, April 2012. Western Pac Surveill Response J. 2012 Oct–Dec, 3(4): 25–28.

UNICEF. 2014. Available from: http://www.unicef.org/cambodia/12681_22270.html.

Vindigni, S.M., A. Srijan, B. Womgstitwilairoong, R. Marcus, J. Meek, P.L. Riley and C. Mason. 2007. Prevalence of foodborne microorganisms in retail foods in Thailand. Foodborne Pathog. Dis., 4: 208–215.

WHO. Food safety, Fact sheet N°399, November 2014. http://www.who.int/mediacentre/ factsheets/fs399/en/(Accessed 6 December 2014).

Worldwide Food Safety Status

Mohammad Aminul Islam[1,]* and Enne de Boer[2]

Introduction

Food safety and foodborne diseases

Foodborne diseases are an important public health, social and economic issue in both the developed world and developing countries. An estimated three million people around the world die each year from food and water-borne disease, with millions more becoming sick. Foodborne diseases are mainly associated with acute, rather mild and self-limiting gastrointestinal symptoms (nausea, vomiting and diarrhoea) caused by microbial contamination of food. However, foodborne diseases may lead to chronic diseases, such as affection of the immune system, multi-organ failure, abortion and neurological disorders. Chemical contamination of foodstuffs, including methyl mercury, lead, arsenic, dioxins and aflatoxins (among others) may cause acute and chronic health effects such as neuro-developmental disorders, cardiovascular disease, cancers and renal disease.

Ensuring safe food, whether domestically produced and consumed, imported or exported, is essential for the protection of human health and for improving the quality of life in all countries. Disease caused by contaminated food and water is a major challenge for health authorities in many countries and communities, especially those that lack basic sanitary services. Food safety policies and actions need to cover the entire food chain, from production to consumption. Despite the implementation of preventive measures, outbreaks of foodborne zoonotic diseases have occurred and will continue to occur. A rapid response to outbreaks of such diseases is essential.

[1] Food Safety Research Group, Centre for Food and Waterborne Diseases, International Centre for Diarrhoeal Diseases Research, Bangladesh (icddr,b), Mohakhali, Dhaka 1212.
[2] Laboratory for Feed and Food Safety, Netherlands Food and Consumer Product Safety Authority (NVWA), Akkermaalsbos 4, 6708 WB, Wageningen, The Netherlands.
* Corresponding author: maislam@icddrb.org

Coordination between all the agencies involved, for example those responsible for public health, food and feed safety, agriculture and fisheries, industry and trade, local authorities, inspection and tourism, is important.

Global organizations for food safety

The World Health Organization (WHO) and the Food and Agriculture Organization of the United Nations (FAO) are in the forefront of the development of risk based approaches for the management of public health hazards in food.

WHO is the directing and coordinating authority for health within the United Nations system.

WHO Member States adopted a resolution in 2000 to recognize food safety as an essential public health function.

The primary aim of FAO is to make sure people have regular access to enough high-quality food to lead active, healthy lives. Related aims include improving the safety and quality of food at all stages of the food chain and providing scientific evidence to solve food safety issues.

Codex Alimentarius is the FAO/WHO body setting food safety and quality standards for consumer protection and facilitation of domestic and international trade.

FAO and WHO have produced guidance documents for use by national authorities on foodborne outbreak investigation, establishing food safety emergency response plans, applying risk analysis principles during food safety emergencies and developing national food recall systems.

Several programmes and networks in relation to food safety have been initiated by WHO and FAO:

- The Emergency Prevention System for Food Safety (EMPRES Food Safety) and the International Food Safety Authorities network (INFOSAN) aim reinforcing preparedness, and recognizing and responding rapidly to food safety emergencies. INFOSAN links together national authorities in Member States responsible for managing food safety emergencies.
- The Global Foodborne Infections Network (GFN) is a capacity-building program that promotes integrated, laboratory based surveillance and intersectional collaboration among human health, veterinary and food-related disciplines.
- FOSCOLLAB is a WHO platform for food safety professionals and enables users to:
  - Access food safety data and information quickly
  - Maximize the utility of already existing sources and minimize duplication of efforts
  - Integrate data and information coming from animal/agriculture, food and human health areas to improve global public health
  - Promote better generation of data
  - Strengthen the underlying sources by promoting awareness and increased utilization

By integrating multiple sources of reliable data, FOSCOLLAB helps overcome the challenges of accessing these key sources in a timely manner. It allows for better risk assessment and decision-making by food safety professionals and authorities.

- The Global Outbreak Alert and Response Network (GOARN) is a technical collaboration of existing institutions and networks who pool human and technical resources for the rapid identification, confirmation and response to outbreaks of international importance. The Network provides an operational framework to link this expertise and skill to keep the international community constantly alert to the threat of outbreaks and ready to respond.

The World Organization for Animal Health (OIE) is working with relevant organizations to reduce foodborne risks to human health due to hazards arising from animal production.

Surveillance of food safety status and foodborne diseases

Investigation of foodborne diseases

The epidemiology of food and waterborne illnesses at the global level is poorly understood in many countries. Little information on disease burden and on pathogens responsible for the illnesses is available and therefore limiting appropriate prevention measures can be taken.

Large numbers of marginalized populations with poor access to health are often at high risk to food, waterborne and zoonotic infections, causing considerable morbidity and largely undetected by routine surveillance systems.

Prompt and thorough investigation of foodborne outbreaks is particularly useful for a timely identification of etiological agents, sources and vehicles, and allows for the prevention of further cases by identifying and eliminating the source of infection.

Moreover, outbreak investigations are critical means of identifying new agents and new vehicles, as well as maintaining awareness of contemporary problems (Pires et al., 2012).

Though foodborne outbreaks receive the most media and political attention, the main part of the burden of foodborne diseases consists of sporadic cases. Thus far, few countries have implemented surveillance of sporadic cases of foodborne disease, particularly in the developing world, where the majority of reported human cases are associated with foodborne outbreaks (Pires et al., 2012).

The burden of foodborne diseases can be defined as the incidence, prevalence of morbidity, disability and mortality associated with acute and chronic manifestations of foodborne disease.

The importance of foodborne disease burden estimates in guiding food safety policy-making has been recognized by the international community and reaffirmed by WHO Member States in the 2010 Resolution of the World Health Assembly on 'Advancing Food Safety Initiatives' (WHA 63.3) (www.who.int/foodsafety/foscollab/).

Burden of disease data are key to enabling food safety decision-makers to appropriately allocate resources to foodborne disease prevention and control efforts and to prioritise specific food safety issues within a country.

The WHO launched the Initiative to Estimate the Global Burden of Foodborne Diseases in 2006 (Hird et al., 2009). The Initiative aims to assemble a quantitative description of the foodborne disease burden globally and has as objectives:

1. To strengthen the capacity of countries in conducting burden of foodborne disease assessments and to increase the number of countries who have undertaken a burden of foodborne disease study.
2. To provide estimates on the global burden of foodborne diseases according to age, sex and regions for a defined list of causative agents of microbial, parasitic, and chemical origin.
3. To increase awareness and commitment among Member States for the implementation of food safety standards.
4. To encourage countries to use burden of foodborne disease estimates (e.g., for cost-effective analyses of prevention, intervention and control measures).

The estimation of the burden of foodborne disease consists of two main components:

1. A population-based component—Population survey based on self-reported cases of gastrointestinal illness and the use of a standardized questionnaire, administered by trained interviewers.
2. A laboratory-based component—Enhanced laboratory testing of patient's samples.

Surveillance of foodborne diseases

Several studies have been conducted to attribute foodborne disease to the responsible food and animal sources in a variety of developed countries, mainly focusing on the most frequent causes of foodborne disease, namely *Salmonella* and *Campylobacter* (Pires et al., 2012).

As opposed to data from sporadic disease surveillance, data from foodborne outbreak investigations are frequently available, both in developed and developing countries. In the absence of other evidence, outbreak data are fundamental source of information to inform food safety policies as well as to indicate the importance of foodborne pathogens and food sources for human disease in a population, i.e., source attribution (Pires et al., 2012).

Traditional surveillance systems tend to capture only a fraction of the existing foodborne and zoonotic disease burden. Communicable disease surveillance in countries and regions has primarily been based on syndrome surveillance and there is limited aetiology (based laboratory-confirmed cases) since stools are not routinely collected for lab testing.

A simple summarization of results of outbreak investigations can be useful for identifying the most common food vehicles involved in outbreaks and thus identifying the most important reported sources for human illness (EFSA, 2014; Pires et al., 2012).

Laboratory testing

Laboratory testing in combination with a population survey is important to estimate the true impact of specific pathogens to food-related disease in the populations. There is a continuous need to strengthen the quality of food analysis and laboratory networks.

PulseNet International is a network of national and regional laboratory networks dedicated to tracking foodborne infections worldwide. Each laboratory utilizes standardized genotyping methods, sharing information in real-time. The standard method is based on fragmentation of chromosomal DNA using rare-cutting restriction endonucleases, followed by the resolution of resultant DNA fragments by pulsed field gel electrophoresis (PFGE). The method is used for subtyping of foodborne pathogens. PFGE for foodborne pathogens such as *Salmonella*, *Listeria monocytogenes* and to a lesser extent *Campylobacter* has been standardized internationally through various PulseNet International networks in most continents to allow for inter-laboratory comparison (Swaminathan et al., 2006). The standardized PFGE protocols are now in use in laboratories worldwide for surveillance of foodborne infections, and for outbreak detection and source tracing of outbreaks of both foodborne and nosocomial bacterial pathogens.

Global Salm–Surv is a passive surveillance system that collects annual Salmonella summary data from member institutions on number and serotypes of Salmonella isolates from human, animal, environmental and food sources.

Food safety status in continents—surveillance systems and data

Latin America and the Caribbean

The region includes the following countries: Argentina, Bahamas, Turks and Caicos Islands, Belize, Bolivia, Brazil, Chile, Colombia, Costa Rica, CPC Barbados, Cuba, Dominican Republic, Ecuador, El Salvador, Guatemala, Guyana, Haiti, Honduras, Jamaica, Nicaragua, Panama, Paraguay, Peru, Puerto Rico, Suriname, Trinidad and Tobago, Uruguay, Venezuela.

The Central and South American Region covers a large biodiversity. The region's different climates set the stage for a wide range of food safety problems, resulting from the prevalence of pathogens, cultural practices that promote pathogen growth and spread, the globalization of trade, and increased travel and tourism.

Approximately 10% of the world's population lives in Latin America and the Caribbean, in total around 600 million people (UNFPA, ECLAC). Owing to the increasing rural exodus to urban areas, only 20% of these on an average live in rural areas (ECLAC). This region is characterized by deep, ingrained inequality (UNFPA).

The region is a major world supplier of an extensive range of agricultural products of importance for food and in industry. It accounts for a significant percentage of world trade in agricultural and livestock products such as soya, sugar, coffee, cereals, fruit, vegetables, meat and dairy products. Brazil is one of the world's leading exporters of chicken meat (FAO, 2014). Food exports from the region are 12% of the world's total food trade, and this figure could increase rapidly over the coming decades if food safety and quality standards are improved.

Food safety organizations

Pan American Health Organization (PAHO), founded in 1902, is the world's oldest international public health agency. It provides technical cooperation and mobilizes partnerships to improve health and quality of life in the countries of the Americas. PAHO serves as the Regional Office for the Americas of the World Health Organization (WHO).

The Strategic Objective of PAHO addresses food safety (ensuring that chemical, microbiological, zoonotic and other hazards do not pose a risk to health) as well as food security (access and availability of appropriate food).

PAHO established the Veterinary Public Health Program. The program's overall objective is to collaborate with member countries in the development, implementation, and evaluation of policies and programs that lead to food safety and protection and to the prevention, control, or eradication of zoonoses.

PAHO's Veterinary Public Health Program established two specialized regional centers:

The Pan American Foot-and-Mouth Disease Center (PANAFTOSA), created in 1951 in Rio de Janeiro (Brazil) and the Pan American Institute for Food Protection and Zoonoses (INPPAZ), established in 1991 in Buenos Aires (Argentina). In 2005, INPPAZ was closed, and a specialized food safety technical team was established in PANAFTOSA.

RIMSA stands for Inter-American Meeting in Health and Agriculture at Ministerial Level. It is the regional forum for collaboration and coordination among the health and agricultural sectors, at the highest political level. PAHO counts on this meeting to development technical cooperation in the area of Veterinary Public Health.

Pan American Centers have been an important modality of PAHO technical cooperation for more than 60 years. In that period, PAHO has created or administered 13 centers (CLATES, ECO, PASCAP, CEPANZO, INPPAZ, INCAP, CEPIS, Regional Program on Bioethics in Chile, CAREC, CFNI, CLAP, PANAFTOSA, and BIREME) and eliminated nine. Current Centers include PANAFTOSA, the Latin American and Caribbean Center on Health Sciences Information (BIREME), the Latin American Center for Perinatology and Human Development/Women's and Reproductive Health (CLAP/SMR); and the Caribbean Epidemiology Center (CAREC) and the Caribbean Food and Nutrition Institute (CFNI), which were transferred at the end of 2012 to the Caribbean Public Health Agency (CARPHA). CARPHA aims the promotion and strengthening of national integrated foodborne disease surveillance and laboratory testing and initiation of the Burden of illness (BOI) studies.

The Caribbean Agricultural Health and Food Safety Agency (CAHFSA) was established in 2010, but activities started in 2014. The Headquarters of CAHFSA are established in Suriname. The primary remit of the Agency is to assist the Caribbean Community (CARICOM) Member States to coordinate and strengthen their infrastructure, institutional and human resource capacity to effectively deliver agricultural goods which achieve the international agricultural health and food safety standards, measures and guidelines in order to safeguard human health and to prevent the introduction of or minimise the incidence of transmission of agricultural pests across national borders.

The creation of the *Pan American Commission for Food Safety (COPAIA)* was suggested by ministers of Health and Agriculture of the Americas, during the RIMSA meeting in 2001.

COPAIA's main purpose is to help improve food safety throughout the entire food chain by sustaining the political will of the countries of the region for adopting food safety programs and promoting coordination and integration with producers and consumers. COPAIA works along the following lines of action: (1) promotion of intersectional coordination; (2) strengthening of food safety systems; (3) development of policies aimed at the modernization of food inspection; (4) promotion of integrated systems for surveillance of foodborne diseases; (5) development of strategic alliances in education and social communication in food safety; (6) promotion of the participation from Region's countries in the Codex Alimentarius work.

Laboratory testing

Agriculture and food exporters must comply with food safety regulations and must perform laboratory testing through the whole supply chain. This requires well equipped laboratories using new technologies (LC-MSMS, GC-MSMS), which are hardly available in South America.

The Inter-American Network of Food Analysis Laboratories (INFAL) or Red Interamericana de Laboratorios de Análisis de Alimentos (RILAA).

INFAL has as mission, to promote the assurance of the safety and food quality in the Region of the Americas, in order to prevent foodborne diseases, protect the health of the consumer and to facilitate the trade, promoting and strengthening the development and interaction of the food analysis laboratories within the framework of integrated national programs of food protection.

Regional Reference laboratories have been established in several countries in South America, Central America and the Caribbean, such as INEI-ANLIS "Carlos G. Malbrán" (Argentina), INCIENSA (Costa Rica), GDD_CDC (Guatemala), CAREC (Trinidad y Tobago).

The PulseNet program was started in Latin America in 2004. Its principal goal is to support the Pan American regional strategy in the Regional Plan for Food Safety in strengthening surveillance of foodborne diseases (FBD) and to reinforce technical cooperation among the member countries in relation to food safety and health.

INEI-ANLIS "Carlos G. Malbrán" is in charge of the technical support regarding PFGE protocols, analysis, certification and quality control programs (Regional Reference Laboratory), PAHO provides all the aspects needed for the communication among members, server development and maintenance for the Regional Databases and project developments.

PulseNet Latin America has established the capability in the participating countries for genotype bacterial pathogens strains with standardized protocols for selected pathogens (*Salmonella* spp., *Vibrio cholerae*, *Escherichia coli* O157 and STEC no-O157, *Shigella* spp., *Campylobacter* spp., *Listeria monocytogenes*) and has initiated the creation of a regional database of the isolates. Current PulseNet Latin America participants include 15 countries and CAREC.

Surveillance systems and surveillance data

The Ministries of Health of each country within Latin America and the Caribbean are responsible for the collection of data from outbreak investigations, which are compiled under the Regional Information System on foodborne disease surveillance, SIRVETA. The network system is promoting exchange of information referring to collection, verification, analysis and systematic interpretation of exact data on food contaminants and diseases transmitted by food. National agencies active in surveillance and monitoring of foodborne diseases exist in some countries in the region, such as Anvisa (Brazil) and Senasa (Argentina).

Between 1993 and 2010, 6313 bacterial outbreaks were reported by 20 countries. In general, the most important sources of bacterial disease were meat, dairy products, water and vegetables in the 1990s, and eggs, vegetables, grains and beans in the 2000s (Pires et al., 2012).

The incidence of illnesses caused by microorganisms that are principally foodborne, such as *Salmonella* and *Campylobacter* spp., continued to increase considerably in many countries.

Fluctuations of the most important sources of disease for each pathogen between decades and countries were observed, which may be a consequence of changes in the control of zoonotic disease over the years, of changes in food consumption habits, or of changes in public health focus and availability of data of different pathogens. The study identified data gaps in the region and highlighted the importance of effective surveillance systems to identify sources of disease (Pires et al., 2012).

Total outbreaks by country in Sirveta database.				
Anguilla (0)	Cayman Islands (0)	French Guiana (0)	Montserrat (0)	St. Lucia (0)
Antigua and Barbuda (0)	Chile (5077)	Grenada (0)	Netherlands Antilles (0)	St. Vincent and the Grenadines (0)
Argentina (69)	Colombia (1)	Guadaloupe (0)	Nicaragua (258)	Suriname (0)
Aruba (0)	Costa Rica (117)	Guatemala (0)	Panama (17)	Trinidad and Tobago (25)
Bahamas (50)	Cuba (3094)	Guyana (0)	Paraguay (41)	Turks and Caicos Islands (0)
Barbados (1)	Dominica (0)	Haiti (0)	Peru (77)	Uruguay (23)
Belize (0)	Dominican Republic (52)	Honduras (20)	Puerto Rico (0)	Venezuela (188)
Bolivia (4)	Ecuador (28)	Jamaica (0)	St. Kitts and Nevis (0)	Virgin Islands (0)
Brazil (292)	El Salvador (17)	Martinique (0)		

Data from PubMed (http://regionalnews.safefoodinternational.org; http://www.ncbi. nlm.nih.gov/pubmed) on food and waterborne illness outbreaks in 2013 include

- Cholera outbreaks in Haiti, Dominican Republic and Cuba. These outbreaks are believed to be part of an ongoing epidemic of cholera in Haiti after that country's 2010 earthquake. Another cholera outbreak was documented in Mexico.
- Salmonellosis in Mexico after eating sushi at a regional Japanese restaurant.
- Norovirus outbreak in Chile caused by contaminated water after insufficient chlorine treatments of municipal water supplies.
- Trichinellosis in Argentina attributed to consumption of raw pork sausage and salami, originating from pigs slaughtered in unlicensed small-scale and home butcheries.
- Listeriosis outbreak in Chile due to contaminated camembert soft cheese.
- Fatal produce-associated *E. coli* outbreak in Guatemala.
- Hepatitis A outbreaks in Belize and Columbia.
- Waterborne *E. coli* outbreak in Guyana (Northwest).

Examples of recent research papers on different aspects of food safety status in Latin America:

- Burden of laboratory-confirmed *Campylobacter* infections in Guatemala 2008–2012: Results from a facility-based surveillance system (Benoit et al., 2014).
- Investigation of food and water microbiological conditions and foodborne disease outbreaks in the Federal District, Brazil (Nunes et al., 2013).
- Microbiological food safety issues in Brazil: Bacterial pathogens (Gomes et al., 2013).
- Risk factors for sporadic Shiga toxin-producing *Escherichia coli* infections in children, Argentina (Rivas et al., 2008).

Food safety developments

International organizations have recognized foodborne diseases as a significant public health issue in the Central and South American Region, giving rise to a number of innovative programs to promote networking and collaboration among countries in Central and South America. Intersectional collaboration in the field of food safety has been established throughout the Region between the public health sector, including epidemiologists and clinical laboratories, and the agriculture and livestock sector and its food analysis laboratories.

The FAO in 2011 urged Latin American countries to enhance their capacity to guarantee food safety. Each country needs specific emergency plans suitable to its development level and a national food control system to guarantee food safety. To reinforce the capacity to prepare for, recognize and react to emergencies is key to guaranteeing the health of consumers. FAO is actively working with Member countries in the region to assure the safety of fresh fruits and vegetable and the safety and quality of coffee.

The annual WHO-GFN and PulseNet workshops play an important role in the process of improving the food safety status. Several countries in the region are implementing integrated projects for antimicrobial resistance surveillance and projects to improve Salmonella control. Some countries are implementing projects to strengthen integrated surveillance of foodborne diseases and the capacity for epidemiological surveillance.

Since the beginning of the 1990s, PAHO/WHO developed a strategy to work with member States to build national capacity for foodborne disease surveillance. Guidelines for the Establishment & Strengthening of Foodborne Disease Surveillance Systems were initiated and PANAFTOSA-PAHO/WHO is working on training courses at country and regional level to create conditions for adequate surveillance of foodborne diseases. Nevertheless, deficiencies in foodborne disease surveillance coverage persist throughout the region, and few cases and outbreaks are detected and reported (Pires et al., 2012).

The existence of scientific and technological institutions with trained personnel facilitates exchange of information and technology. However, scientific and technological institutions work in isolation and duplicate activities, and it is well known that there is a lack of ongoing networked studies and research at regional level. Moreover, there is a lack of continuity in research into and spread of technologies owing to frequent leadership changes in research programmes and in national health and plant health services.

Measures to be taken include assuring safety throughout the entire food chain, by enforcing science-based regulations and applying risk analysis focussing on those critical points in the food chain which require closer monitoring by the relevant government authorities; and improving coordination both among countries in the Region, and between food regulators and food producers in each country. Despite some relative success in some countries, multiple agencies with fragmented responsibilities, combined with a lack of human and financial resources, are making efforts to achieve uniform high food standards across the region difficult. Not all countries in the Region have adequate food safety regulatory frameworks and enforcing capabilities and food legislation has not always been harmonised with Codex Alimentarius standards. This makes exporting food to the rest of the world, more difficult. Moreover, food safety and quality reference laboratories do not exist in all countries of the Region. All countries in the region must have its own food-testing facilities, or have access to suitable ones in the region. Many challenges remain to improving food safety in the region. The countries of the region recognize the importance of developing practical actions for capacity building and promotion of food safety in the region.

Europe

Food safety organizations

The European Food Safety Authority (EFSA), located in Parma, Italy, was established and funded by the European Union (EU) as an independent agency in 2002. EFSA provides objective scientific advice on all matters, in close collaboration with national authorities and in open consultation with its stakeholders, with a direct or indirect impact on food and feed safety, including animal health and welfare and plant protection. EFSA's work falls into two areas: risk assessment and risk communication.

In particular, EFSA's risk assessments provide risk managers (EU institutions with political accountability, i.e., the European Commission, the European Parliament and the Council) with a sound scientific basis for defining policy-driven legislative or regulatory measures required to ensure a high level of consumer protection with regard to food and feed safety. EFSA communicates to the public in an open and transparent way on all matters within its remit. Collection and analysis of scientific data, identification of emerging risks and scientific support to the Commission, particularly in the case of a food crisis, are also part of EFSA's mandate, as laid down in the founding Regulation (EC) No. 178/2002.

EFSA has a broad remit, covering the entire food chain from farm to fork, including topics related to animal health and welfare, biological hazards, pesticides and contaminants, genetically modified organisms, nutrition and food and feed additives, as well as plant health.

The European Centre for Disease Prevention and Control (ECDC), an EU agency based in Stockholm, Sweden, was established in 2005. The objective of ECDC is to strengthen Europe's defences against infectious diseases. ECDC's mission is to identify, assess and communicate current and emerging threats to human health posed by infectious diseases. In order to achieve this mission, ECDC works in partnership with national public health bodies across Europe to strengthen and develop EU-wide disease surveillance and early warning systems. By working with experts throughout Europe, ECDC pools Europe's knowledge on health so as to develop authoritative scientific opinions about the risks posed by current and emerging infectious diseases.

Laboratory testing

According to EC Regulation 882/2004 the designation of Community (European) and national reference laboratories should contribute to a high quality and uniformity of analytical results. The activities of reference laboratories should cover all the areas of feed and food law and animal health, in particular those areas where there is a need for precise analytical and diagnostic results.

The Community (European) reference laboratories (EU-RL) for feed and food are responsible for:

a) providing national reference laboratories with details of analytical methods, including reference methods;

b) coordinating application by the national reference laboratories of these methods, in particular by organising comparative testing and by ensuring an appropriate follow-up of such comparative testing in accordance with internationally accepted protocols;

c) coordinating, within their area of competence, practical arrangements needed to apply new analytical methods and informing national reference laboratories of advances in this field;

d) conducting initial and further training courses for the benefit of staff from national reference laboratories and of experts from developing countries;

e) providing scientific and technical assistance to the Commission, especially in cases where Member States contest the results of analyses;

f) collaborating with laboratories responsible for analysing feed and food in third countries.

Member States shall arrange for the designation of one or more national reference laboratories for each European reference laboratory.

PulseNet Europe, as part of the PulseNet International network started in 2004. The Salmonella, Listeria and STEC PFGE profiles from the PulseNet Europe database are now incorporated into the pilot Molecular Surveillance System (MSS) established by ECDC in 2012 as part of TESSy (EFSA, 2013).

Surveillance systems and surveillance data

Collecting data that are relevant to food safety is central to EFSA, because such data are an integral part of risk assessment. These data are particularly valuable for quantitatively estimating risks and/or for identifying to what extent a given control measure or intervention strategy can reduce the burden of a zoonotic disease in humans. Annual monitoring data provide updates on the current situation and help to inform risk managers and Member States of recent developments.

In the field of biological risks for human health, Directive 2003/99/EC2 lays down the requirement for an EU system for monitoring and reporting information, which obliges EU Member States to collect relevant and comparable data on zoonoses, zoonotic agents, antimicrobial resistance and foodborne outbreaks and to report these data annually to the EC. EFSA is assigned the tasks of examining the collected data and preparing the annual EU summary reports, in collaboration with the ECDC, which collects and analyses corresponding data on human cases. In addition, EFSA runs the data collection applications on behalf of the EC.

According to Directive 2003/99/EC2 on zoonoses, monitoring is based on the systems already in place in Member States. However, the Directive also foresees that detailed rules for monitoring may, where necessary, be laid down in EU legislation, to make the data easier to compile and compare. In addition, EFSA issues technical specifications and submits external reports for monitoring and reporting certain zoonoses, antimicrobial resistance and foodborne outbreaks, to improve analyses and make it easier to compare the data between Member States.

A yearly report of the distribution and other epidemiological characteristics of zoonoses in EU Member States is an essential component of assessing the impact of these diseases and potential preventive measures. However, it has also been noted that the reports do not always clearly identify the reference population, the data sources, and data collection approaches (surveillance methodology) used for the various diseases. EFSA regularly reviews its data requirements to improve its preparedness to answer risk assessment questions, by ensuring that it continues to collect readily available, stable data and has a good knowledge of ad hoc data sources throughout the EU.

EFSA and ECDC analysed information submitted by 27 European Union Member States on the occurrence of zoonoses and foodborne outbreaks in 2012 (EFSA, 2014).

Campylobacteriosis was the most commonly reported zoonosis, with 214,268 confirmed human cases. The occurrence of *Campylobacter* continued to be high in broiler meat at EU level. The decreasing trend in confirmed salmonellosis cases in humans continued with a total of 91,034 cases reported in 2012. Most Member States met their *Salmonella* reduction targets for poultry. In foodstuffs, Salmonella was most often detected in meat and products thereof. The number of confirmed human listeriosis cases increased to 1,642. *Listeria* was seldom detected above the legal safety limit from ready-to-eat foods. A total of 5,671 confirmed verocytotoxigenic *Escherichia coli* (VTEC) infections were reported. VTEC was also reported from food and animals. The number of human tuberculosis cases due to *Mycobacterium bovis* was 125 cases, and 328 cases of brucellosis in humans were reported. *Trichinella* caused 301 human cases and was mainly detected in wildlife. A total of 643 confirmed human cases of Q fever were reported. A total of 232 cases of West Nile fever in humans were reported. Most of the 5,363 reported foodborne outbreaks were caused by *Salmonella*, bacterial toxins, viruses and *Campylobacter*, and the main food sources were eggs, mixed foods and fish and fishery products.

Conclusions and future developments

The globalization of food supply impacts patterns of foodborne disease outbreaks worldwide, and consumers are having increased concern about microbiological food safety. In this sense, the assessment of epidemiological data of foodborne diseases in different countries has not only local impact, but it can also be of general interest, especially in the case of major global producers and exporters of several agricultural food products (Gomes et al., 2013).

Although attempts have been made in most countries and regions to coordinate control systems across the animal health, food safety and human health sectors, many systems are generally poorly coordinated.

A veterinary infrastructure that detects and effectively eradicates outbreaks of animal diseases and the control of plant health and produce is important to allow the export of agricultural products to countries and regions where stringent food safety regulations on pesticide residues, antibiotic residues, contaminants, banned substances, mycotoxins and allergens, among others are enforced.

Possible obstacles to horizontal cooperation among countries to improve health and plant health conditions at regional or sub regional level, because countries view themselves as competitors in the same international market for agricultural products.

Most countries apply SPS risk management measures forced by international market demands, for entering food trade or for maintaining their markets. Most countries don't have SPS measures as a priority for internal priority for internal markets. SPS issues become politically important when related to trade and economic development.

Improving food safety for export markets has a positive effect on local markets and the health of the population.

There is a continuous need to build capacity related to food safety and risk analysis. Food safety should be a priority in relation to other areas requiring increased government support.

Most national food control systems involve several ministries, making coordination among different agencies challenging.

Food safety activities are best undertaken using an integrated, multidisciplinary approach that considers the whole of the food chain. Experts, representing government food regulatory authorities, food producers, consumer associations and international organizations, must agree on a comprehensive plan which includes measures to improve the safety and quality of food both for the peoples of the region and for export.

Most countries are making efforts to align regulatory frameworks with the requirements of the WTO SPS/TBT Agreements.

The most important component of an improved food safety system is surveillance of foodborne disease

Surveillance systems for foodborne disease vary in capacity by country. Generally, the more developed the country is, the more funding that is put into its surveillance programs, but no country has an outstanding system that could serve as a model for all others (Todd, 2006).

Most countries have some passive system that allows data on foodborne illness to be sent to centralized authorities where summaries are generated.

Active surveillance systems collect data targeted to answer specific epidemiological questions more efficiently, but at such a high cost that most countries do not have the resources, except on a occasional basis (Todd, 2006).

Surveillance over the years has generated much interesting information on how disease agents are transmitted through the food supply and where contamination and growth by pathogens in the food production and preparation chain typically occur (Todd, 2006).

One of the major problems in trying to estimate the number of cases of foodborne diseases, and in identifying their causes, is under-reporting—many of those affected do not contact the medical services and, even if they are reported, many disease outbreaks are not investigated or the investigations do not result in identification of the source of the problem.

There is a recognized need for better coordination of surveillance policies for animal health, food pathogens and foodborne diseases.

References

Benoit, S.R., B. Lopez, W. Arvelo, O. Henao, M.B. Parsons, L. Reyes, J.C. Moir and K. Lindblade. 2014. Burden of laboratory-confirmed Campylobacter infections in Guatemala 2008–2012. Results from a facility-based surveillance system. J. Epidemiol. Glob. Health, 4: 51–59.

Directive 2003/99/EC of the European Parliament and of the Council of 17 November 2003 on the monitoring of zoonoses and zoonotic agents.

EFSA. 2013. Scientific Opinion on the evaluation of molecular typing methods for major foodborne microbiological hazards and their use for attribution modelling, outbreak investigation and scanning surveillance: Part 1 (evaluation of methods and applications). EFSA Journal 11(12): 3502.

EFSA (European Food Safety Authority), ECDC (European Centre for Disease Prevention and Control). 2014. The European Union Summary Report on Trends and Sources of Zoonoses, Zoonotic Agents and Foodborne Outbreaks in 2012; EFSA Journal, 12: 3547, 312.

FAO. 2014. FAOSTAT (http://faostat3.fao.org/faostat-gateway/go/to/home/E).

Gomes, B.C., B.D. Franco and E.C. De Martinis. 2013. Microbiological food safety issues in Brazil: bacterial pathogens. Foodborne Pathog. Dis., 10: 197–205.

Hird, S., C. Stein, T. Kuchenmüller and R. Green. 2009. Meeting report: Second annual meeting of the World Health Organization initiative to estimate the global burden of foodborne disease. Int. J. Food Microbiol., 133: 210–212.

Nunes, M.C., A.L. Arrais de Alencar Mota Caldas and E. Dutra. 2013. Investigation of food and water microbiological conditions and foodborne disease outbreaks in the Federal District, Brazil. Food Control, 34: 235–240.

Pires, S.M., A.R. Vieira, E. Perez, D. Lo Fo Wong and T. Hald. 2012. Attributing human foodborne illness to food sources and water in Latin America and the Caribbean using data from outbreak investigations. Int. J. Food Microbiol., 152: 129–138.

REGULATION (EC) No 178/2002 OF THE EUROPEAN PARLIAMENT AND OF THE COUNCIL of 28 January 2002 laying down the general principles and requirements of food law, establishing the European Food Safety Authority and laying down procedures in matters of food safety.

REGULATION (EC) No 882/2004 OF THE EUROPEAN PARLIAMENT AND OF THE COUNCIL of 29 April 2004 on official controls performed to ensure the verification of compliance with feed and food law, animal health and animal welfare rules.

Rivas, M., S. Sosa-Estani, J. Rangel, M.G. Caletti, P. Vallés, C.D. Roldán, L. Balbi, M.C. Marsano de Mollar, D. Amoedo, E. Miliwebsky, I. Chinen, R.M. Hoekstra, P. Mead and P.M. Griffin. 2008. Risk factors for sporadic Shiga toxin-producing *Escherichia coli* infections in children, Argentina. Emerg. Infect. Dis., 14: 763–71.

Swaminathan, B., P. Gerner-Smidt, L.K. Ng, S. Lukinmaa, K.M. Kam, S. Rolando, E.P. Gutierrez and N. Binsztein. 2006. Building PulseNet International: an interconnected system of laboratory networks to facilitate timely public health recognition and response to foodborne disease outbreaks and emerging foodborne diseases. Foodborne Pathog. Dis., 3: 36–50.

Todd, E.C.D. 2006. Challenges to global surveillance of disease patterns. Marine Pollution Bulletin, 53: 569–578.

10

Impact of Changing Lifestyles and Consumer Demands on Food Safety

Abdul Khaleque,[1,]* *Mohammad Mustafizur Rahman,*[2]
Kazi Selim Anwar[3] *and Md. Nazim Uddin*[4]

Introduction

Changing demographics with rising purchasing power of consumers all over the world is shifting preferences in global food production. Modern technological innovations are allowing consumers to be aware of food production process, packaging and distribution information. Thus helping consumers make informed decision about their food choices. Better access to diet information linking food quality, safety, and nutrition is contributing to a demand for structural change in large-scale food production, processing, and retailing worldwide. Family food budget and expenditure are linked to income, and it greatly varies between developing and developed countries. In developing countries, higher household income is linked to increased demand for meat products, often leading to increased import of livestock. In developing countries changing lifestyle associated with increased purchasing power of the consumer is slowly shifting demand for processed food however price hinders widespread adoption.

[1] Department of Biology and Chemistry, North South University, Dhaka 1229, Bangladesh.
[2] National Institute of Diabetes and Digestive and Kidney Diseases (NIDDK), National Institutes of Health (NIH) Bethesda, Maryland 20892, USA.
[3] Department of Microbiology, Faculty of Medicine, AIMST University, Semeling 08100 Bedong, Kedah, Malaysia.
[4] Bangladesh Agriculture Research Institute (BARI), Gazipur 1701, Bangladesh.
* Corresponding author: akhaleque@northsouth.edu; abdul.khaleque@northsouth.edu

In developed countries diversification of diet has increased demand for both quality and safety but laborsaving food choices results in an increased import of processed food. Consumers in developed countries are also inclined to organic food for quality and safely; and also paying more attention to better animal welfare. In developed countries both public and private sectors have responded to consumer demand shift in these quality attributes by implementing changes in agricultural food production, processing, quality control, management of livestock, and assurance schemes.

Global food market is constantly evolving, driven not only by changes in lifestyle and consumer preferences but also due to a shifting nature of food supply chains, public policies, and demographic trends. Changes in these preferences and perceptions impacted heavily on food quality and safety in recent times. New technology in association with social media changed the way consumers learn about agricultural food production, processing, livestock management, animal handing, pesticide and fertilizer application, and associated impact on global environment. In both developed and developing countries consumers are now concerned for food safety when making their food choices. As border disappeared with the globalization of food chains also increased foodborne illnesses at significant level. Consumer concerns are affecting the global network of agriculture practice. Consumers are increasingly expecting transparency, and more reliable information concerning agricultural production processes. Consequently, livestock markets and big-name retailers are attempting to find new avenues to regain consumer confidence in their labels. Despite being well informed about safe products, consumers ultimately prefer to choose and enjoy the tasty alternatives. It appears that sensations related to taste beat the guarantee of food safety in consumer behavior subsequently standards, labels and quality signal no link between the different attributes of the products and their production processes.

Consumer Lifestyles and Food Choices

Consumption of food reflects the identity of the consumer. Lindeman and Stark (1999) suggested that eating healthy, non-fattening diet rich in vegetable components is the basis of good health and longevity. Diversified diet is the manifestation of a positive lifestyle, and it also simplifies our food choices. Today food choice is complex due to the same reason, an array of processed food products although diverse in nature is available in the supermarket (Amarra et al., 2008). Today consumers are kind of confused while trying to satisfy their preferences, and achieve their goals. The large number of food products available on the supermarket separates the consumer into two categories, active choosers and passive pickers (Mäkelä, 2000; Schwarz, 2005; Sun, 2008). Active choosers choose their food in part that cooking and eating is segment of an enjoyable lifestyle. On the other hand passive pickers choose their food out of necessity. According to Schwarz (2005), food choices are mostly based on consumer habits irrespective of the type, active or passive. Social customs and religious restrictions also play a role in food choice for both active choosers and passive pickers. In most instances food choice decisions are made more or less routinely and thus save time, and effort. The overwhelming number of food products available in supermarkets, the decision-making require complicated evaluation which many consumers try to avoid. Consumer food choices are made primarily based on: (a) personal goals, (b) importance of each goal, (c) choices available, (d) how likely each choice options would help meet (Schwarz, 2005; Bisogni et al., 2012; Franchi, 2012).

Every consumer weighs food-related goals consciously or unconsciously, and values that are meaningful to them. In addition, available resources, personal and social factors and consumption affect food choices. Consumers are flexible in their food choices depending on the situation or context making investigation of food choice motivators as a challenge to researchers (Cohen and Warlop, 2001; Niva, 2008; Connors et al., 2001; Järvelä et al., 2006; Johansen et al., 2011). The consumer-developed categories simply meant to make a distinction of foods as good or bad based on their perception of quality of food. For instance, in judging the healthfulness of a food, consumers first and foremost pay attention to the fat content whereas freshness and nutrition values are overlooked (Oakes and Slotterback, 2002). Järvelä et al. (2006) found that avoidance is a common strategy observed in consumers paying attention to chemical risk in food products, for instance, avoiding liver or kidney as food or foods known to be high in additives. Healthfulness of a food is often approached via a choice strategy, such as favoring low fat or finished food products. However, the food choice task is not as simple as that. Connors et al. (2001) argued that benefits from the process of categorizing food products led to the evolution of multi-dimensional category of foods. Food categories are reliant on our social behavior also: eating alone, eating with children, eating with a spouse, eating out, eating as a guest at someone's home, with guests in one's own home, etc. (Herman and Higgs, 2015).

Healthy food choices dominate consumers under certain situations, where taste is the most relevant issue to be considered. Convenience dominates a situation where a person is hungry, or reach home late always at the end of a long day of work. The preferences of food choice and health of children in many such families vary greatly (Patrick and Nicklas, 2005). Further, pleasing one's desires is not possible in many situations, so consumers prioritize the benefits sought in a given situation. For instance, the taste of all the family members is not always the same, sometimes tasty, and or a favorite food or staple diet does not fit in the healthy food category. Finished foods that are convenient and quick-to-prepare are never the healthiest. The strategy becomes just to simplify food choices by eliminating one or more benefits from consideration. The benefit prioritization schemes commonly used by consumers are relatively stable (Roberto et al., 2015). When consumers are faced with new information, and situations, they always attempt hard to maintain their food choices that had worked previously. If consumers are forced to reconsider their choice strategies, they try to collect information from health professionals, close friends, and family members first; and then turn to media to learn about current trends in the society (Connors et al., 2001; Connors and Rozell, 2004). It is often seen that some food choice strategies are more unwavering than others. Taste and price often rule against healthfulness in a consumer's mind (Järvelä et al., 2006). Sensory motives like appearance, taste, and smell, have been found to be the most influential determinants of consumer food choice (Honkanen and Frewer, 2009). Important factors are also health, convenience, cost, and ethical/religious concerns. In addition, natural content, familiarity, mood, and weight control may influence the dietary choices of consumers (Steptoe and Pollard,

1995; Spence et al., 2013). Some of these motivating factors can be grouped into attributes, and some into benefits that attributes bring. Over a period of long time, food choice strategies have a tendency to become routine and thus simplify the food choice. Consumers devote their time, money, physical and emotional energy for novel food products. For example, pizza was a new experience in the Finnish food culture in the 1970s, but today ready-to-eat pizzas can be bought in the supermarkets. Routine consumption and refined consumption certainly varies, but not only in a particular place but also to a great extent of consumers with different socio-economic perspectives (See also Mäkelä, 2000). The division of the products from routine consumption to more refined consumption is illustrated in Fig. 1.

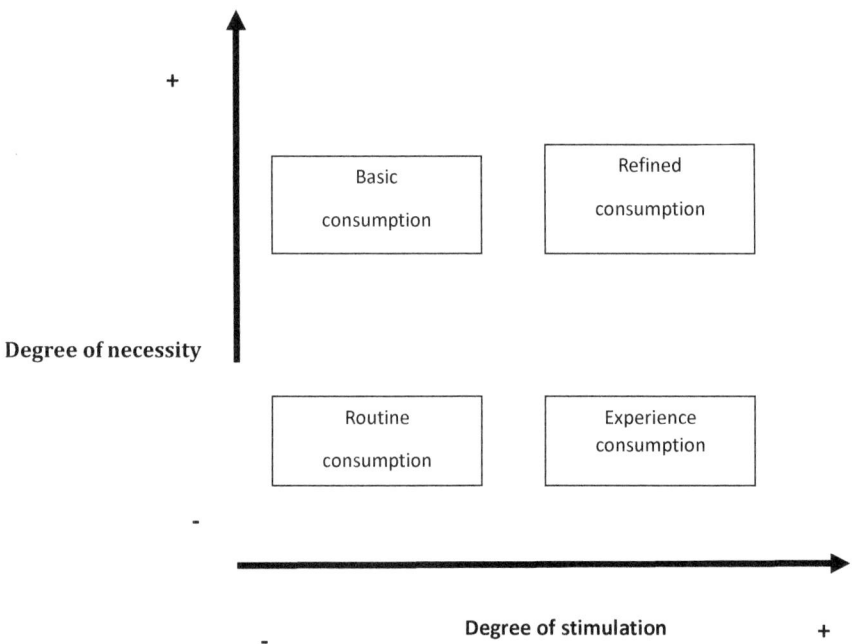

Figure 1. Two-dimensional consumptions (Adopted from Wikstrom, 2000; Zukin and Maguire, 2004).

Zukin and Maguire (2004) divided food consumption in four large groups based on necessity. They also emphasized on how much stimulation is required to bring a particular consumer group (see Fig. 2.3). Food choices and consumption provide a little or no personal stimulation, as basic food consumption is a routine process. However a refined food consumption model characterizes necessity which consumers pay interest with stimulation, for example, a specially cooked meal for a festival or family occasion. The routine food consumption does not give much pleasure to consumers. Examples of such routines are passive TV viewing or the habit of buying certain snack on their way to home after a long workday. However refined food consumption choices provide stimulation, and leads to personal satisfaction. Also trying out new and exciting food products, which is not a necessity, brings new experience and stimulation for the consumer (Hollywood et al., 2013; Cannuscio et al., 2013).

Food-Related Lifestyle Affects Consumers' Attitudes

Food-Related Lifestyle (FRL) is an application of means-end theory (MET). Lifestyle is stated as a dominant system connecting situation-specific product opinions that allow adding personal values. There are five domains of lifestyle, which mediate between values and product attributes that motivate behavior intentions and which can be used to explain food purchases: ways of shopping, quality aspects, cooking methods, consumption situations, and purchasing motives (Scholderer et al., 2004; Szakaly et al., 2012). Among the five FRL domains described, the ways of shopping, cooking methods, and consumption situations measure differences in the habitual use of scripts and skills (Perez-Cueto et al., 2010). The importance of quality aspects measures a schema for the evaluation of product attributes. The purchasing motives domain measures differences in the significance associated with food-related personal values. The FRL model is a questionnaire consisting of 69 Likert-type statements (Krosnick and Presser, 2010). These statements measure 23 lifestyle dimensions in five domains, which are:

1. The ways of shopping (importance of product information, positive attitude towards advertising, enjoyment from shopping, specialty shops, the price criterion, and shopping list),
2. The cooking methods (interest in cooking, looking for new ways, convenience, the whole family, planning, and woman's task),
3. The quality aspects (health, price/quality relation, novelty, organic product, taste, and freshness),
4. The consumption situation (snacks versus meals, and eating out/social event) and
5. The purchasing motives (self-fulfillment in food, security, and social relationships).

Attitude Behavior Context (ABC) model (Fig. 2)

This is a model of socio-environmental behavior that incorporates contextual factors limiting one's ability to act on their intentions. Contextual factors included socioeconomic and demographic variables. It is modulated by community characteristics that may constraint access to organic and local foods.

Means-End Chain (MEC) theory

This is a model of consumer assumption based choice of food products whose attributes and consequences reflect consumers' goals that is value of the products they buy for the functional and psychological benefits it carries.

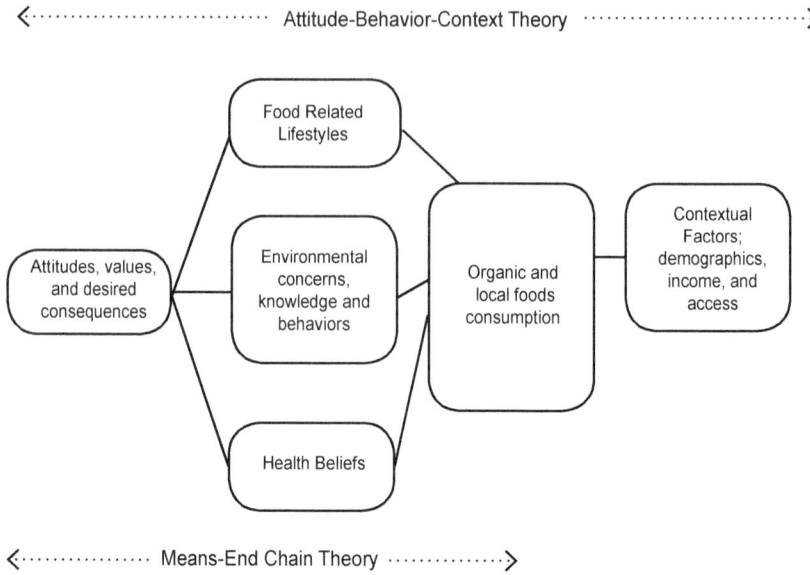

Figure 2. Attitude-Behavior-Context (ABC) theory as an overall framework, and containing Means-end chain (MEC) theory, Health Belief (HB), and Food-related lifestyle (FRL) models (Brunso et al., 2004; Nie and Zepeda, 2011).

Food-Related Lifestyle (FRL) model

In this model FRL is an application of MEC theory. FRL have five components of lifestyle, which arbitrate between value and product attribute that motivate food purchases as well as ways of shopping, consideration of quality aspects, cooking methods, consumption situations, and finally purchasing motives of consumers.

Health Belief (HB) model

In this model consumers only focus on desired health outcomes of their food choice or preferences.

In FRL model ways of shopping refers to consumers' food shopping behavior, and how they use available information, such as labels. Food quality aspects refer to the attributes consumers seek from products. Choice of food cooking methods refers to effort and time is put into meal preparation; and who is responsible for daily meal preparation in a family. Food consumption situations tackle when and how food is eaten. Consumer purchasing motives incorporates desired consequences of a meal that is satisfaction and pleasure (Amarra et al., 2008; Hollywood et al., 2013).

In Europe the validity of the FRL-instrument has been tested across food cultures (Candel, 2001; Kesic and Piri-Rajh, 2003; Scholderer et al., 2004; O'Sullivan et al., 2005). The studies revealed various consumer segments associated with different food-related lifestyles. In Great Britain, six FRL segments were found: the adventurous food consumer (17%), the careless consumer (14%), the conservative consumer (9%), the rational consumer (26%), the snacking consumer (20%), and the uninvolved consumer (14%) (Wycherley et al., 2008). In another study Bae et al. (2010) described the food-related lifestyle of undergraduates in Korea. The relationship between types of FRL and restaurant selection presented a canonical correlation. Both the taste-seeking and safety-seeking types of consumers have significant positive relationships with the quality of the food, taste and service provided in the restaurants. Recently, the attitudes of consumer towards different food categories have been further investigated, for instance, towards convenience food (Buckley et al., 2007; Ryan et al., 2002) and specialty food (Wycherley et al., 2008). Recently FRL differences were examined between obese and non-obese consumers in five European countries (Pérez-Cueto et al., 2010). A body mass index (BMI) of 30 kg/m was used to divide the respondents into obese or non-obese groups. In all countries examined, obese respondents scored lower on most FRL quality dimensions, particularly on novelty, freshness, consumption of organic products, healthy food choice, food security, and eating socially. On the other hand, non-obese respondents were found to be more interested in food price (Pérez-Cueto et al., 2010).

Conceptual model of global consumer food choice (Fig. 3)

The conceptual model of global consumer food choice represents three main factors involved in food choice: life course, influences, and personal preferences. It incorporates personal roles and social atmospheres to which the consumer is exposed. Religion, moral values, ideals, available resources, and social agenda can shape consumer food choice preferences. A conscious dialogue and unconscious social policies influence a food-related consumer choice situation. The funnel shape of conceptual model of food choice indicates social attributes of the process. A single food choice event results from the diverse set of personal and environmental inputs. It influences a food choice situation to a extent where the social settings affect how people construct and execute personal food choice preferences, a process that may either be deliberate or automatic (Perez-Cueto et al., 2010).

Life course perspective on food decision-making includes personal experiences and current involvement of consumers in food trends and transitions, e.g., upbringing characteristics of a generation, relationship with food culture, and the social demand of current food choices (Sobal and Bisogni, 2009). Influences mutually shape one another as well as interact and compete with one another. Food choice expectations, standards, and social beliefs compared by consumers who judge and evaluate food choices. Following are the many variables that greatly influence consumer food choices: Individual factors, likes/dislikes, individual food styles, food centeredness (pleasure, health, safety, or symbolism) and emotions (emotional cues, moods, and feelings) as well as characteristics like gender, age, health status, sensory preferences (or taste sensitivities) and state of hunger. It also incorporates cravings, preferences for particular foods or food types, and aversions; physiological factors such as allergic response, hunger. Families and households provided one of the most important sets of interpersonal relationships influencing consumer food choices (e.g., the mother or the husband or the wife who sacrifices their own priorities to meet the family's needs).

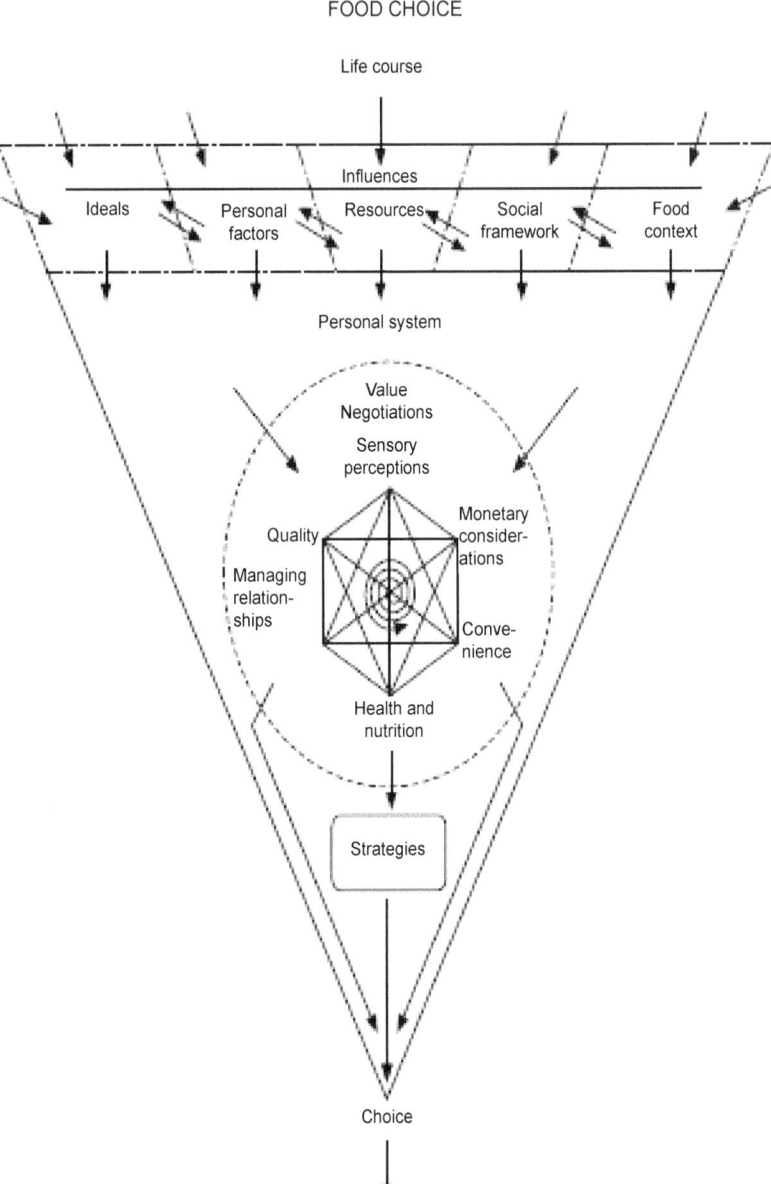

Figure 3. A conceptual model of the components in the food choice process (Adapted from Furst et al., 1996).

Socio-economical environment and demography influence food choices, for example, physical surroundings, social climate of the choice setting, food supply chains, types of food and food sources availability including seasonal or market factors. A particular market can offer expanded or constrained food choice possibilities to establish a tone or ambiance that influences the food choice process in consumers (Jang et al., 2009).

Personal preferences in food choice conceptual model value negotiations that involved weighing of different considerations in making choices and strategies that involve choice based on habitual patterns. Sensory perceptions dominate value that is driven mostly by taste, and vary widely in consumers. In this model the limiting factor is consumer food choice that is less negotiable, for example, included dimensions of texture, odor, or appearance. Sensory perceptions, especially taste, and price considerations are frequently in conflict during food choice by consumers. Taste is weighed against convenience in most instances whereas tolerance for food aversions and willingness to accept particular food choices are generally influenced by availability and the social situation (Jang et al., 2009). Monetary considerations, in other words price and perceived value of food is a very striking point as price more often conflict with taste, quality, and food safety. Convenience is often spoken of as a commodity that is how much time to be spent or saved. Health and nutrition factors relating to bodily well being or disease avoidance is weighted in terms of healthfulness of consumer. Finally strategies or well-established habits or rules (heuristics) are key component of consumer food choice (Dibsdall et al., 2003; Maillot et al., 2007).

The consumer food choice model (Fig. 4)

In this model life course events and experiences dominate personal food choice decisions. According to Sobal and Bisogni (2009) expectations about the future courses are a key concept in a consumer's determined opinion, taste, and food

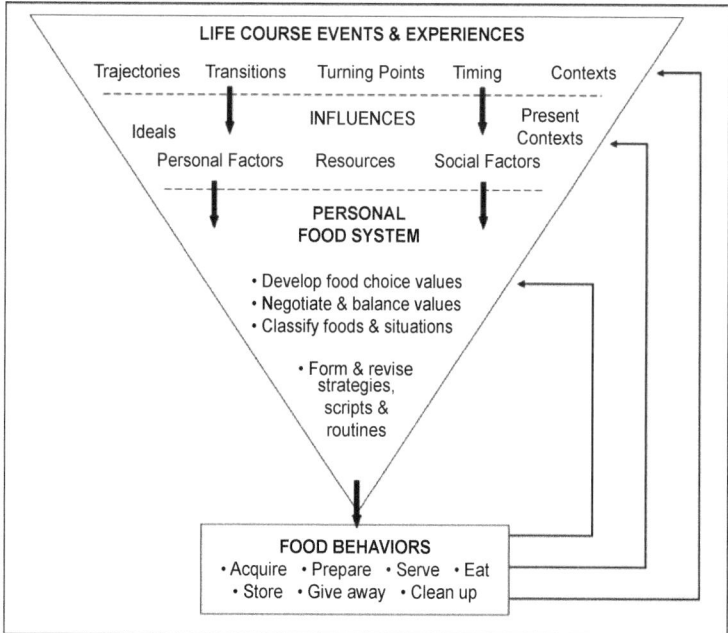

Figure 4. Consumer food choice process model (Adapted from Sobal and Bisogni, 2009).

choice strategies to influence food preferences. Any major shift in life course leads to a change in food choice preferences (e.g., change in residence through migration, change in family type and nature through marriage, change in workplace, change in health due to disease conditions). Any major transitions lead to sweeping reforms of consumer food choices (e.g., a shift from eating unhealthy food to a fat restricted diet post-heart surgery). Timing of transitions and contexts are the environments in which consumers live, for example, social, cultural, political, economic, environment, and other conditions that either facilitate or constrain changes in the food choice trajectories. At microlevel contexts families, friends, schools, workplaces, churches, communities, and many other social and physical assemblies shape consumer food choice trajectories (Scheffer, 2009; Wethington and Johnson-Askew, 2009).

Contributions of life course perspective model on food choice greatly depend on influences by various factors. Influences are grouped into five major categories: *cultural ideals, personal factors, resources, social factors,* and *present contexts.* Cultural ideals include the knowledgeable system of rules, and value plans shared by a group of consumer. Cultural ideals provide the standards that can be used as reference points by consumers to weigh food choice related behaviors. Personal factors are characteristics of persons that effect their food choice behaviors, inclusive genetic predisposition to disease, sensory sensitivity to food tastes, personality, gender, role of parents in making food choices. Resources are the properties that consumers contemplate in crafting their food choice decisions, for example, household income, knowledge of food, relationships, values, traditions, etc. Social factors are the relationships among individuals that can limit or assist their food choice decisions, such as, eating alone or with coworkers, family, friends, outdoor, indoor, etc. could support to eat healthy. Contexts are larger settings influencing consumer food choice decisions that include social and physical environments. Social establishments, mass media, and government policies create economic circumstances whereas physical conditions include climate, and other material objects that either aid or restrict consumer food choice decisions (Wethington and Johnson-Askew, 2009; Larson and Story, 2009).

The development of consumer food choice values is achieved by balancing foods and situations. Personal systems of cognitive processes for food dictate development of strategies, and routines for recurring food choice decisions. Consumer food choice preferences are considerations of taste, cost, health, convenience, managing relationships, and feelings that people would attach to the term "healthy eating". In our society, to simplify food choice decisions consumers classify foods and situations according to categories they develop based on characteristics of the food value. Personal systems of cognitive processes help negotiate and balance competing values using heuristics like prioritizing values, taste, cost, convenience, and health (Sobal and Bisogni, 2009; Gillespie and Johnson-Askew, 2009). Strategies for food choice decisions include elimination, limitation, substitution, addition, and modification to make food choice decisions either more automatic or reflective. Food choice expectations and plans are ceremonial awareness of consumer held responsible for food choice related behaviors in situations that are familiar to them. Food choice expectations and plans provide predictability and comfort to consumer. Such personal systems are the cognitive processes involved in food choice decisions and are close to actual individual food behavior in comparison to distal influences and prior life course experiences. Personal system model indicates that through food choice behaviors consumer shape their life course experiences, influences, and personal food system. It also shapes nutritional status, healthfulness, and disease avoidance. Individual involvement in food acquisition, preparation, cooking, after-meal cleanup provides acquaintance and skills for imminent resource management. It also stimulates trying out new ways of shopping, cooking and eating behavior to discover new strategies and revise their food choice expectations (Sobal and Bisogni, 2009; Rothman et al., 2009).

Model of food consumption: Socio-economic effects

The demand of food and food choice decisions are functions of income and prices, as well as tastes and preferences. Cost is in the consumer's mind instead of price because the actual price that a consumer pays is a function of food price, travel costs to the store, and the in-store availability of specific foods. Even though a small grocery might be very close to an individual, if there is no in-store availability of fresh fruits, e.g., a consumer might have to travel to a distant supermarket. Car ownership could lower overall travel costs if it shortens travel time to stores. A detailed demographic characteristics, including age, race, ethnicity, schooling, and other variables, is useful for capturing unobserved information on consumers' tastes and preferences. Tastes and food preferences are based on cultural food habits associated with particular ethnic groups, or they might be based on knowledge and concern of the consumer regarding diet and health outcomes (Boylan et al., 2010; Rose et al. 2010; Franchi, 2012) (Fig. 5).

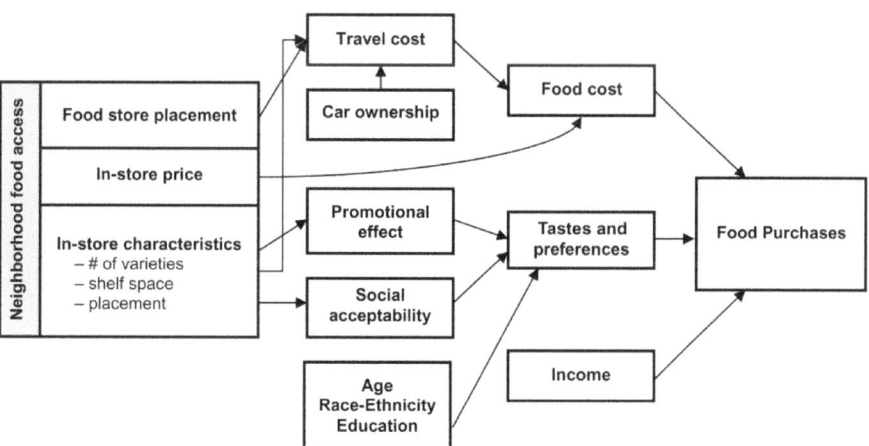

Figure 5. Model of food consumption: Socio-economic effect (Adapted from Rose et al., 2010).

Conceptual model for food systems and health disparities

The conceptual model for food choice decision demonstrates comprehensive food system situations that interplay with the environment in participating communities. The model argues that retail stores, restaurants, schools, workplaces, and public policy in turn have mutual interactions with other issues in the social environment that changes over time. All these factors affect individual inclination to a healthy and sustainable diet. In the conceptual model, existing disparities in the social environment that include race, ethnicity, demography, gender, and purchasing power alter the entire relationship so that a diverse society may vary widely in the types, qualities, and quantities of food. Consumers also consider availability, affordability, and cultural acceptance of food choices. In "feedback" loops shown in grey (Fig. 6) depicts disparities in individual and community likelihood of obtaining good food. It can affect the extent to which such food choices are made available either in certain communities or society wide. For example, knowledge about food and nutrition in college students has led to a level of demand for more sustainable campus food. A demand for a sustainable food

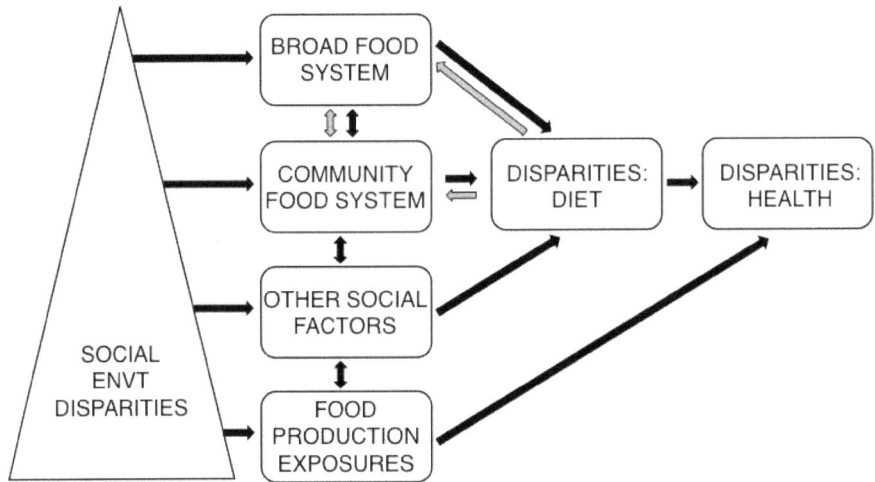

Figure 6. Conceptual model for food systems and health disparities (Adapted from Furst, 1996; Neff et al., 2009).

in college campuses has caused large institutional food providers to change their offerings (Kolodinsky, 2007; Peterson et al., 2010). Personal food choice decisions and associated socio-economic factors affect the course that vary within and between communities, and may contribute to disparities in healthfulness. Interventions to change the food choice, preferences and health relationship fall into two broad categories using population-based strategies, (a) aimed at changing factors affecting the entire population; and (b) aimed at changing the food system exposures or food demand within specific sectors of the population. It is always critical to acknowledge that population-based public health interventions may improve conditions for all. Any efforts that use new information technologies or social media may have this effect. The group that starts out healthier is often better educated with more resources and is therefore more likely to use new technology and information more rapidly and effectively. Better eaters thus have a greater gain in healthfulness than those with lesser education and fewer resources (Oakes and Slotterback, 2002; Kolodinsky et al., 2007; Neff, 2009).

Conceptual model for factors influencing consumer food choice decisions

The conceptual model represents the factors influencing consumer food choice decisions including many categorizations from production to consumption. Krebs-Smith and Kantor (2001) summarized the methods of assessment pertaining to food supply, foods acquired and foods consumed that ultimately affect food choice, especially as they pertain to fruits and vegetables by individuals (Irala-Estevez et al., 2000) (Fig. 7). The current model offers a portrayal of the interplay among agricultural, economic and social forces. The current model also weighs the role of food supply, acquisition and consumption. Present valuation processes such as data on food supply, food acquired and food consumption help develop an understanding of the prime factors and their relationships to the food production and consumption sequence (Zukin and Maguire, 2004; Sun, 2008).

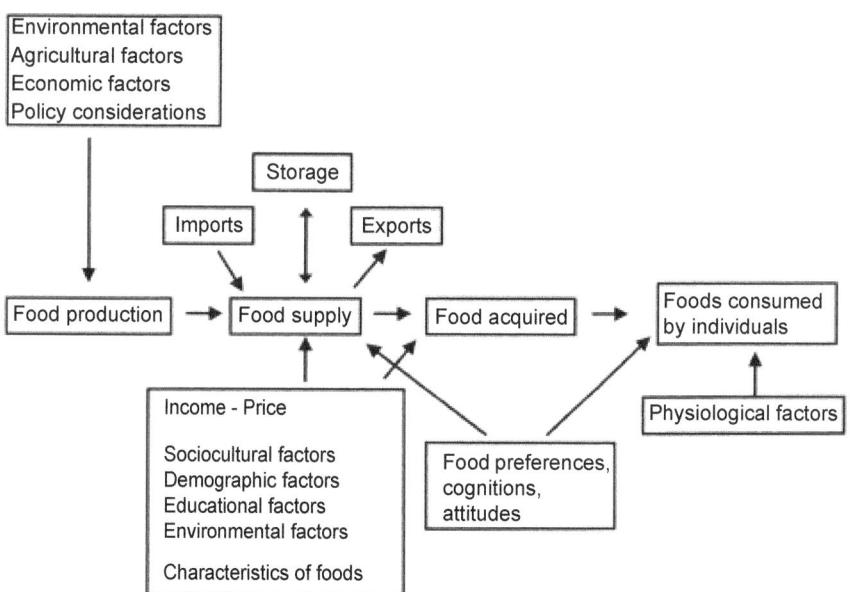

Figure 7. Conceptual model for factors influencing consumer food choice decisions (Adapted from Krebs-Smith and Kantor, 2001).

The Impact of Social Conditions that Affect Consumer Food Choice

Consumer food intake is largely dependent on social influences specially referring to one or more persons having an impact on the eating behavior of others either consciously or subconsciously (Gillespie and Johnson-Askew, 2009; Hermans et al., 2012; Herman and Higgs, 2015). Food choice decisions are influenced by social factors even when eating alone due to mindsets and lifestyles developed through social interaction and family influences. We tend to eat more with our friends and family in comparison to when eating alone. In most instances the quality and the quantity of food increase as the number of friends for diner grows larger (Herman et al., 2003; Hermans et al., 2012). Low socio-economic status and poor health is subjective to gender, age, culture, environment, and community networks. Healthy food choices are also influenced by individual lifestyle factors and behaviors. Many studies reported that there are differences in social welfare associated with food habits and nutrient values (Bisogni et al., 2012). In low-income groups there is a greater tendency to consume unbalanced diets with low intakes of fresh fruit and vegetables leading to under-nutrition (micronutrients deficiency) and over-nutrition (energy overconsumption) resulting in overweight and obesity (Sunguya et al., 2014). Within the members of such low-income community, depending on the age group, gender and level of deprivation also led to chronic diseases at an earlier age compared with higher socio-economic groups; usually identified by educational and occupational levels (Neff et al., 2009; Mayen et al., 2014).

The Impact of Income on Food Choice Decisions

Individual income provides the ability to purchase food. Income is a key element to determine the types of food, and other goods are being purchased by consumers. In the last 25 years, there has been a significant increase in individual income levels worldwide. The World Bank reported that during the period 2000 to 2014, individual income in most areas of the world grew, with the exception of East Asia (Mayen et al., 2014). Higher consumer income allowed individuals to spend more on food, especially on meat, poultry, and specialized food products. Income inequalities affect consumer preferences in large part to the differences in food choice decision across the globe, and to changes in diet and nutrition of a population over time. Rise in food price outpaced income growth, and the percent of average household income spent on food has fallen. Individual income is also associated with changes in the value of foods consumed. High-earner consumers have diets that are more diversified, and the primary source of calories changes from root crops, cereals and other staples to animal products when compared to low-earner consumers. A shift from cereals and pulses to animal products is associated with increasing household income. Recent studies across a wide range of countries show that the share of protein from animal sources increases as incomes (GDP). High-earner consumers also stress on other food attributes, such as convenience (Boylan et al., 2011; Popkin, 2014).

The consumer food demand pyramid, as illustrated in Fig. 8, is a model of the consumer choice process (Kinsey, 2000). The food demand pyramid implies that low- and middle-income families emphasis on meeting survival needs first that is obtaining sufficient calories. Food price and food safety are basic concerns in low- and middle-income. At a lower purchasing levels food safety triumphs information about food quality. High-earner consumers expect foods to be safe from microbial contamination and other health hazards similar to low- and middle-income consumers but they also focus to health and nutrition value of food when making food choice decisions (Kennedy et al., 2005; Kaur and Hegde, 2008; Popkin, 2014). High-earner consumers can purchase products that satisfy preferences above and beyond basic nutritional needs, such as better taste, variety and convenience. In developing countries increased meat demand is associated with the improvement in economy. When safety and affordability, lower on the food pyramid have been met, consumers want to satisfy taste, variation and convenience. High-earner consumers demand expanded information about the food they consume. They also want to know how food products affect health and lifestyle, and most often become concerned about the impact of food choice decisions on other people, on the environment and livestock. Thus, with rising income demand for agricultural food products with diverse features evolves, presenting both opportunities and risks to current food chains (Wikstrom, 2000; USDA-ERS, 2005; Neff, 2009). Low- to moderate-income families in developed countries and in developing nations still demand an affordable livestock for a convenient source of animal proteins. In midcentury, the entry of women into the labor force contributed to the rise in consumer economics as consumers devote less time for food preparation, and have moved away from raw food products to food choices that are easy to prepare or requires little preparation (Candel, 2001; Jang et al., 2009).

Consumers with low household income find it difficult to achieve a balanced diet. Low-income consumers are often referred to as experiencing food poverty (Riches, 1997). There are many facets to food poverty but the core impediments to achieve a balanced healthy diet demands high price, availability and nutrition

Figure 8. Consumer Food Demand Pyramid (Adapted from Kinsey, 2000).

knowledge (Dibsdall et al., 2003). These factors have led to the development of areas identified as food deserts. A diet dependent on energy-rich but nutrient-poor foods is a consequence of lack of purchasing power to buy wholesome foods. Moreover, a lack of proper cooking facilities in the home increases the need to eat convenient fast food, which has a potentially higher energy density. Sometimes the low-income consumer groups also face with logistical obstacles to eat well due to a lack of transportation. Public transport is not always viable solution for many, particularly those with young children or various mobility difficulties. Finally, a lack of knowledge or conflicting knowledge on diet and nutrition along with lack of motivation and the loss of cooking skills can inhibit a practice of healthy eating. Low education and low-income levels determine food choices and behavior that can ultimately lead to diet-related diseases. The origins of many of the problems faced by low-income consumers emphasize the need for a multidisciplinary approach to improve health inequalities (Kinsey, 1994; Irala-Estevez, et al., 2000; Hollywood et al., 2013).

Impact of Changing Food Markets

Demand for increased food quality and safety in global markets are affecting the world's food supply chain. Impact of increased competition, trade barriers, and changing regulations is changing the nature of all suppliers in the global food supply chain. Agricultural suppliers account for 25% of the economy of developed countries; and maybe more for developing countries (Cheek, 2006). The food market chain contains a number of sub-sectors, for example, the input supply sector, the production sector, the post-farm marketing sector, the primary processing sector, the secondary processing sector of more complex goods, and the final distribution sector that includes both the retail and restaurant sectors. The organization of food market chain has been changing fast, and the relationship among the sectors is ever changing. The primary changes lead to the emergence of relatively larger agricultural firms. The big box retailers or food supermarkets started dominating the retail distribution (Pinstrup-Anderson, 2014). In developed countries food retailing by the food-away-from-home method has also grown full-fledged to meet the needs of high-income and time-starved consumers, who demand more convenience. In the USA, about 50% expenditures are for food eaten away in comparison to Canada where food expenditure (spent on meals away from home) has risen nearly to 34% (USDA-ERS, 2005; Zafiriou, 2005). In developing nations like Mexico, consumption of local and traditional foods, such as tacos and tortas, declined recently. Similar to other developing countries, in Mexico consumption of meal away from home has risen sharply, especially in the rapidly growing urban areas (Barkley and Barkley, 2015). While the general change in retailing has been the supremacy of big-box retail stores in the global food chain market, a new segment of food market is imitated in small units that are becoming a vital part of the food chain market. Small mom-pop stores, web-based stores, smaller-sized food markets, and direct food marketing by producers have reinforced the niche of global food chain market. It is important to mention here that organic food products have become one of the emergent segments of the food chain market and, organic meat, poultry, dairy and eggs are being sold widely in conventional markets globally (Zafiriou, 2005; Hughner et al., 2007).

Impact of Consumer's Perceptions and Trust

Consumer education on risk promotes safer handling of food. Risk communication can be the best way to reduce the risk of developing foodborne illness of consumer who is at the end of the food chain. In the global context, each country has its own health concerns and risk management priorities although studies assessing food safety concerns are limited. In both developed and developing countries consumer concerns include foodborne bacterial contamination, excess hormones, unnecessary antibiotic in poultry and livestock, irradiated foods, etc. Ranking of the consumer concerns and priorities vary among nations. Consumer concerns stem mainly from country-wide differences in perceptions about food safety. Consumer perceptions are the outcome of a multifaceted function of factors, for example, differences in each country's baseline of food safety guideline, risks from imported food, access to information about food safety, extent and nature of risk, experience of food safety incidents, and finally trust in the sources of information (Macintyre et al., 1998; Medeiros et al., 2004; Patil et al., 2005; Hollywood, 2013). There may even be basic differences, like, some societies consider diarrheal diseases as a natural/normal occurrence due to factors such as teething, eating hot/spicy foods, indigestion instead of perceiving diarrhea as a symptom of disease that can be transmitted easily to cause an outbreak leading to a national disaster. The food safety risks are the same across countries; different countries may perceive and handle food safety risks differently due to disparities in access to advanced detection, and mitigation technology. For example, the United States has a zero-tolerance policy for *Listeria monocytogenes* organism in all ready-to-eat foods compared to any other nations of the world. Factors that contributed the most to foodborne illness are improper holding temperatures, inadequate cooking, contaminated equipment, food from unsafe sources, and poor personal hygiene (Motarjemi and Käferstein, 1997; Buzby and Roberts, 1999; Kennedy et al., 2005; McEntire, 2013).

Consumer considerations and practices associated with food choice, healthfulness, and safety concerns are always balanced against consumer habit and realism (Lindenman and Stark, 1999; Sun, 2008). Berg (2004) reported four main categories of consumer trust about food choice preferences, and supported by other studies (de jonge et al., 2007; Leikas et al., 2007) as follows:

(a) Sensible consumers who attempt to choose healthy foods for their diets so they can feel assured that the foods they consume are beneficial.
(b) Skeptical consumers who are anxious about choosing foods for their diets but are abysmal that the foods they consume could be harmful to healthfulness.
(c) Naïve consumers who are not worried about choosing foods for their regular diets.
(d) Denying consumers who panic that the foods they consume are risky to themselves or their families but never attempt healthy foods.

Current literature reported that the trust-in-food-safety that consumers are concerned to their FRL accommodated in the main domain of consumption situations that are quite similar globally despite their different degrees of trust in food safety (Patil et al., 2005; Kaur and Hegde, 2008). Additionally, taste, and convenience play very important role in food choice decisions in all four types of consumer groups. It is from a different perspective that four consumer types vary in level of trust in food safety. Consumer groups emphasize differently to the taste of foods, the convenience of cooking, and the security. Based on current literature findings, the big-box retailers are making efforts to enhance the food taste, and reinforce food safety as well as provide convenience for cooking to adapt to changing trend in food-related lifestyle globally (Lindenman and Stark, 1999; Leikas et al., 2007; Papadopoulos et al., 2012).

Influence of Foodborne Incidents and Publicity on Consumer Behavior

Highly publicized international food safety incidents may lead to lasting changes in consumer perceptions about food safety. A change in consumer perception always leads to a substantial change in their food purchasing patterns. In many instances a strong public outcry bring changes in government regulations affecting domestic and imported livestock and food products (Berg et al., 2005; Chen, 2008; de Jonge et al., 2008). So, the current hypothesis is that subsequent resolution of a problem that instigated a major food safety incident, consumer perceptions about the associated food and about the producer country's ability to manufacture safe food may be changed, and these perception changes have a long-lasting impact on global demand and trade. To explore this hypothesis, three international food safety incidents are mentioned here. First, the 1996 outbreaks from the pathogen, *Cyclospora*, on Guatemalan raspberries in the United States and Canada (Manuel et al., 2000); second, a bovine spongiform encephalopathy (BSE) crisis in the United Kingdom (Beck et al., 2005), and finally the 1999 contamination of feed in Belgium by cancer-causing dioxin (Buzby and Chandran, 2003). Case studies demonstrated a global change in consumer perceptions and how that related to changes in consumer behavior, which affected global trade greatly. Economic impact data illustrated a severe implication to agricultural industry, particularly during periods when food exports were reduced, suspended, or denied entry in to affected food market (Manuel et al., 2000; Satcher, 2000; Buzby and Chandran, 2003; Beck et al., 2005).

Influence of Media Publicity in Consumer Food Choice

Consumers now get many kinds of information that may affect their decision-making. Beside traditional print and TV advertisements, there are social utilities on the internet where consumers communicate with each other, there are informative TV and radio programs like those featuring chefs (Holt, 2002; Harris et al., 2009). In addition, consumers really need aid in decision-making because new products and brands are constantly introduced into the consumer market. Food related decision-making has become more complicated in recent years. Consumers do not have the time, knowledge or skills required to be experts in every field in the consumer market. Therefore, they seek advice from other people, use filtering devices such as restaurant guides on the social media or in the internet follow a particular lifestyle trend (Kaur and Hegde, 2008; Wangberg et al., 2009).

Impact of New Technologies on Consumer Perception about Food

Consumers now demand greater food variety and the availability of these foods year round. Improved information and transportation technologies have significantly changed the way food is produced, processed, transported and delivered to consumers. The "new food economy" incorporates information, production and distribution technologies that have reorganized distribution channels (Hammoudi et al., 2009). New technologies allowed increased integration of various market activities and increased use of private contracting in global supply networks. Buyers are now associated with large retail food networks, where reputation, quality and delivery are important attributes of the transactions. These retailers traditionally used importers, brokers, distributors and wholesalers as suppliers, but, increasingly, they are contacting directly with farmers and processors, or integrating vertically. For example, most major retail food store chains in North America have acquired their own agricultural firms to supply perishable products, such as dairy, produce, meat, poultry and value-added items (Zafiriou, 2005; Ploeg et al., 2015). Rapid developments in food production methods and food processing choices created new opportunities for food attributes. New developments can be categorized into the following three models: (a) food production methods to improve nutrient levels; (b) achieve better efficiencies in production; and (c) improvement in food safety. Technological advances in agricultural methods that enhanced nutrition, production and safety (foodborne illness) also created public concerns about safety of genetically modified (GM) crops, livestock, and poultry; social inequality, impact on environment, and finally unethical treatment of livestock and poultry. Getting consumer acceptance of complex, and occasionally controversial agricultural technologies, and often misunderstanding large-scale commercial practices demanded more information to sway consumer perceptions which could be achieved only by effective communication programs (Wilcock et al., 2004; Grunert, 2005).

Impact of Dietary Guidelines and Consumer Health

The dietary guidelines released by the U.S. Department of Agriculture (USDA) influence public education on nutrition, policy and food programs run by different federal and state agencies. The guidelines released in 2005, and updated in 2015 provided guidance on food choice decisions with recommendations for body weight management and necessary physical activity. The current guidelines inspire increased consumption of fresh fruits and a variety of vegetables, whole grain diet, and discourage use of highly processed, high-calorie, high-fat equivalents. Recommendations are made to limit fats, especially saturated fats and trans-fats. The current guidelines encourage choosing carbohydrates from fiber-rich fruits, vegetables; and limit intake of added sugars in various food products. Similarly, in many developing countries, the government is preparing its first set of dietary guidelines that will include recommendations on the "plate of good eating" (Kennedy, 2004). The government's public health guidance is expected to advise consumers to balance their diet by a portion from each food groups: fresh fruits and vegetables; cereals and beans; and animal products, i.e., eating a balanced meal (Kennedy, 2005; Ferruzzi et al., 2012).

Conclusion

Food safety is a public concern across the world, as food production, processing, and distribution become more integrated, food safety problems in one country can quickly pose problems in another country. Therefore, harmonization of standards and certification program should be developed across the world. Alternatively, the growth of strong retail chains can support private systems for food safety and quality control through internal mechanisms, e.g., vertically integrated food supply chains or private mechanisms such as brand names, contracting arrangements, animal identification and tracking systems. New technologies have allowed more rapid measurement of product attributes (e.g., fat content, drug residue) allowing the buyer to specify attributes of interest and better match preferences.

There is increased competition in various products and process of food attributes, therefore nutritional attributes, food handling and warnings, and product attributes such as country of origin should be clearly labeled. New methods and technologies (e.g., electronic information in the retail store environment) may provide alternatives to traditional media for educating consumers and allow highly motivated consumers to move from summary information on the label to more complete information available through the internet. Public agencies may serve the role of deciding what type of information to provide to the general consumer. Lack of information can lead to consumer misconceptions about production methods and techniques. At the same time, agricultural production is under increasing scrutiny from consumer groups. Both sources may threaten continued growth in processed food consumption and perpetuate lack of understanding about issues surrounding production. Educating consumers about commercial production and enhancing the public's knowledge and awareness of food production methods may have long-term benefits in maintaining consumer confidence and demand for food products. Different food retailing environments exist within the world market. The dominance of four or five large firms characterizes both the Canadian and U.S. markets, and nontraditional retailers are having a significant effect on retailing. This type of environment provides increased consumer product choice at low prices; however, it may reduce consumer choice over other products that may serve smaller consumer segments. In some markets, the presence of large merchandisers co-exists with smaller, niche segments. In other cases, the presence of large firms may limit the ability of smaller market segments, such as specialty meat markets with store-based operations, to survive. Some suggest that government's intervention in food retailing; and/or labeling is needed to preserve consumer choice globally. In the 21st century nutritional translation needs to blend with food science for the betterment of global food safety and healthfulness.

References

Amarra, M.S., Y.B. Yee and A. Drewnowski. 2008. Symposium on understanding and influencing consumer food behaviours for health: executive summary report. Asia Pac. J. Clin. Nutr., 17: 530–9.

Bae, H.J., M.J. Chae and K. Ryu. 2010. Consumer behaviors towards ready-to-eat foods based on food-related lifestyles in Korea. Nutrition Research and Practice, 4: 332–8.

Barkley, A. and P.W. Barkley. 2015. The economic approach to polarization. *In*: Depolarizing Food and Agriculture: An economic approach. 1st Edition. Routledge Publishers, Oxon, Canada.

Beck, M., D. Asenova and G. Dickson. 2005. Public administration, science, and risk assessment: A case study of the U.K. bovine spongiform encephalopathy crisis. Public. Admin. Rev., 65: 396–408.

Berg, L. 2004. Trust in food in the age of mad cow disease: A comparative study of consumers' evaluation of food safety in Belgium, Britain and Norway. Appetite, 42: 21–32.

Berg, L., U. Kjaernes, E. Ganskau, V. Minina, L. Voltchkova, B. Halkier and L. Holm. 2005. Trust in food safety in Russia, Denmark and Norway. European Societies, 7: 103–129.

Bisogni, C.A., M. Jastran, M. Seligson and A. Thompson. 2012. How people interpret healthy eating: contributions of qualitative research. J. Nutr. Educ. Behav., 44: 282–301.

Boylan, S., T. Lallukka, E. Lahelma, H. Pikhart, S. Malyutina, A. Pajak, R. Kubinova, O. Bragina, U. Stepaniak, A. Gillis-Januszewska, G. Simonova, A. Peasey and M. Bobak. 2011. Public Health Nutrition, 14: 678–687.

Brunso, K., J. Scholderer and K.G. Grunert. 2004. Testing relationship between values and food-related lifestyle: results from two European countries. Apetite, 43: 195–205.

Buckley, M., C. Cowan and M. McCarthy. 2007. The convenience food market in Great Britain: Convenience food lifestyle (CFL) segments. Appetite, 49: 600–617.

Buzby, J.C. and R. Chandran. 2003. The Belgian dioxin crisis and its effects on agricultural production and exports. *In*: International Trade and Food Safety, published by Economic Research Service, USDA. AER-828: 125–136.

Candel, M.J.J.M. 2001. Consumers' convenience orientation toward meal preparation: conceptualization and measurement. Appetite, 36: 15–28.

Cannuscio, C.C., K. Tappe, A. Hillier, A. Buttenheim, A. Karpyn and K. Glanz. 2013. Urban food environments and residents' shopping behaviors. Am. J. Prev. Med., 45: 606–614.

Cheek, P. 2006. Factors impacting the acceptance of traceability in the food supply chain in the United States of America. Rev. Sci. Tech., 25: 313–319.

Chen, M.F. 2008. Consumer trust in food safety—a multidisciplinary approach and empericial evidence from Taiwan. Risk Anal., 28: 1553–1569.

Cohen, J.B. and L. Warlop. 2001. Motivational perspective on means-end chain. London and New York: Lawrence Erlbaum Associates.

Connors, M., C.A. Bisogni, J. Sobal and C.M. Devine. 2001. Managing values in personal food Systems. Appetite, 36: 189–200.

Connors, P.L. and S.B. Rozell. 2004. Using a visual plate waste study to monitor menu performance. J. Am. Diet. Assoc., 104: 94–96.

de Jonge, J., H. van Trijp, R.J. Jan Renes and L. Frewer. 2007. Understanding consumer confidence in the safety of food: its two-dimentional structure and determinants. Risk Anal., 27: 729–40.

de Jonge, J., J.C. van Trijp, I.A. van der Lans, R.J. Jan Renes and L. Frewer. 2008. How trust in institutions and organizations builds general consumer confidence in the safety of food: a decomposition of effects. Appetite, 51: 311–7.

Dibsdall, L.A., N. Lambert, R.F. Bobbin and L.J. Frewer. 2003. Low-income consumers' attitudes and behaviour towards access, availability and motivation to eat fruit and vegetables. Public Health Nutrition, 6: 159–68.

Ferruzzi, M.G., D.G. Peterson, R.P. Singh, S.J. Schwartz and M.R. Freedman. 2012. Nutritional translation blended with food science: 21st century applications.

Franchi, M. 2012. Food choice: beyond the chemical content. Int. J. Food Sci. Nutr., 63: 17–28.

Furst, T., M. Connors, C.A. Bisogni, J. Sobal and L.W. Falk. 1996. Food Choice: A conceptual Model of the Process. Appetite, 26: 247–66.

Gillespie, A.M.H. and W.L. Johnson-Askew. 2009. Ann. Behav. Med., 38: S31–S36.

Grunert, K.G. 2005. Food quality and safety: consumer perception and demand. Eur. Rev. Agric. Econ., 32: 369–910.

Hammoudi, A., R. Hoffmann and Y. Surry. 2009. Food safety standards and agri-food supply chains: an introductory overview. Eur. Rev. Agric. Econ., 36: 469–78.

Harris, J.L., J.A. Bargh and K.D. Brownell. 2009. Priming effects of television food advertising on eating behavior. Health Psychology, 28: 404–13.

Herman, C.P. and S. Higgs. 2015. Social influences on eating. An introduction to the special issue. Appetite, 86: 1–2.

Herman, C.P., D.A. Roth and J. Polivy. 2003. Effects of the presence of others on food intake. A normative interpretation. Phycol. Bull., 129: 873–86.

Hermans, R.C.J., A. Lichtwarck-Aschoff, K.E. Bevelander, C.P. Herman, J.K. Larsen and R.C.M.E. Engels. 2012. Mimicry of Food Intake: The Dynamic Interplay between Eating Companions. PLOS ONE, 7: e31027.

Hollywood, L.E., G.J. Cuskelly, M. O'Brien, A. McConnon, J. Barnett, M.M. Raats and M. Dean. 2013. Healthful grocery shopping. Perceptions and barriers. Appetite, 70: 119–26.

Holt, D.B. 2002. Why Do Brands Cause Trouble? A dialectical theory of consumer culture and branding. Journal of Consumer Research, 29: 70–90.

Honkanen, P. and L. Frewer. 2009. Russian Consumers' Motives for Food Choice. Appetite, 52: 363–71.

Hughner, R.S., P. McDonagh, A. Prothero, C.J. Shultz and J. Stanton. 2007. Who are organic food consumers? A compilation and review of why people purchase organic food. J. Consumer Behaviour, 6: 94–110.

Irala-Estevez, J.D., M. Groth, L. Johansson, U. Oltersdorf, R. Prattala and M.A. Martinez-Gonzalez. 2000. A systematic review of socioeconomic differences in food habits in Europe: consumption of fruit and vegetables. European Journal of Clinical Nutrition, 54: 706–714.

Jang, Y.J., W.G. Kim and I.S. Yang. 2009. Food-related Lifestyle Segments and Mature Consumers' Attitudes to Home Meal Replacement. Intl CHRIE Conference-Refereed Track http://scholarworks.umass.edu/refereed/Sessions/Saturday/12.

Järvelä, K., J. Mäkelä and S. Piiroinen. 2006. Consumers' everyday food choice strategies in Finland. International Journal of Consumer Studies, 30: 309–316.

Johansen, S.B., T. Næs and M. Hersleth. 2011. Motivation for choice and healthiness perception of calorie-reduced dairy products. A cross-cultural study. Appetite, 56: 15–24.

Kaur, M. and A.M. Hegde. 2008. Are we aware of what we are, we are what we Eat—An Epidemiological Survey. International Journal of Clinical Pediatric Dentistry, 1: 13–6.

Kennedy, E.T. 2004. Dietary diversity, diet quality, and body weight regulation. Nutr. Rev., 62: S78–81.

Kennedy, E.T. 2005. The global face of nutrition: What can governments and industry do? J. Nutr., 135: 913–5.

Kennedy, J., V. Jackson, C. Cowan, I.B. David and M.D. Bolton. 2005. Consumer food safety knowledge: segmentation of Irish home food preparers based on food safety knowledge and practice. British Food Journal, 107: 441–52.

Kesic, T. and S. Piri-Rajh. 2003. Market segmentation on the basis of food-related lifestyles of Croatian families. British Food Journal, 105: 162–74.

Kinsey, J.D. 1994. Food and families' socioeconomic status. J. Nutr., 124: 1878S–1885S.

Kinsey, J. 2000. The Changing Global Consumer. Presented at the 2000 IAMA World Food and Agribusiness Congress. Chicago, IL.

Kolodinsky, J., J.R. Harvey-Berino, L. Berlin, R.K. Johnson and T.W. Reynolds. 2007. Knowledge of current dietary guidelines and food choice by college students: better eaters have higher knowledge of dietary guidance. J. Am. Diet. Assoc., 107: 1409–13.

Krebs-Smith, S.M. and L.S. Kantor. 2001. Choose a variety of fruits and vegetables daily: understanding the complexities. The American Society for Nutritional Sciences, 131: 4875–5015.

Krosnick, J.A. and S. Presser. 2010. Questions and questionnaire design. In: Handbook of Survey Research. 2nd Edition. Chapter 9, pp. 263–313. Emerald Publishers, New York.

Larson, N. and M. Story. 2009. A review of environmental influences on food choices. Ann. Behav. Med., 38: S56–S73.

Leikas, S., M. Lindeman, K. Roininen and L. Lahteenmaki. 2007. Food risk perceptions, gender, and individual differences in avoidance and approach motivation, intuitive and analytic thinking styles, and anxiety. Appetite, 48: 232–240.

Lindeman, M. and K. Stark. 1999. Pleasure, pursuit of health or negotiation of identity? Personality correlates of food choice motives among young and middle-aged women. Appetite, 33: 141–161.

Macintyre, S., J. Reilly, D. Miller and J. Eldridge. 1998. Food choice, food scares and health: the role of the media. In: A. Murcott (ed.). The Nation's Diet: The Social Science of Food Choice. London: Longman Publishers.

Maillot, M., N. Darmon, M. Darmon, L. Lafay and A. Drewnowski. 2007. Nutrient-Dense food groups have high energy costs: an econometric approach to nutrient profiling. J. Nutr., 137: 1815–1820.

Mäkelä, J. 2000. Cultural definitions of the meal. pp. 7–18. In: H.L. Meiselman (ed.). Dimensions of the Meal. The Science, Culture, Business, and Art of Eating. Amsterdam: Aspen Publishers.

Manuel, D.G., S. Neamatullah, R. Shahin, D. Reymond, J. Keystone, J. Carison, C. Le Bar, B.L. Herwaldt and D.H. Werker. 2000. An outbreak of cyclosporiasis in 1996 associated with consumption of fresh berries-Ontario. Can J. Infect. Dis., 11: 86–92.

Mayen, A.L., P. Marques-Vidal, F. Paccaud, P. Bovet and S. Stringhini. 2014. American Journal of Clinical Nutrition, 100: 1520–1531.

McEntire, J. 2013. Foodborne disease: the global movement of food and people. Infect. Dis. Clin. North Am., 3: 687–93.

Medeiros, L.C., V.N. Hillers, G. Chen, V. Bergmann, P. Kendall and M. Schroeder. 2004. Design and development of food safety knowledge and attitude scales for consumer food safety education. J. Am. Diet. Assoc., 104: 1671–1677.

Motarjemi, Y. and F.K. Kaferstein. 1997. Global estimation of foodborne diseases. World Health Stat Q, 50: 5–11.

Neff, R.A., A.M. Palmer, S.E. Mckenzie and R.S. Lawrence. 2009. Food systems and public health disparities. Journal of Hunger & Environmental Nutrition, 4: 282–314.

Nie, C. and L. Zepeda. 2011. Lifestyle segmentation of US food shoppers to examine organic and local food consumption. Appetite, 57: 28–37.

Niva, M. 2008. Consumers and the Conceptual and Practical Appropriation of Functional Foods. National Consumer Research Centre, Helsinki.

O'Sullivan, C., J. Scholderer and C. Cowan. 2005. Measurement equivalence of the food related lifestyle instrument (FRL) in Ireland and Great Britain. Food Quality and Preference, 16: 1–12.

Oakes, M.E. and C.S. Slotterback. 2002. The good, the bad, and the ugly: characteristics used by young, middle-aged, and older men and women, dieters and non-dieters to judge healthfulness of Foods. Appetite, 38: 91–97.

Papadopoulos, A., J.M. Sargeant, S.E. Majowicz, B. Sheldrick, C. McKeen, J. Wilson and C.E. Dewey. 2012. Enhancing public trust in the food safety regulatory system. Health Policy, 107: 98–103.

Patil, S.R., S. Cates and R. Morales. 2005. Consumer food safety knowledge, practices, and demographic differences; finding from meta-analysis. J. Food Prot., 68: 1884–1894.

Patrick, H. and T.A. Nicklas. 2005. A review of family and social determinants of children's eating patterns and diet quality. J. Am. Coll. Nutr., 24: 83–92.

Pérez-Cueto, F.J.A., W. Verbeke, M. Dutra de Barcellos, O. Kehagia, G. Chryssochoidis, J. Scholderer and K.G. Grunert. 2010. Food-related lifestyles and their association to obesity in five european countries. appetite, 54: 156–162.

Peterson, S., D.P. Duncan, D.B. Null, S.L. Roth and L. Gill. 2010. Positive changes in perceptions and selections of healthful foods by college students after a short-term point-of-selection intervention at a dining hall. J. Am. Coll. Health, 58: 425–431.

Pinstrup-Anderson, P. 2014. Food systems and human nutrition: relationships and policy interventions. *In*: Brian Thompson and Leslie Amoroso (eds.). Improving Diets and Nutrition: Food-based Approaches, 1st Edition. Food and Agriculture Organization of the United Nations (FAO), Rome, Italy.

Ploeg, M.V., L. Mancino, J.E. Todd, D.M. Clay and B. Scharadin. 2015. Where do americans usually shop for food and how do they travel to get there? Initial findings from the national household food acquisition and purchase survey, USDA. www.ers.usda.gov/ publications/eib-eco-nomic-information-bulletin/eib138.

Popkin, B.M. 2014. Synthesis and implications: China's nutrition transition in the context of changes across other low- and middle-income countries. Obes. Rev., 1: 60–7.

Riches, G. 1997. Hunger, food security and welfare policies: issues and debates in First World societies. Proceedings of Nutrition Society, 56: 63–74.

Roberto, C.A., B. Swinburn, C. Hawkes, T.T. Huang, S.A. Costa, M. Ashe, L. Zwicker, J.H. Cawley and K.D. Brownell. 2015. Patchy progress on obesity prevention: emerging examples, entrenched barriers, and new thinking. Lancet, 385: 2400–2409.

Rose, D., J.N. Bodor, P.L. Hutchinson and C.M. Swalm. 2010. The importance of a multi-dimensional approach for studying the links between food access and consumption. The Journal of Nutrition, 140: 1170–1174.

Rothman, A.J., P. Sheeran and W. Wood. 2009. Reflective and automatic processes in the initiation and maintenance of dietary change. Ann. Behav. Med., 38: S4–17.

Ryan, I., C. Cowan, M. McCarthy and C. O'Sullivan. 2002. Food-related lifestyle segments in ireland with a convenience orientation. Journal of International Food & Agribusiness Marketing, 14: 29–47.

Scheffer, M. 2009. Critical Transitions in Nature and Society. *In*: Princeton Studies in Complexity. 1st Edition. Princeton University Press, New Jersey.

Scholderer, J., K. Brunsø, L. Bredahl and K.G. Grunert. 2004. Cross-cultural validity of the foodrelated lifestyles instrument (FRL) within Western Europe. Appetite, 42: 197–211.

Schwarz, B. 2005. The Paradox of Choice. Why More is Less? New York: HarperCollins Publishers.

Sobal, J. and C.A. Bisogni. 2009. Constructing Food Choice Decisions. Ann. Behav. Med., 38: S37–46.

Spence, M., M.B. Livingstone, L.E. Hollywood, E.R. Gibney, S.A. O'Brien, L.K. Pourshahidi and M. Dean. 2013. A qualitative study of psychological, social and behavioral barriers to appropriate food portion size control. Int. J. Behav. Nutr. Phys. Act., 10: 92.

Steptoe, A. and T.M. Pollard. 1995. Development of a Measure of the Motives Underlying the Selection of Food: the Food Choice Questionnaire. Appetite, 25: 267–284.

Sun, Y.C. 2008. Health concern, food choice motives, and altitudes toward healthy eating: The mediating role of food choice motives. Appetite, 51: 42–49.

Sunguya, B.F., K.I. Ong, S. Dhakal, L.B. Mlunde, A. Shibanuma, J. Yasuoka and M. Jimba. 2014. Nutr. J. 13: 65.

Szakaly, Z., V. Szente, G. Kover, Z. Polereczki and O. Szigeti. 2012. 58: 406–413.

U.S. Department of Agriculture—Economic Research Service (USDA-ERS). 2005. ERS briefing room on food cpi, prices and expenditures. http://ers.usda.gov/Briefing/CPIFoodAndExpenditures/.

U.S. Department of Agriculture. Scientific report of the 2015 dietary guidelines advisory committee. http://www.health.gov/dietaryguidelines/2015-scientific-report/.

U.S. Department of Labor, Bureau of Labor Statistics. 2005. American time use survey—2004 results announced by BLS. http://www.bls.gov/tus/home.htm#news.

Wangberg, S., H. Andreassen, P. Kummervold, R. Wynn and T. Sorensen. 2009. Use of the internet for health purposes: trends in norway 2000–2010. Scandinavian Journal of Caring Sciences, 23: 691–696.

Wethington, E. and W.L. Johnson-Askew. 2009. Contributions of the life course perspective to research on food decision making. Ann. Behav. Med., 38: S74–S80.

Wikström, S. 2000. Consumers and consumption in transition. *In*: Academic Perspectives on the Future of the Consumer Goods Industry, ECR Europe Academic Partnership.

Wilcock, A., M. Pun, J. Khanona and M. Aung. 2004. Consumer attitudes, knowledge and behavior: a review of food safety issues. Trends in Food Sci. Tech., 15: 55–66.

Wycherley, A., M. McCarthy and C. Cowan. 2008. Speciality food orientation of food related lifestyle (FRL) segments in Great Britain. Food Quality and Preference, 19: 498–510.

Zafiriou, M. 2005. Food retailing in Canada: Trends, dynamics and consequences. Agriculture and Agri-Food Canada. Paper presented at the Pacific Economic Cooperation Meetings, Pacific Food Systems Outlook. Kunming, China.

Zukin, S. and J.S. Maguire. 2004. Consumers and consumption. Annual Review of Sociology, 30: 173–197.

11

Risk and Hazard Analysis of Foods

Malik Altaf Hussain

Introduction

Food safety is an emerging global challenge. There are many challenges to achieve supply of safe food for a domestic or an international market. One of the biggest problem is continuously changing global patterns of food production and consumption. These problems may include climate change, international trade, new technology development and growing population. These factors are also closely linked to ensure supply of nutritionally balanced and safe food to the world population. In other words 'food security' is similarly an emerging global problem.

A wide range of hazards associated with food are posing risks to health and obstacles to international food trade. Different food items may carry risks from potential hazards that may come with them, therefore, the risks must be assessed and managed to ensure consumer's safety. Risk analysis is a science-based systematic and disciplined approach to make food safety decisions easier. Hazard analysis and risk analysis are two important steps of the risk assessment process (Table 1). Risk analysis is a powerful tool that helps carrying out analysis to solve complex food safety problems. Risk analysis is generally conducted around a specific hazard or a combination of hazards.

Hazard is a situation or an agent that can cause a threat to life, health and environment. Normally hazards exist in dormant form and could create to an emergency situation only when they become active. Physical hazards, chemical hazards and biological hazards are three major types of hazards. Examples of physical hazards

Centre for Food Research and Innovation & Department of Wine, Food and Molecular Biosciences, Lincoln University, Christchurch, New Zealand.
Email: Malik.Hussain@lincoln.ac.nz

Table 1. Difference between risk analysis and hazard analysis.

Hazard analysis

Hazard analysis is the identification all possible hazards potentially created by a product, process or application. It is the first and necessary step to conduct risk assessment.

Risk analysis

Risk analysis is a systematic science-based approach to assess the probability and severity of the potential risks of an identified hazard becoming reality. It is the step after the identification of potential hazards in the risk assessment process.

---

are rocks, metal shavings and pieces of glass in foods. Benzoic acid, acrylamide and mercury are common chemical hazards that can be present in food products. Biological hazards are viruses, parasites, bacteria, and fungi. Temperature and time are critical in controlling biological hazards due to the growth of a pathogenic microorganism as well as its toxin production. Salt, sugar and brine solution are commonly used to preserve food by altering the moisture and regulating the acidity of certain foods. Addition of these compounds make foods more shelf stable by limiting microbial growth.

Presence of one or more hazards could create a hazardous situation that is also known as a foodborne outbreak or a food scare. Routine product testing or consumer's complaints can become the first source of information whether a hazard is present in the finished product. Epidemiological and historical data as well as past foodborne outbreaks are also sources of important information. Cross-contamination through unhygienic hands, raw foods and surfaces like cutting boards and cleaning cloths can transport harmful substances to ready-to-eat (RTE) foods. Therefore, each element of the food supply chain—raw material, processing, packaging, storage, and distribution—has the potential to become a source of different hazards. There are numerous food safety management systems like HACCP (Hazards Analysis and Critical Control Points), FSP (Food Safety Plan), RMP (Risk Management Programme), and WSMP (Wine Standards Management Plans) to manage these food hazards.

Hazard analysis is an important process to protect consumers from any adverse health effect due to consumption of food. Identification of a significant hazard in a food means control measures are needed to manage the hazard. Control measure is a term referred to any activity that is used to eliminate a food hazard or at least reduce it to an acceptable level. Modification of product formulation, better control on processing steps and improvement in plant operations are the examples of the manipulation that can be used as a part of control measure. Thermal processing is a control measure in killing vegetative cells of pathogens in the high risk foods. Higher temperature may kill the microorganisms whereas lower temperature normally prevents germination and growth of microbial cells. Control measures like irradiation, acidification, fermentation, refrigeration, filtering, metal detectors, X-ray devices and many other methods are available to eliminate or reduce hazards in food. Rejecting ingredients with high concentration of a harmful chemical is another way to apply control measures (Food Safety Cooperation Forum, 2012).

Identification and detection of a hazard is a primary step toward risk analysis. In food safety, risk analysis plays two important roles; helping in the production of the highest quality food products and ensuring the safety and protection of public health (Renwick et al., 2003). It also assists to comply with international and national standards and market regulations. An effective risk analysis should strengthen the food safety systems and reduce the foodborne illnesses. Risk analysis has three distinct but closely connected components: risk assessment, risk management, and risk communication (see Section on Risk Management and Communication in this chapter).

Different Types of Hazards

Hazards are broadly classified into three groups—biological hazards, chemical hazards and physical hazards (Table 2).

A detailed description of these groups is given in the following sections.

Table 2. Basic definitions of biological, chemical and physical hazards.

Biological hazard

A biological hazard is an agent of biological nature (viruses, bacteria and parasites) that can transmit through food or water and can harm human health or even cause death.

Chemical hazard

A chemical hazard is a toxic compound that can enter into food through production system and could pose acute or chronic health risk to the consumer.

Physical hazard

A physical hazard is an object that is foreign in nature and present in food with the potential to cause injury or health risk to the consumer.

Biological hazards

A diverse range of biological agents such as viruses, parasites, bacteria, protozoa and nematodes could be associated with different foods (Table 3). Any biological agent that is capable of growing in food or may be carried through food and can cause illness is generally regarded as a 'biological hazard'. Common examples of biological hazards are the detection of pathogenic microorganisms like *Campylobacter*, *Listeria* and *Salmonella* in foods. This specific type of biological hazard is also known as 'microbiological hazard'. Norwalk type viruses are the most common example of foodborne viruses. *Giardia intestinalis* is an example of a protozoan that may be a biological food hazard. It is the most common cause of parasitic gastrointestinal disease; estimates show that 20,000 cases of giardiasis occur each year in the USA alone and its prevalence in the world's population is between 20 and 30%. It is important to have some understanding of the risks associated with different types of microbiological hazards.

Viruses

They represent the simplest and the smallest form of life on our planet. One of the key features of viruses is the lack of ability to reproduce outside a living cell. However, viruses are quite able to survive for a period of time in inanimate objects like door

Table 3. Common biological hazards and their association with specific foods.

Biological agent	Organism	Associated foods or food products
Bacteria	*Campylobacter*	Chicken
	Clostridium botulinum	Pork, beef, chicken
	Clostridium difficile	Seafood, coconut milk
	Clostridium perfringens	Egg, tuna, chicken
	Cronobacter	Chicken, egg, red meat
	Escherichia coli	Beef, milk, egg
	Listeria monocytogenes	Crab, shrimp, fish
	Salmonella	Chicken, egg, red meat, fish, vegetables
	Shigella	Meat, poultry, fish, egg
	Staphylcoccus aureus	Beef, pork, turkey, egg
	Vibrio parahaemolyticus	Pork sausage, chicken
	Vibrio vulnificus	Pork, veal, lamb, beef
	Yersinia enterocolitica	Poultry, lamb, beef
Parasite	Anisakids	Fish
	Cryptosporidium	Chicken, pork, mince
	Cyclospora	Beef, pork, lamb
	Entamoeba	Beef, pork, chicken
	Giardia	Beef, pork, chicken
	Toxoplasma	Beef, pork
	Trichinella	Pork
Viruses	Hepatitis A	Shellfish
	Hepatitis E	Pork, deer meat
	Norovirus	Shellfish, RTE foods
Fungus	Mycotoxin	Corn, Peanut, Wheat, Barley
Others	Prions	Beef

handles. When a virus enters into a host cell, it reproduces quickly and causes disease. Some viruses are resistant to heat and cold. Food can act as a transmission mode for viruses from one host to another; however, viruses do not multiply in food. The most common viruses in food are hepatitis A and norovirus.

Viruses are responsible for the large number of foodborne illnesses and deaths throughout the world. Centers for Disease Control and Prevention (CDC) reported that norovirus caused 58% of foodborne cases in 2011 in the USA (Guiterrez, 2013). Norovirus is highly contagious and has very low infectious dose, i.e., only 18 virus particles are enough to infect a person. It can cause symptoms like diarrhea, vomiting and stomach pain. Typical transmission route for norovirus is from stool or vomit of an infected person or from food products that are harvested from contaminated water. Oyster, fruit and vegetables are common food sources of norovirus. This virus can survive to 60°C and quick steaming process commonly used for cooking shellfish may not be efficient to destroy the viral particles. CDC recommends use of chlorine bleach solution with 1000–5000 ppm (5.25% per gallon of water) concentration to disinfect contaminated area (CDC—Norovirus Illness: Facts for Food Handlers).

Bacteria

Food is an ideal medium for the growth of harmful bacteria. The bacterial population can become double in every 10–30 minutes under favorable conditions. Several sources (human, utensil, air, soil, surfaces, etc.) can contaminate food with pathogenic bacteria easily. A single cell can become billions in 10–12 hours. Consumption of harmful bacteria in large numbers can cause symptoms like diarrhea, abdominal cramps, headache, chills, vomiting, and jaundice. *Campylobacter, Clostridium, Escherichia coli, Salmonella* and *Staphylococcus* are the most common pathogens in food. Bacteria also produce toxin and the toxin may still be present even though the bacterial cells killed during cooking, freezing, smoking or curing. Good personal hygiene is a very effective way to reduce contamination levels of harmful microorganisms.

    Salmonella is one of the most dangerous known foodborne pathogen. According to CDC *Salmonella* is responsible for 11% cases of foodborne illnesses in the USA—the second leading cause after norovirus (CDC, 2011). A vast majority of bacteria prefer to grow on food with a pH range of 4.6 to 7.5, under aerobic condition, in the presence of moisture and at temperatures ranging from 5 to 57°C (Guiterrez, 2013). *Salmonella* can be eliminated through cooking and pasteurization. Eggs, poultry, meat, unpasteurized milk, unpasteurized juice, cheese, raw fruits, raw vegetables, spices, and nuts are the most common sources. An infection with *Salmonella* is clinically known salmonellosis with typical symptoms such as diarrhea, fever, abdominal cramps and vomiting. One of the species *Salmonella enterica* subsp. *enterica* contains over 60% of the total number of serovars and 99% of the serovars that are capable of infecting cold and warm blooded animals as well as humans.

Parasites

They need a host to live and survive. Examples of parasites are *Trichinella spiralis* in pork and larval Anisakids roundworm in fish. Refrigeration, the most commonly used preservation technique for meat and seafood, does not kill parasites. On the other hand, freezing can be effectively used to kill parasites present in meat or seafood. For example parasites are killed in frozen seafood (blast frozen to –35°C in 15 hours or regular frozen to –20°C in seven days). Besides freezing the food product, anisakiasis can be prevented by cooking higher above 60°C but salting and marinating do not eliminate the parasite (John and William, 2006). *Asisakis simplex* is a common parasite found in raw and undercooked seafood and can cause tingling in the throat and cause the person to cough up worms (Guiterrez, 2013). Fish, squid and crustaceans, generally perform as a transport host. This means the larva grows in their bodies but do not mature into adults. When larger marine mammals like seals, sea lions, dolphins and whales consume the smaller fish or crustacean, the larva then grows into adult worms. This cycle will continue through the eggs of the worm passed from the marine mammal's feces, hatched into larva and consumed by crustacean or fish and so on. In humans, however, the parasites cannot mature and reproduce thus causing pain in the body. Japan, Netherland and Spain have the most human infections with around 14,000 cases reported since 2000.

Food contaminations occur through various channels and most of the time due to contact with people, pests, dirty equipment, or raw foods. Raw meat, milk, fish and poultry are grouped as 'high risk foods' and thus have higher probability to be contaminated with harmful bacteria. One way to maintain lower microbial counts is to buy these foods from reputable suppliers and keeping them at refrigerated temperature. The outer surface of the meat cut generally carries high microbial loads as compared to inner tissues. However, in the case of minced meat product like burger patties, the microbial population is uniformly distributed throughout the product. Whole chicken and fish on the other hand have higher number of bacteria in the center part near the intestine. This is the reason why cooking them thoroughly is crucial and the center temperature needs to reach 75°C for destruction of harmful bacteria. Some bacteria have the ability to form spores to cope with harsh environmental conditions such as heat treatment. A spore is a seed-like coated structure that can germinate when the surrounding environment becomes favorable for growth. Therefore, special measures must be taken to manage spore-forming bacteria.

Chemical hazards

Chemical hazards are diverse and complicated and could cause irreversible damage to the human body. They can enter into the food chain through different sources, so it is really important to test foods for the presence of potential chemical hazards. Examples of toxic chemicals include excessive amounts of heavy metals, pesticides, herbicides, insecticides, vitamins, minerals, preservatives, disinfectants, detergents and cleaning compounds. Chemical contamination can occur when cleaning chemicals, rodent baits or insecticides gets into the food. The most common cause of chemical contamination is poor manufacturing practices, i.e., storing chemicals near food ingredients, spillage in food areas, cleaning residues on cleaned surfaces and cleaning equipment that is too near to food preparation areas. Table 4 shows different types of chemical hazards routinely detected in specific foods.

Chemical hazards can also be divided into three groups:

1. **Naturally occurring:** Mycotoxins (e.g., aflatoxin), scombrotoxin, ciguatoxin
2. **Unintentionally added:** Agricultural chemicals, heavy metals (e.g., lead, zinc)
3. **Intentionally added:** Preservatives (e.g., nitrite and sulfiting agents), nutritional additives (e.g., niacin, vitamin A)

Ciguatoxins

This is an excellent example of naturally found toxins in a food product. A specific seafood toxin from pufferfish, moray eels and freshwater minnows is called ciguatoxin. Its detection through smell and taste is not possible. It cannot be destroyed as well, thus the only way to avoid the toxin is by buying seafood from regularly inspected

Table 4. Examples of chemical hazards in foods.

Chemical hazard type	Specific example	Common food products
Food additives	Sodium chloride (preservatives)	Shrimp, lobster
	Benzoic acid (preservatives)	Fruit juice, soft drink, pickle
	Ascorbic acid (colour retention agents)	Fruit juice, beverage
	Sodium metabisulphate (preservatives)	Shrimp, lobster
Environmental pollutants	Arsenic (heavy metals)	
	Mercury (heavy metals)	Canned tuna, shellfish
	Lead (heavy metals)	Tomato sauce, juice, soda
	Dioxin (organic pollutants)	Livestock, fish, shellfish
Chemical from food processing	Acrylamide	Potato chips, french fries
	Ethyl carbamate	Wine, beer, fermented food
	Furan	Canned food
Pesticides and veterinary drug residues	Azoxystrobin	Peaches
	Clethodim	Beans
	Thiocarbamate	Apples
	Drug residues	Animal products
Natural toxin	Oxalic acid	Rhubarb
	Aflatoxin	Peanut, treenut, fig, rice
	Ciguatoxins	Baraccuda, grouper, eel
	Saxitoxin	Oysters, clams, scallops kles, whelks
	Tetrodotoxin	Pufferfish, octopus
Others	Phytohaemagglutinin	Red kidney bean
	Allergen (peanut, egg, milk, gluten)	Biscuit, bread, protein bar
	Biotechnological hazards	GM foods

supplier (Guiterrez, 2013). An alga, the marine dinoflagellate *Gambierdiscus toxicus*, is the principal known source of ciguatoxins. *Gambierdiscus toxicus* is associated with seaweeds, sediments and dead coral. This group of biotoxins accumulates in certain fish species that feed on toxic algae, or prey on toxic herbivorous fish species.

Sodium chloride and sodium metabisulphite

These are examples of intentionally added chemicals in foods where they are not expected to be present. This can also be regarded as food adulteration and a common problem in many food products. Seafood products like shrimp and lobster often contain chemical additives like sodium chloride and sodium metabisulphite that can delay post-mortem changes and appeals to consumers as fresh product. Food processing companies should test some of the raw samples to ensure the additives amount is not higher than 100 ppm.

Acrylamide

It is an example of unintentionally added chemical hazard. There are certain chemicals (like acrylamide) that are produced during normally processing of some food products. For example, there could be formation of high amounts of acrylamide and trans fatty acids during frying of potato chips or barbequing meat. Both acrylamide and trans fat acids have shown adverse health effectives; acrylamide is carcinogenic (Tareke et al., 2002) and trans fatty acids can lead to coronary heart disease (Benatar, 2010).

Physical hazards

Physical hazards are the foreign objects that get into food, or are already present in food, and may cause illness, injury or distress to the person eating it. Extraneous material, usually non-toxic but represents unsanitary conditions, may be found in a food that are foreign to that particular food. Physical contamination in food by foreign bodies is a common occurrence through sources such as packaging, broken equipment, storage area and people. For example, glass, wood, hair, jewelry, insects and metals can accidently dropped or entered into food during processing (see Table 5). Some

Table 5. Physical hazards identified in different foods.

Extraneous material	Common foods or food product
Plastic	Bread, fish ball, crab stick
Glass pieces, fragments, particles	Bottled beer, jam, chutney
Wood	Nuts, peanut, canned fruit
Industrial rubber	Processed food
Metal pieces *(needles, knife blades and mesh gloves)*	Minced beef, pork, chicken
Stones	Rice, whole-wheat
Shell fragments	Crab, shrimp, oyster
Cleaning equipment *(sponges, cloth, bristles, and tissue)*	Processed food
Packaging material	Processed food
Metal twist ties, elastic bands, string	Processed food
Infestations, insect and rodent dropping	Fresh fruit and vegetables
Medications	Livestock, fruit, vegetables
Band-aid, finger cots or glove	Processed food
Tartrate crystals	Wine, grape juice
Struvite (mineral deposits)	Canned fish product
Life infestations	Fresh fruit and vegetables
Bone fragments	Chicken, fish, turkey
Others *(pen, keys, paper clips, staples, and jewelery)*	Cereal, potato chips

objects like a piece of glass, metal fragments or bone could cause serious injuries. For instance, hard and sharp objects can cause damage to the mouth, teeth, gums, throat, and intestine. Other contaminants such as hair or insects may be offensive but not necessarily a danger to health hence should be considered seriously and controlled.

Though health consequences of physical hazards are far less dangerous than microbiological hazards or some extremely toxic chemicals but occurrence of physical hazards in food can trigger food recall that can affect a brand image and cost huge financial loss. Manufactured products like infant formulas and beverage for specific or vulnerable consumers (infants, elderly and pregnant women) are at high-risk level and physical hazard must be managed strictly in this case. It requires very simple measures to prevent physical contamination. Some examples are given below:

- Raw materials and food ingredients should be inspected thoroughly for field contaminants like stones in cereals.
- Food handling should be done according to good manufacturing practices (GMPs).
- Food handlers should not use any jewelry, artificial nails or hairnets in processing plants.
- Potential sources of physical hazards should be minimized in the processing and storage area.
- Using the protective acrylic bulb or lamp covers to prevent breakable glass contamination.
- Installing a metal detector where potential of metal contamination is higher.

Physical hazard can cause very serious health problems indirectly. Classical example can be hair, urine and droppings left by a rodent in the food can carry disease like hantavirus (Lee et al., 1982). Therefore, an effective pest control programme is essential in preventing such infestation. Vermins like cockroaches, flies and other insects can cause hazards by falling into the food and leaving behind droppings that contain million of pathogens. The vermin normally moves around dirty places like garbage and then moves to sanitized area like the food preparation table and thus are a source of cross-contamination of pathogens. Other examples are non-typical where people were hospitalized because they accidentally ingested bristles from wire brushes used for cleaning grills. In some extreme cases the patients had gone through emergency surgery to repair internal bleeding. A bristle can puncture the lining of the intestinal wall and allow bacteria to enter the bloodstream and cause infection.

Prevention of hazards

It is important to know the hazards and their possible sources in foods; however, effective detection and elimination systems are far more crucial to minimize any hazard occurrence and protect the consumer's health. Table 6 lists examples of different prevention techniques to eliminate or reduce hazards. Using optical system can offer several benefits in detecting a hazard. This system recognizes object's color, size, shape, structural properties and chemical composition (Sun, 2007). Optical system can also detect pathogen microorganisms in several hours whereas plating system takes several days to produce results. Faster result enables the manufacturer to ship

Table 6. Different types of hazard prevention.

Prevention	Hazard
Physical Exclusion	
Heat: boiling, frying, baking, roasting, steaming	Biological
Pest control	Physical
Sanitizing equipment	Biological, chemical
Chilling or freezing	Biological
Irradiation	Biological
Filter/sieve	Biological, physical
Brine solution	Biological
Detection Methods	
Metal detectors	Physical
Optical systems	Physical, chemical, biological
Magnet	Physical
X-ray machines	Physical
Screening of raw materials, conveyor belt, end product	Physical
Others	
Proper equipment design and calibration	Physical, chemical, biological
Employee training program	Physical, chemical, biological
Facility maintenance	Physical, chemical, biological
Good manufacturing practices	Physical, chemical, biological

their product straightaway thus increasing profit from low storage cost and speedier production. There are optical systems that can be used for allergen detection through optical thin-film biosensor chips and results are seen through color change. The equipment can detect food allergens like soybean, wheat, peanut, cashew, shrimp, fish, beef, and chicken (Wang et al., 2011). Elimination of potential sources of extraneous material within the establishment is also important, especially to prevent physical hazard contamination.

Use of durable materials (such as aluminum or plastic) could help to prevent glass shattering. Food could be rejected if metal is detected, therefore use of magnets and metal detectors is recommended. The equipment for detection should be properly maintained to ensure accurate functioning. X-ray machines and food radar system can be used to identify hazards like stones, bones, hard plastic and metal.

Hazard Analysis Approaches

A diverse range of food safety management systems including HACCP, FSP, RMP and WSMP are designed to assess hazards using different methodological approaches. Ultimate objective of all these systems is same, i.e., to eliminate or reduce occurrence of a hazard. One of the most widely used and universally accepted systems is HACCP (or a food safety system based on HACCP principles). The purpose of HACCP is to find, correct and prevent hazards throughout the food production process. This system uses an approach to identify critical control points (CCPs) in a food production or processing system as a whole or part of it. A CCP is generally referred to any step that is essential to prevent or eliminate a food hazard to an acceptable level. In reality accurate determination of CCPs is the backbone of a HACCP plan and normally done through the help of a decision tree (Fig. 1) (Hellier, 2000). The HACCP team needs to regularly evaluate the hazard analysis as well as the flow diagram (Schmidt and Newslow, 2013). The CCP decision tree questions whether there are control measures available for the hazard, whether control measures can eliminate hazard to an acceptable level, whether hazard can increase to an unacceptable level, and whether subsequent steps will eliminate the hazard to an acceptable level.

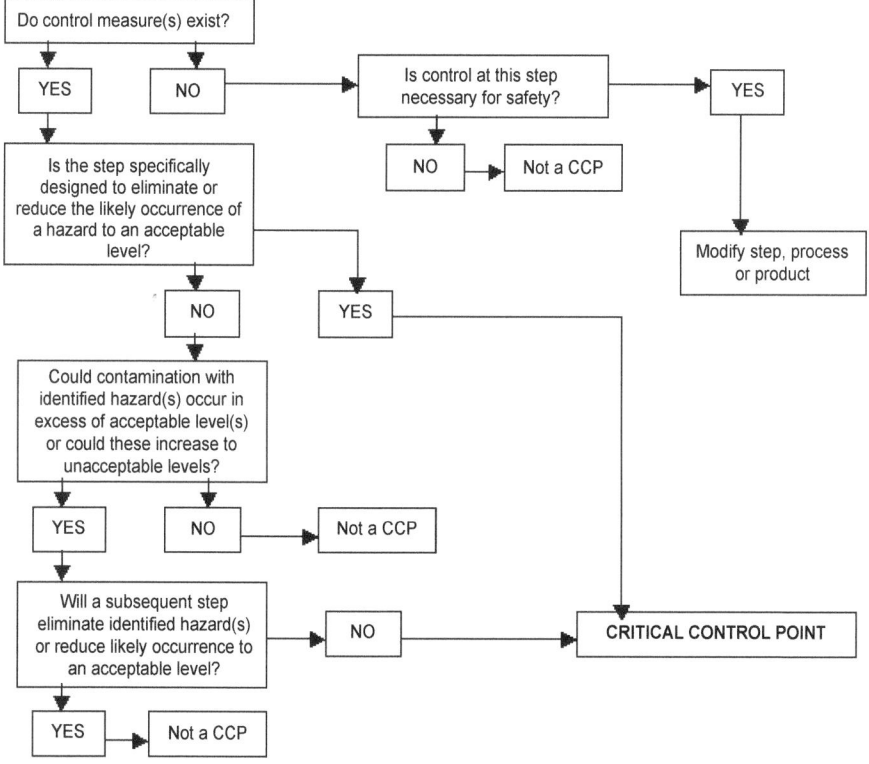

Figure 1. Critical control point (CCP) determination decision tree (Adapted from Hellier, 2000).

Risk Analysis

Risk analysis is the science-based assessment of risks associated with an identified hazard. By definition it is a process of identifying the potential risks and analyzing what could happen if a hazard occurs. Risk analysis in food industry is meant to be the scientific evaluation of known or potential adverse health effects resulting from human exposure to foodborne hazards. Risk assessments of a hazard can be qualitative or quantitative in order to estimate the probability and severity of illness from consuming food carrying a biological, chemical or physical hazard (Magnússon et al., 2012). Risk assessment helps to identify the possible entry points of food hazards in the supply chain and inform the interventions that will have the greatest impact on reducing risk. The process of risk analysis of a hazard has four steps as described in Fig. 2.

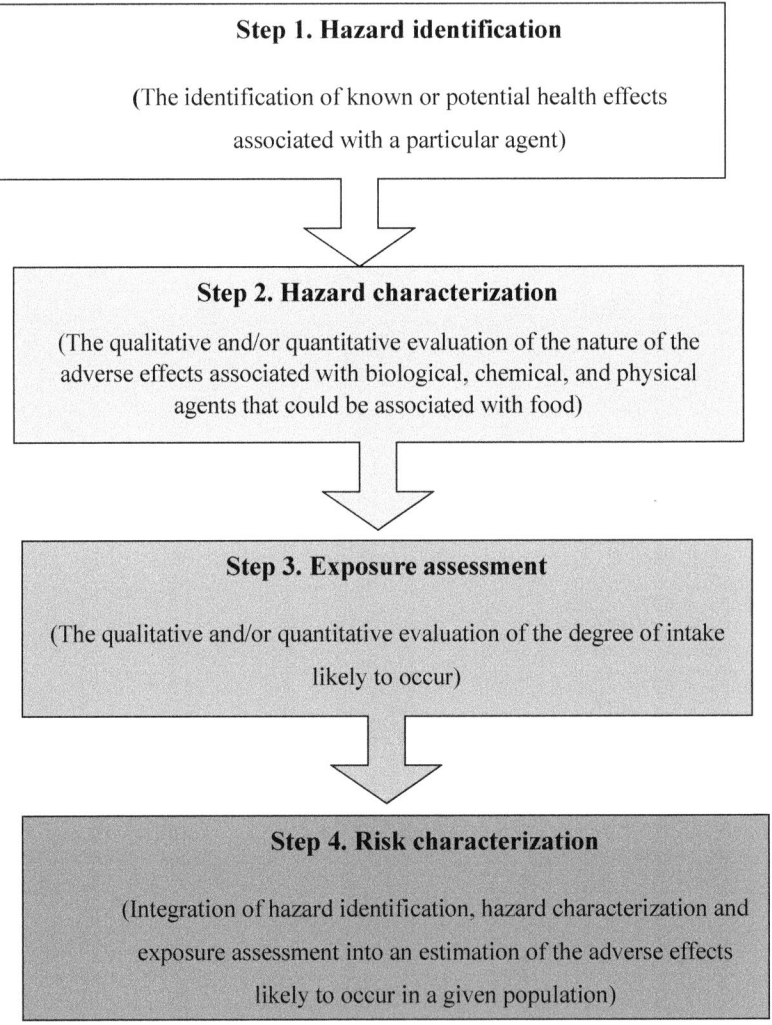

Figure 2. Four major steps involved in risk analysis of a food hazard (Whitehead and Field, 1992).

Hazard identification

This is the first and an important step in risk analysis of food or food products. This step helps to identify the possible microbiological, chemical and physical hazards that can occur at any stage of the food supply chain. Hazard identification covers entire food system, i.e., production, processing, manufacture, storage and distribution, and until fork. It is vital to consider how the customer might handle the food before consumption (FAO, 2006). Different types of hazards in food are discussed in the next section.

Hazard characterization

Hazard characterization is the second step in the process of risk analysis. This step looks at the characteristics of an identified hazard to provide information on the potential adverse health effects attributable to a specific hazard, the mechanisms by which hazards exert their harmful effects, and the associated dose, route of transmission, duration, timing of exposure, and their respective attendant uncertainties (Buchanan et al., 2000). Qualitative and, wherever possible, quantitative description of the inherent properties and the nature of the hazard (biological, chemical, or physical) associated with food systems can be available as a result of successful completion of hazard characterization step. Generally, a dose-response assessment must be performed for chemical agents; however, in case of biological or physical agents a dose-response assessment should be performed if enough data is available.

Hazard characterization is a step that can be a stand-alone process to understand the nature of a specific hazard or a component of risk analysis process of an entire food system. A hazard characterization done for a particular pathogen may serve as a common module or building block to conduct risk assessments for a variety of purposes and in an assortment of commodities. Once hazard characterization of a specific agent or situation developed in a part of the world, it could be used as base to conduct hazard characterization by combining an exposure assessment specific to a country or region. For an instance, a hazard characterization developed for water exposure may be adapted to a food exposure scenario by taking into consideration the food matrix effects. Adaptability is one of the key features of hazard characterizations. For example, the potential adverse health effects of *Listeria monocytogenes* as a hazard will stay quite similar in risk assessments conducted in different parts of the world. On other hand, exposure assessments are highly specific to the production, processing and consumption patterns within a country or region.

Exposure assessment

Exposure assessment is the third step in the risk analysis process and vital in a sense to determine rationale based possibility of potential adverse health effects. Exposure occurs through contact with a hazard by inhaling air, drinking water, eating food, or touching a variety of products that contain the hazard. Exposure assessment is widely used for risk assessment of chemical hazards.

This step attempts to answer some key questions about a hazard:

- Who or what is exposed (e.g., people, aquatic ecosystems)?
- What is the mode of exposure (e.g., breathing air, drinking water, skin contact or any other routes)?

- How much exposure occurs?
- How often and for how long does exposure occur, that is, what is its frequency and duration?

Exposure assessment has two important components, i.e., the concentration of the hazard and the extent of the contact. For risk analysis purposes, a quantitative exposure assessment approach is needed and exposure information must be clearly linked to the hazard identification and dose-response relationship. The information of an exposure assessment and a hazard assessment (characterization) of a chemical hazard are combined to determine the actual threat to health in risk analysis process. It is clear now that a hazard characterization provides an understanding of the potential for the chemical to cause adverse effects to humans and plant and animal life (EPA, 2003). Therefore, exposure assessment and hazard assessment are needed together in order to make conclusions about the likelihood of adverse effects in the exposed population.

Different population groups may need different levels of exposure assessment. Therefore, exposure assessment of a hazard should be carried out for different population groups for estimating the intake of the target users compared to the intake of the general population. It is recommended to calculate the respective prevalence of subjects with an exposure above and below the health-based guidance level (Bottex et al., 2008). Assessments for screening purpose are based primarily on readily available monitoring data and other data, conservative assumptions and simple models. These assessments are good to quickly prioritize further research work.

However, advanced assessments focus on higher priority exposures and attempt to represent actual environmental conditions and exposures. These types of assessments require well-designed monitoring studies, more data and more sophisticated models. Lack of high quality exposure data for certain subgroups of the population such as children, infants and neonates is challenging to conduct comprehensive risk exposure assessment.

Risk characterization

Risk characterization is the fourth and final step in the risk analysis process. A risk characterization reports in a comprehensive manner the nature and presence or absence of risks of a hazard (an agent or a condition). It also provides detailed information about how the risk was assessed, where assumptions and uncertainties still exist, and where policy choices will need to be made. Risk characterization covers both human health risk assessment and ecological risk assessment. According to EPA (2000) report "the risk characterization integrates information from the preceding components of the risk assessment and synthesizes an overall conclusion about risk that is complete, informative, and useful for decision makers".

In real situation where risk analysis is being conducted, each component of the risk assessment for example hazard assessment, dose-response assessment, exposure assessment has an individual risk characterization that should include findings, assumptions, limitations, and uncertainties. Then these sub-sets of individual risk characterizations provide the information basis to prepare an integrative risk characterization analysis (Magnússon et al., 2012). The final risk analysis document should contain overall risk characterization (the individual risk characterizations and an integrative analysis).

Principles of Conducting Risk Characterization

Risk characterization should be fully and systematically documented, and it is the basic principle. A good risk characterization covers the scope of the assessment, expresses results clearly, articulates major assumptions and uncertainties, identifies reasonable alternative interpretations, and separates scientific conclusions from policy judgments (Knudsen, 2010). Environmental Protection Agency (EPA) developed the following four principles for conducting risk characterizations (as listed in EPA, 2000):

1. **Transparency:** The characterization should fully and explicitly disclose the risk assessment methods, default assumptions, logic, rationale, extrapolations, uncertainties, and overall strength of each step in the assessment.
2. **Clarity:** The products from the risk assessment should be readily understood by readers inside and outside of the risk assessment process. Documents should be concise, free of jargon, and should use understandable tables, graphs, and equations as needed.
3. **Consistent:** The risk assessment should be conducted and presented in a manner, which is consistent with EPA policy, and consistent with other risk characterizations of similar scope prepared across programs within the EPA.
4. **Reasonable:** The risk assessment should be based on sound judgment, with methods and assumptions consistent with the current state-of-the-science and conveyed in a manner that is complete, balanced and informative.

The above described four principles (Transparency, Clarity, Consistency, and Reasonableness) are collectively referred to as TCCR. It is important to achieve TCCR in each risk characterization process. A process should have a formal record and summary of any constraints, uncertainties, and assumptions and their impact on the risk assessment. Therefore, the risk assessment step should be based on the same four principles as the risk characterization.

Risk Management and Communication

Risk management

Risk management needs development of procedures and protocols to address or handle risks of a hazard. Risk management is an integral part of overall risk analysis frame work. Risk analysis frame work has three tiers; risk assessment, risk management, and risk communication (see Fig. 3). A complete chapter is dedicated to discuss risk assessment and pathogen management in food (see Chapter 10). A documented plan called "Risk Management Programme (RMP)" is implemented for this purpose. An RMP is a written programme designed to manage the hazards (biological, chemical or physical) and risks linked to them. A well-designed RMP should assist food processing to meet the requirements of national standards and ensure the safety of the products. An RMP, which is developed for food safety risk management, must be functional

Figure 3. Risk analysis frame work.

in both strategic, long-term situations and in the shorter term work of national food safety authorities. The RMP implementation has two phases:

Phase 1: This phase of the RMP involves preliminary risk management activities such as identification of a food safety issue, risk profiling using available scientific information and initiation of an appropriate further action.

Phase 2: This phase of the RMP consists of identifying and evaluating a variety of possible options (e.g., controlling, preventing, reducing, eliminating or in some other manner mitigating) to manage the risk. Once risk management strategies have been developed, the relevant stakeholders must ensure that these are implemented, monitored and activities reviewed in an effective manner.

The objective of RMP is to determine whether the measures that were selected and implemented are in fact achieving the risk management goals they were meant to achieve, and whether they are having any other unintended effects (FAO, 2006).

Risk communication

The definition of risk communication given by FAO (2006) states "the interactive exchange of information and opinions throughout the risk analysis process concerning risk, risk-related factors and risk perceptions, among risk assessors, risk managers, consumers, industry, the academic community and other interested parties, including the explanation of risk assessment findings and the basis of risk management decisions". Risk communication is another vital component of the overall risk analysis frame work (Fig. 3). In fact, it is the most challenging part of the framework. It is not an easy task to develop an effective risk communication strategy. Over the last three decades, academics and practitioners have worked to develop a formula for risk communication and a practical tool for predicting how the public perceives risk. Many government agencies in different countries and the food industry have identified the need for effective risk communication to reduce risks in foods. Risk communication is a process that starts from the beginning and continues throughout the process of risk analysis. There are several communication challenges related to explaining scientific concepts, describing the concepts of variability and uncertainty and explaining why food safety decisions cannot be based on "zero risk" (Buchanan, 2011).

Importance of Risk and Hazard Analysis

Hazard analysis and risk analysis are activities of a food system to monitor and assess them on an ongoing basis. It is extremely important due to the fact that hazards are dynamic in nature or potential adverse effect of a risk on health may change. For example, food habits and food composition are changing widely and rapidly and may have an impact on the current hazard and risk analyses. The hidden risks present in each food should be identified, characterized and communicated to the consumer for public health. A few examples are given below to highlight the importance of risk and hazard analysis:

- There is a group of population who are really sensitive to food risks like infants, immune-compromised (pregnant women and chronically ill patients) and elderly peoples, so updated risk analysis is really important for these fragments of the society.

Another best example for the need of risk analysis can be demonstrated using the case of probiotics. Microbiological contaminants (i.e., pathogenic bacteria) in food are well-known for adverse effects on human health. However, many microorganisms are integral to various food productions and some of them are believed to have positive health effects. This specific group of microorganisms is termed as 'probiotics' and predominantly belong to genera *Lactobacillus* and *Bifidobacteria*. Probiotics are rarely associated with disease and are thought to have low pathogenicity (Bernardeau et al., 2008). Yet, there are reports that a small portion of the population who may be at risk of adverse effects following probiotic administration, i.e., through over-stimulation of the immune system (Sanders et al., 2010) and transfer of antibiotic resistance to pathogens (Courvalin, 2006). Therefore, evaluation of the safety status of foods or products generally perceived healthy for the general population should also have risk analysis done, especially for the vulnerable section of the population.

Conclusion

Increase in food safety scares and crisis over the last few decades has highlighted the importance of good hazard and risk analysis. A hazard-free or risk-free food is not available yet, which means that risk analysis should be a major part of food safety management. Though there are several limitations for risk analysis at the moment but overcoming the lack of good quality and comprehensive data on different type of hazards (biological, chemical and physical) is important for safe food production in future. Evaluation of any possible hazard to distinguish critical control point from mere control point will bring greater understanding of processing steps and control measures. CCPs determination is crucial to ensure that the hazard is eliminated to an acceptable level. In conclusion, hazard analysis and risk analysis provide the foundation of creating a proper food safety management system.

References

Benatar, J.R. 2010. Trans fatty acids and coronary artery disease. Open Access J. Clin. Trials, 2: 9–13.

Bernardeau, M., J.P. Vernoux, S. Henri-Dubernet and M. Gueguen. 2008. Safety assessment of dairy microorganisms: the *Lactobacillus* genus. Int. J. Food Microbiol., 126: 278–285.

Bottex, B., J.L.C.M. Dorne, D. Carlander, D. Benford, H. Przyrembel, C. Heppner, J. Kleiner and A. Cockburn. 2008. Risk-benefit health assessment of food—Food fortification and nitrate in vegetables. Trends Food Sci. Technol., 19: S113–S119.

Buchanan, R.L. 2011. Understanding and managing food safety risks. Food Safety Magazine, December 2010/January 2011 issue (cover story).

Buchanan, R.L., J.L. Smith and W. Long. 2000. Microbial risk assessment: dose–response relations and risk characterization. Int. J. Food Microbiol., 58: 159–172.

[CDC] Centre for Disease Control and Prevention. Norovirus Illness: Facts for Food Handlers CS234745-B. (www.cdc.gov/norovirus/downloads/foodhandlers.pdf). Accessed 2015.

[CDC] Centre for Disease Control and Prevention. 2011. CDC 2011 estimates: findings (http://www.cdc.gov/foodborneburden/2011-foodborne-estimates.html). Accessed 2013.

Courvalin, P. 2006. Antibiotic resistance: the pros and cons of probiotics. Dig. Liver Dis., 38: 261–265.

[EPA] Environment Protection Agency. 2000. Risk Characterization Handbook. EPA 100-B-00 002 (http://www.epa.gov/spc/pdfs/rchandbk.pdf). Accessed 2014.

[EPA] Environment Protection Agency. 2003. Understanding PCB Risks (http://www.epa.gov/housatonic/understandingpcbrisks.html). Accessed 2014.

[FAO] Food and Agriculture Organization. 2006. Food Safety Risk Analysis: A guide for national food safety authorities. FAO Food and Nutrition Paper 87, Rome, Italy, pp. 1–77.

Food Safety Cooperation Forum. 2012. Training Modules on General Food Safety Plans for the Food Industry (FSCF-PTIN: http://fscf ptin.apec.org/docs/APEC%20Food%20Safety%20Modules%202012/English%20Module%20PDF SCM_11_Section_3-3_HACCP_Principle_1-Hazard_Analysis_6-2012 English.pdf). Accessed 2014.

Guiterrez, J. 2013. Food hazards—Learn how to avoid them and the foodborne illnesses they cause. Today's Dietician., 15(11): 50.

Hellier, K. 2000. Hazard Analysis and Critical Control Points for water supplies. Conference paper in 63rd Annual Water Industry Engineers and Operators' Conference Civic Centre Warrnambool, Australia, pp. 101–109.

John, D. and A. William. 2006. Markell and Voge's Medical Parasitology. St. Louis: Saunders.

Knudsen, I.B. 2010. The SAFE FOODS framework for integrated risk analysis of food: An approach designed for science-based, transparent, open and participatory management of food safety. Food Cont., 21: 1653–1661.

Lee, H.W., L.J. Baek and K.M. Johnson. 1982. Isolation of Hantaan virus, the etiologic agent of Korean hemorrhagic fever, from wild urban rats. J. Infect. Dis., 146: 638–644.

Magnússon, S.H., H. Gunnlaugsdóttir, H. van Loveren, F. Holm, N. Kalogeras, O. Leino, J.M. Luteijn, G. Odekerken, M.V. Pohjola, M.J. Tijhuis, J.T. Tuomisto, O. Ueland, B.C. White and H. Verhagen. 2012. State of the art in benefit–risk analysis: Food microbiology. Food Chem. Toxicol., 50: 33–39.

Renwick, A.G., S.M. Barlow, I. Hertz-Picciotto, A.R. Boobis, E. Dybing, L. Edler, G. Eisenbrand, J.B. Greig, J. Kleiner, J. Lambe, D.J. Muller, M.R. Smith, A. Tritscher, S. Tuijtelaars, P.A. van den Brandt, R. Walker and R. Kroes. 2003. Risk characterisation of chemicals in food and diet. Food Chem. Toxicol., 41: 1211–1271.

Sanders, M.E., L. Akkermans, D. Haller, C. Hammerman, J. Heimbach, G. Huys, D. Levy, D. Mack, P. Phothirath, A. Constable, G. Solano-Aguilar and E. Vaughan. 2010. Assessment of probiotic safety for human use. Gut Microbes, 1: 1–22.

Schmidt, R. and D. Newslow. 2013. Hazard Analysis Critical Control Points (HACCP) Principle 2: Determine Critical Control Points (CCPs). University of Florida IFAS Extension Document FSHN 07-04 (http://edis.ifas.ufl.edu/fs140). Accessed 2014.

Sun, D. 2007. Computer Vision Technology for Food Quality Evaluation. Massachusetts: Academic Press.

Tareke, E., P. Rydberg, P. Karlsson, S. Eriksson and M. Törnqvist. 2002. Analysis of acrylamide, a carcinogen formed in heated foodstuffs. Agric. Food Chem., 50(17): 4998–5006.

Wang, W., J. Han, Y. Wu, F. Yuan, Y. Chen and Y. Ge. 2011. Simultaneous detection of eight food allergens using optical thin-film biosensor chips. J. Agr. Food Chem., 59(13): 6889–6894.

Whitehead, A.J. and C.G. Field. 1992. Risk analysis and food: The experts' view. *In*: Food, Nutrition and Agriculture–15- Food Safety and Trade. FAO Corporate Document Repository online (http://www.fao.org/docrep/v9723t/v9723t08.htm). Accessed 2014.

12

Innovative Measures for Ensuring Food Safety in the Food Value Chain

Richard Bonne,[1] *Didier Montet*[2] *and Nadine Zakhia-Rozis*[2,*]

Introduction

Food safety is a major research topic due to the increasing concern of consumers with public health related issues, along with the stronger sanitary standards set for international trade. To face these constraints and ensure prevention and control of food contaminants, it is essential to implement Food Quality Management Systems (FQMS) and Food Safety Management Systems (FSMS) based on the application of the Hazard Analysis and Critical Control Points (HACCP) method and its prerequisites (Good Practices) throughout the food value chain. However, considering only the technical or analytical features is not enough for this implementation. There is an urgent need for a systematic/proactive, cost-effective approach towards the control of contaminants and phytosanitary residues along the agri-food chains. Efficient FQMS should take into account the socio-economic context along with the organizational and technological capabilities of the chain stakeholders, including farmers, seed producers, cooperatives, storage and transportation infrastructures, SMEs and big enterprises, as well as market intermediaries and actors. On the other hand, this approach of FQMS establishment could be coupled with a collective scientific expertise, at national or regional level, in order to give insight and recommendations to policy-makers, public health and regulatory authorities. The coupling of these two approaches is an innovative way to make food safety a reality, especially in the specific context of less developed countries. This chapter summarises these innovative measures for food safety implementation in agri-food chains, with illustrations in African, Latin American and Asian countries.

[1] Food Safety Senior Expert, 13 rue Caumont, 32000 Auch, France.
[2] Cirad, UMR Qualisud, TA B-95/16, 73 rue Jean-François Breton, 34398 Montpellier cedex 5, France.
* Corresponding author: nadine.zakhia-rozis@cirad.fr

The Food Safety Management System

The implementation of a food safety management system (FSMS) turns out to be a difficult process for many food businesses (FB) managers because of the complexity of the available documents and methods. This chapter describes two original facilitating methods for the implementation of the *Codex Alimentarius* food safety requirements: (1) Good Hygiene Practices (GHP) and Good Manufacturing Practices (GMP) and (2) the HACCP method.

- The method known as "comprehensive hygiene management in food industries" is a rational system of organization for the implementation of GHP/GMP.
- The "alternative method to the Codex decision tree" makes it possible to avoid frequent failures in the use of "the Codex decision tree" for the Critical Control Points (CCPs) determination.

These two innovative methods were developed by applying a deductive reasoning procedure on a validated scheme of occurrence of health and/or food accidents (Bonne et al., 2005).

Logical Analysis of the *Codex alimentarius* Requirements

The situation set up by the *Codex alimentarius* is based on the following principles:

- The goal of the current regulations, GHP/GMP/HACCP guidelines and standards based on the principles of *Codex alimentarius* is to control health or economic food accidents.
- To effectively and efficiently control any phenomenon, it is necessary, first, to establish its exact description.
- The common origin for the encountered difficulties is a major logical flaw: none of the documents mentioned above, gives a description of the mechanism of occurrence of food accidents (health and/or economic ones).

From this logical analysis a possible strategy has been defined to give to FB managers facilitation methods for the Codex principles implementation:

- Establishing a validated description of the mechanism of outbreak of food accidents (health and/or economic ones).
- and using it to:
  ○ Establishing a comprehensive approach for GHP/GMP implementation.
  ○ Designing an alternative method to the "Codex decision tree".

Several key deductions can be deduced from this scheme:

- The concomitant intervention of the contamination and the multiplication is essential to the appearance of a food accident of biological origin.
- Figure 1 explains the mechanism of action of all the preservation methods, which reciprocally establishes its validity.
  ○ total control of contamination or multiplication induces a long lasting preservation (canning, freezing);
  ○ partial control of only one factor or of the both, induces a short lasting preservation (refrigeration, pasteurization).

- The contamination stage of the three types (physical, chemical and biological) is a passive mechanism from the contaminants point of view, not depending on measurable physicochemical parameters. The contamination is the only mechanism intervening in the occurrence of the physical and chemical food accidents.
- Multiplication and survival are active mechanisms which are specific to biological contaminants and depending on measurable physicochemical parameters.

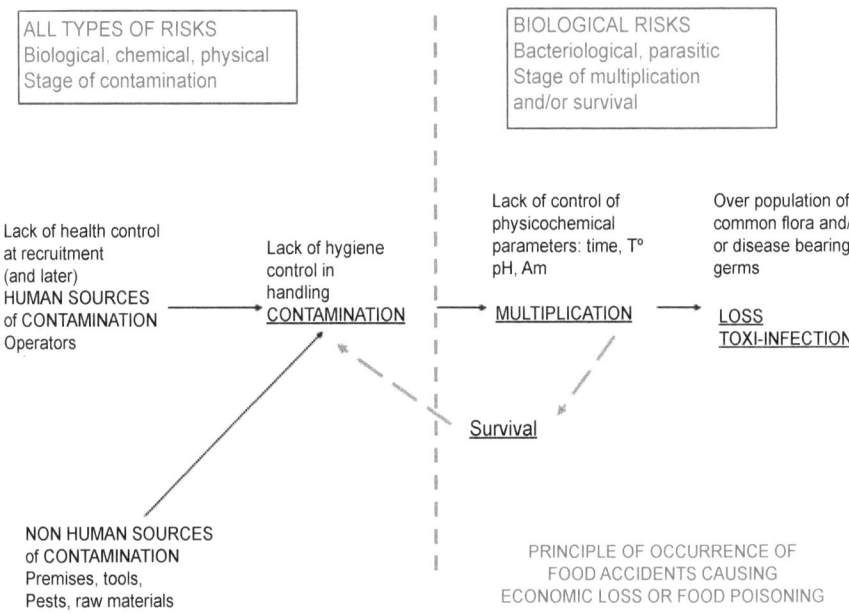

Figure 1. Scheme of occurrence of health/food accidents.

The Comprehensive Hygiene Management Method in Food Industries

The use and specifications of the comprehensive hygiene management method (Fig. 2), which is fully compliant with the *Codex alimentarius* requirements and covers all of them, can be presented as follows:

- The two goals of the comprehensive hygiene management method is to combine alternately the assessment and improvement of the food safety management system in food businesses.
- The assessment stage consists in checking that all specifications (chapters of the manual of conditions of hygiene symbolized as colored boxes) are taken into account in the establishment.
- The improvement stage consists in designing modes of action to satisfy the *Codex alimentarius* requirements ignored (or not fully implemented) in the existing system of hygiene management.

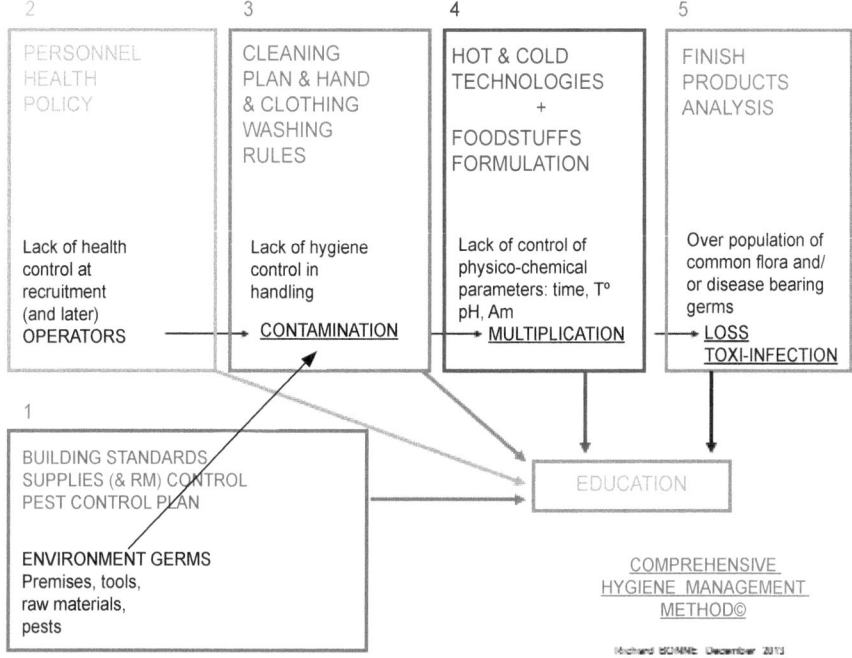

Figure 2. Table of the comprehensive management method.

The "Alternative Method to the *Codex alimentarius* Decision Tree" for Critical Control Points (CCP) Determination

Anyone having frequently used the "Decision Tree" knows it does not always conclude definitively if a particular operation of a manufacturing process is a CCP or not. This difficulty often comes from the fact that an operation identified as a CCP by the decision tree at HACCP step No. 7 does not match all the step No. 8 requirements which state: "Critical limits must be specified and validated if possible for each Critical Control Point" … «criteria often used, include measurements».

This innovative method is applicable to:

- The food safety control means determination on a process (GHP/GMP or CCPs) by food business managers.
- The quick assessment by inspectors/auditors of an FSMS applied on a process.

Its logical leading thread is based on the characteristics of the three types of risks and on the differences observed between the contamination on one hand, and multiplication and survival on the other hand (Fig. 3).

In order to apply the alternative method, five columns should be drawn on the flow diagram of the manufacturing process (two to the left and three to the right).

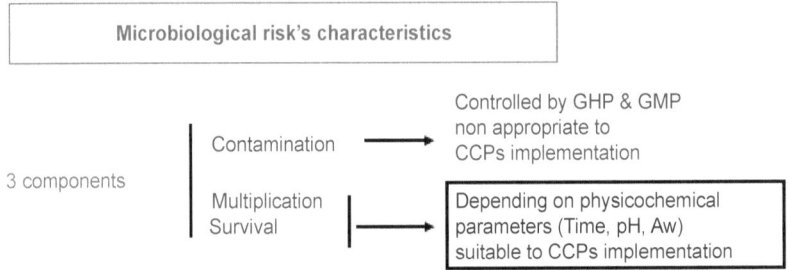

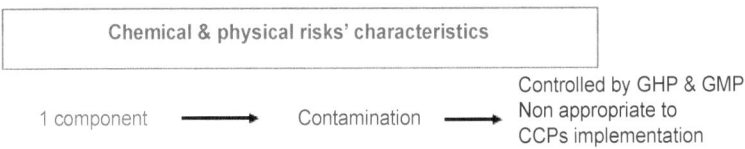

Figure 3. Table of risk comparison.

First of all, the sources of contamination should be entered on the left side of the diagram:

- Inputs (raw materials and packaging) in column 1.
- Points of contact involved in the production process (worker's hands, tools, equipment, work schedules, etc.) in column 2 (Fig. 4).
- Then, the parameters should be entered for each manufacturing stage (if there are some) on the left side of the flow diagram in column 3 (Fig. 5).
- Using the elements in the first three columns, the risks that may be deduced from them (Contamination, Multiplication, Survival), are entered on the right side of the diagram in column 4 (Fig. 6).
- If for any given procedure, elements have been listed (input or contacts) in columns 1 and 2, then a possible risk of contamination must be entered in the column 4 that corresponds to one or more of the three types: chemical, physical, or microbiological.
- If elements of a given manufacturing process have been listed in the column 3 (significant parameters), the eventual risk of multiplication or survival of microbes must be entered, in column 4.

Finally, appropriate means of control and monitoring will be entered in column 5, according to the logistics that enabled us to construct the diagram of the mode of appearance of food-related accidents (Fig. 7):

- for each operation involved in the production procedure, risks of contamination may be controlled by implementing the Good Hygiene and Manufacturing

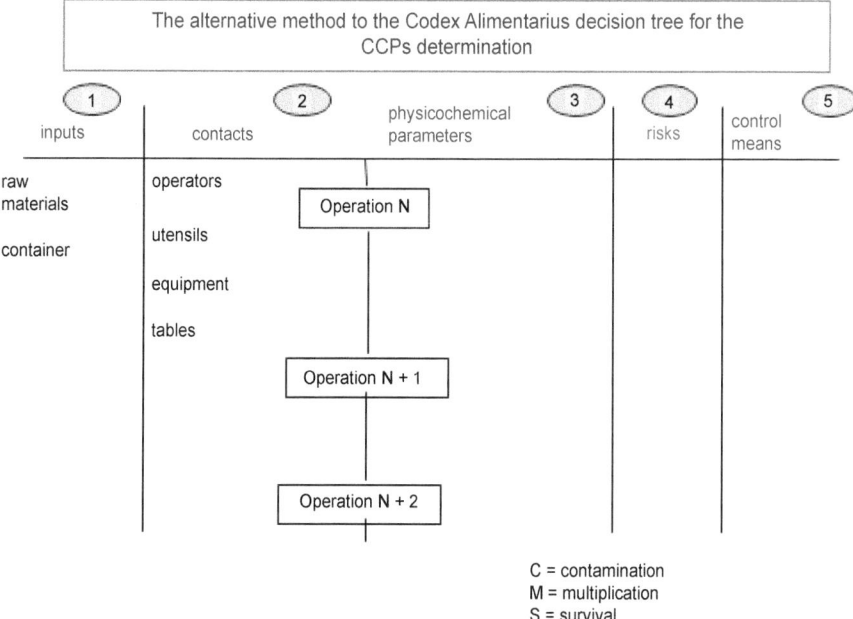

Figure 4. "Alternative Method" with columns 1 & 2 indicating the sources of contamination.

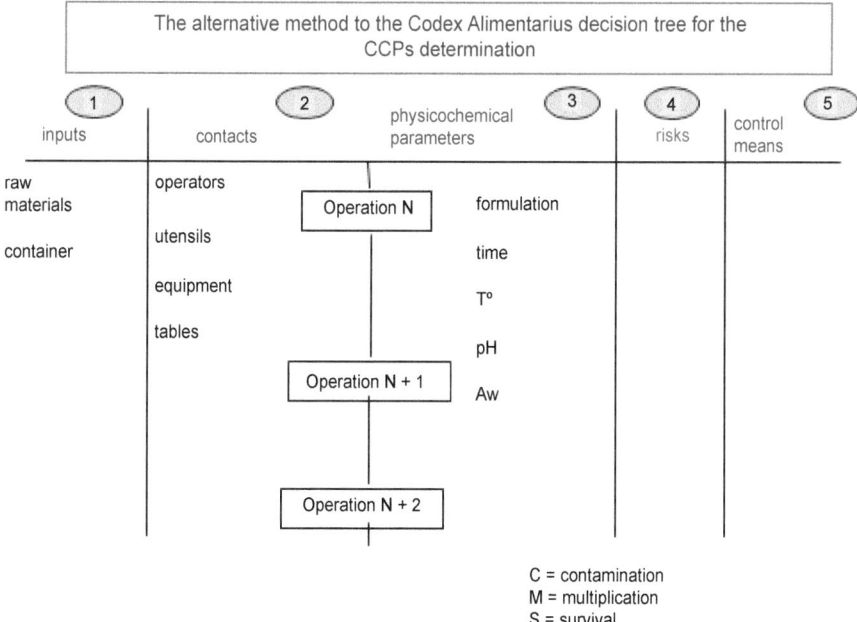

Figure 5. "Alternative Method" with column 3 indicating the parameters.

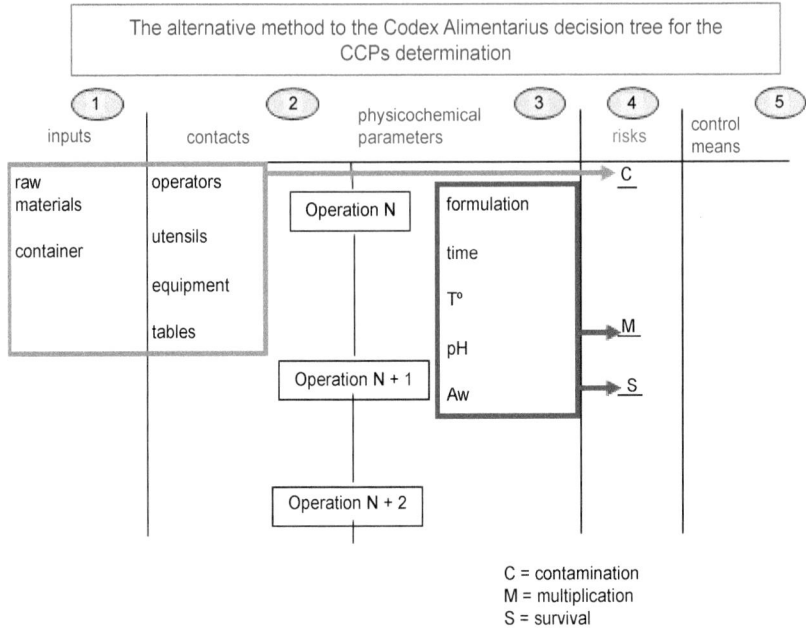

Figure 6. "Alternative Method" with column 4 indicating the deduction of risks.

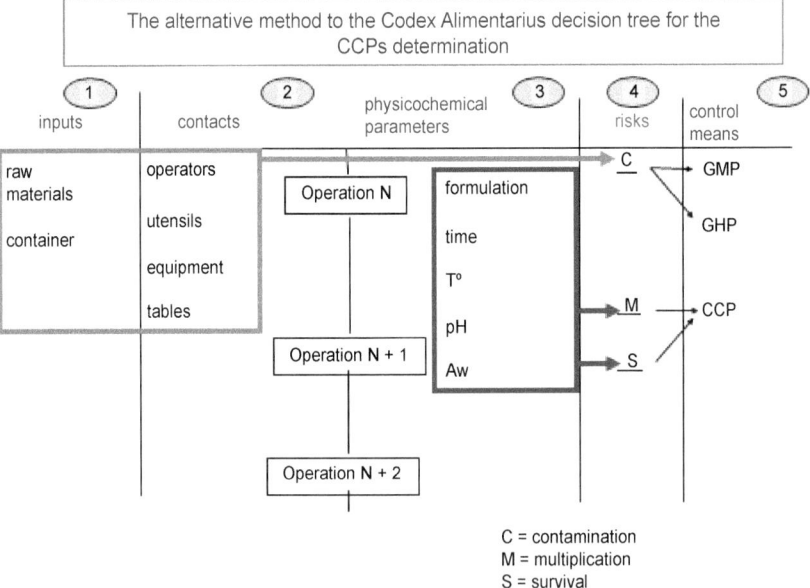

Figure 7. "Alternative Method" with column 5 indicating the deduction of control means (CCPs and/or GHP/GMP).

Practices (GHP and GMP) that were applied beforehand as prerequisites for the application of the HACCP method, thanks to the "Comprehensive Hygiene Management Method" in FB.

- for operations where risks of multiplication or survival have been identified, it will be necessary to monitor and control the parameters that these active mechanisms, characteristic of living organisms, depend on. The implementation of monitoring and control of the relevant parameters takes the form of an established Critical Control Point.

This method can be applied by the FB managers to determine CCPs. It can be also used by the inspectors/auditors to carry out a quick assessment of any FSMS applied in a FB, by drafting their own FSMS on the process flow chart and checking if the proper food safety control means (GHP/GMP or CCPs) are implemented on the right location.

Food Quality Management System: A Tool for Controlling Mycotoxin Contamination

Mycotoxin contamination of food and feed stuffs is among the top priority issues regarding human and animal safety, along with the economic losses they are responsible for due to rejection of contaminated food consignments. Mycotoxins have been defined as "fungal metabolites which, when ingested, inhaled or absorbed through the skin, cause lowered performance, sickness or death in man or animals". Exposure to mycotoxins can produce both acute and chronic toxic effects ranging from death to deleterious effects on the immunity system, hepatic and kidney functions, central nervous, the alimentary tract and so on. Mycotoxins can be carcinogenic, mutagenic, teratogenic and immunosuppressive. The ability of some mycotoxins to compromise the immune system and, consequently, to reduce resistance to infectious disease, is now widely considered to be their most important effect.

The main interventions that may be employed for the control of mycotoxins are prevention of contamination, identification and segregation of contaminated material (quality control, monitoring and legislation), and detoxification. Alternatively a modification of a food processing operation can also help to control the level of mycotoxins in foods. Mold and mycotoxin contamination may occur in any point of the supply chain, both in production or post harvest stages, the latter being the stages where field or earlier stage contamination may be magnified. Mycotoxin contamination then needs to be neutralized through the use of proper pre and post-harvest processes, of adequate equipment and of sound handling practices.

The Latin America Southern Cone (MERCOSUR) countries export cereals to EU but the new European standards create the need for improving cereal chain quality management to achieve those levels and to have a better knowledge of contamination reduction to acceptable levels through further processing. Although several mycotoxin surveys have been performed on cereals in the Latin America Southern Cone countries, the surveillance data are incomplete and widely dispersed. There is an urgent need for a systematic/proactive, cost-effective approach towards the control of mycotoxins throughout the cereal chains in these countries, rather than relying entirely on expensive, wasteful end-point testing/segregation. The MYCOTOX project (http://mycotox.cirad.fr) (ref ICA4-CT-2002-10043) entitled *The Development of a Food Quality Management System for the Control of Mycotoxins in Cereal Production and Processing Chains in Latin America South Cone Countries* was developed with partners from France, UK, Argentina, Brazil, Chile and Uruguay and funded by the European Commission. The overall objective of the project was to improve the competitiveness of domestically and internationally traded cereals by controlling the occurrence of mycotoxins in maize and wheat products used as human food and animal feed. The main output of the project consisted of an efficient Food Quality Management System (FQMS) taking into account the socio-economic context along with the organisational and technological capabilities of the chain stakeholders, focusing on farmers, cooperatives and SMEs.

The innovative challenges tackled by the project were:

- The exploitation of an HACCP approach, involving a multidisciplinary HACCP team, to identify the constraints and opportunities associated with the establishment of a successful domestic and international market for cereal-based foods and feeds.
- The development of novel analytical tools (sampling, sample preparation and analysis) for the determination of mycotoxins.
- The completion of data generation describing the risks associated with mycotoxins in maize and wheat in Latin America South Cone countries.
- The development of novel, validated proactive procedures (control measures, critical limits and monitoring methods) for the control of mycotoxins in a variety of cereal production and processing chains.
- The co-ordination of food quality management activities, for the control of mycotoxins, in Latin America and Europe.
- The development of a Food Quality Management System for the control of mycotoxins in both domestic and international cereal production and processing chains.

The HACCP Application Along the Food Chain: A Great Challenge for Food Safety

The HACCP concept was developed for improving the control of food safety and quality. It was initially applied in developed countries, with well-established and demanding codes to be followed. However, the initial use of HACCP was often limited to the processing stages in the industries which leads to the lack of understanding of potential sources of mycotoxin contamination within the production and distribution components of the commodity system. There is an urgent need for the adaptation of the HACCP approach, so that it may be applied throughout the whole agri-food chain in developing countries, especially those with important export markets. The implementation of HACCP principles in the case of MYCOTOX project enabled the introduction of a harmonised approach to mycotoxin control throughout international commodity systems:

- *It introduced a different approach to the management of food safety within commodity systems, involving Hazard Analysis and Critical Control Point*

(HACCP) principles. HACCP is a multidisciplinary, proactive approach that focuses upon the control of the process, at specific critical points, rather than focusing only upon the finished product. The HACCP approach directly encourages an increase in the proportion of safe material, produced within the commodity system, by maintaining control of the production, processing and marketing activities throughout the system (Fig. 8).

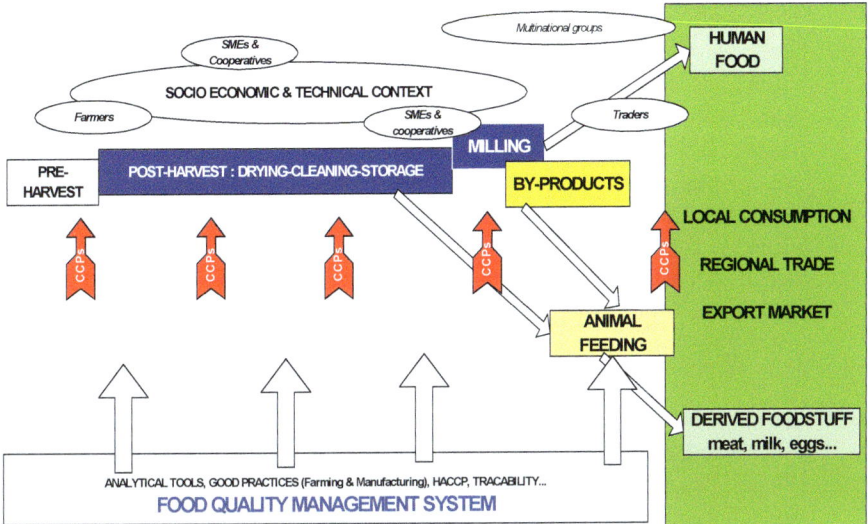

Figure 8. Implementation of a Food Quality Management System along cereal value chains.

- *It employed a multidisciplinary, international team to develop HACCP plans and a Food Quality Management System for the control of mycotoxins in international maize and wheat commodity systems involving Argentina, Brazil, Uruguay, Chile and Europe. This was a challenge of the project as it was pioneer in extending the HACCP application to the whole cereal chain, involving all key players and including socio-economic studies.* A multi-disciplinary HACCP team was assembled in each country, including agronomists, socioeconomists, HACCP specialists, analysts, and representatives of the key players of the selected cereal chain (private sector, seed and grain producers). Confidentiality agreements were signed with the key players in order to officialise their involvement in the project and define the mutual commitment of the consortium and the cereal stakeholders to join efforts and share field activities and outputs.

- The HACCP team *constructed a commodity flow diagram (CFD)*, based on the knowledge of team members and selected specialists, for each participating country that produces significant quantities of the selected commodity. The collected data allowed a detailed overview on the agricultural practices, the raw material supply, the processing and transport channels and actors, the quality control procedures, the socioeconomic context (i.e., trade issues such as volumes, prices, fluxes, seasonal fluctuations, supply chain organization). The team verified the CFD by observing activities at each step, and interviewing those performing the operations. A detailed description of the equipments used and the operating procedures were documented. This multidisciplinary team conducted the mycotoxin hazard analysis to identify commodity (maize and wheat)/mycotoxin combinations that presented an unacceptable risk to human or animal health, and/or which were likely to exceed national or international regulations. Hazard and incidence data were largely drawn from published sources, but when adequate data were lacking, surveillance studies were conducted to establish at which steps in the CFD the mycotoxin hazard originated or at which steps concentrations increased to unacceptable levels (Brabet et al., 2005; Cea, 2006).

- The HACCP team developed and validated *control measures* that prevented, eliminated or reduced mycotoxin content to an acceptable level, when applied to a specific step in the CFD, according to the socio-economic, cultural and institutional issues for each country (i.e., willingness of consumers to pay higher price for mycotoxin-free cereal-based products, potential incentives and costs/benefits for implementation of quality management systems, regulation procedures, local constraints and/or needed services for this implementation, specific institutional and cultural features in the context of each country (Henry et al., 2006).

- The location of Critical Control Points (CCPs) was confirmed, as well as their critical limits and the corrective actions for monitoring these critical limits. Manuals describing *Good Agricultural Practices (GAP)* were developed for the production of safe, high quality maize and wheat, in the Southern Cone region. These manuals compiled advice, recommendations and guidelines on the good practices that should be followed in order to prevent and control efficiently mycotoxin contamination in the Southern Cone region. Recommendations concerned either agricultural practices (choice of resistant variety to *Fusarium*, tillage, fungicide treatment, harvesting date, weather forecasts) or manufacture/storage practices (grain segregation at reception, silo cleaning and emptying, transportation). As the Good Practices are pre-requisite for HACCP plans and Quality Management Systems, the partners agreed upon the need for considering them in an holistic way, and stressed the need for external interventions driven by authorities and regulatory bodies for ensuring their application by all cereal chain stakeholders.

- Verification, documentation and record keeping procedures were established. The resulting *Food Quality Management System (FQMS)* was implemented and provided a means of alleviating the occurrence of the mycotoxins, from the plough to the plate, and enabled key players within the maize and wheat systems to positively work towards a high level of adherence to EU regulations. It also made a significant contribution towards the alleviation of economic losses, amongst European traders and processors, caused by the rejection of contaminated food consignments; and brought about a commensurate reduction in the detrimental environmental effects of product disposal, together with a reduction in the wastage of energy associated with the processing and transportation of material which is to be ultimately rejected.

The concerned official authorities were approached, either in health area (to stress the need for sensitization campaigns oriented towards the consumers upon the negative impact of mycotoxin-contaminated food on human health) or agriculture area (to stress the need for application of Good Agricultural Practices and integrated management systems to solve mycotoxin problems), as well as regulatory bodies (to stress the need for implementation of adequate regional standards and compliance with international requirements).

The MYCOTOX project has used specific case studies to pioneer the application of a number of methodologies and approaches to the control of mycotoxins in cereal chains in the Southern Cone. The application of HACCP (Hazard Analysis and Critical Control Points), which was initially focusing on technical aspects at lab level, to the whole agrifood chain including stakeholders, and the full integration of socio-economic and technical inputs, has been successfully demonstrated and can therefore be recommended for use for the control of mycotoxins (and by extension to other types of contaminants) in all cereals and cereal products (and by extension to other types of commodities).

This approach was essential according to the expressed needs for considering mycotoxin prevention and control in a holistic way. The project's outputs served for external interventions driven by mandatory regulations, policy makers and official authorities to ensure the application of integrated and global Quality Assurance and Management Systems for controlling mycotoxin contamination along the cereal chains in the Latin America Southern Cone region.

The Collective Scientific Expertise: A Tool for Food Safety Implementation

The collective expertise helps in identifying health risks, especially in developing countries. It consists in bringing together a number of experts in a specific field to give a collective opinion on a problem. This type of expertise is applied for many years in the European Union to appraise the health hazards of food. European expertise is framed by the European Regulation 178/2002 which requires expert assessment based on scientific available evidence. This expertise must be conducted in an independent, objective and transparent way (Article 6). This expertise allows providing a strong scientific basis (Article 1), provisions and effective organizational procedures to underpin decision-making in the field of safety of food and feed. The various European agencies use the collective expertise and EFSA (European Food Safety Authority) has the following committees or panels to achieve it:

- Additives and products or substances used in animal feed (FEEDAP).
- Animal health and welfare (AHAW).
- Biological hazards (BIOHAZ), including BSE-TSE-related risks.
- Contaminants in the food chain (CONTAM).
- Dietetic products, nutrition and allergies (NDA).
- Food additives and nutrient sources added to food (ANS).
- Food contact materials, enzymes, flavorings and processing aids (CEF).
- Genetically modified organisms (GMO).

- Plant health (PLH).
- Plant protection products and their residues (PPR).
- Scientific Committee & Emerging Risks (SCER).

This system is extremely expensive to implement and is reproduced in all European countries with different administrative sizes. In France for example, about 250 experts meet monthly in Paris to treat only food health problems.

It is of course out of the question to reproduce these agencies in developing countries because no politician would want to implement such an expensive system to deal with non-apparent causes.

We therefore propose to implement the collective expertise in developing countries to identify local health risks without having to implement a large and expensive battery of analytical means. Be aware that no analytical tool will permit to identify accurately a health risk. In fact, the lab analysis will provide expert information on a number of samples that represent most of the time only a small fraction of the samples of a country. We rely on the assumption that a group of experts on a given topic will bring, in an adversarial discussion, very specific information to a question.

For this purpose and for example, a committee of experts was held in 2009 in Bangkok (Thailand) to identify risks on food exports from Asia to the European Union in the frame of the European project Era-Net. 33 experts identified these risks in 3 working days (Montet et al., 2010). More recently, during the EC-funded Europ-Aid project (entitled 3C Ivoire project), an Ivorian expert committee was formed to identify and discuss food safety risk in Côte d'Ivoire. In 2014, the University of Mansoura in Egypt organized an Egypt-Europe committee of experts to identify the hazards related to food in Egypt (Montet et al., 2015). Two groups of experts representing a total of 33 scientists have worked three days on the biological and chemical hazards and made a report that refers to the national level.

Now we give you a working example from the Ivorian experts. The panel chose to work on slaughterhouses in Abidjan that is what we call a self-referral.

At the beginning of the meeting, all experts agreed to say that Abidjan abattoirs were very dirty and very contaminated. None of the experts possessed analysis report. The objective was to write a brief report to the Ivorian government proposing a course of action that would improve the situation. The president of the expert group then asked the experts what they offer. Almost all offered to perform microbial testing in order to know the state of hygiene of slaughterhouses. In fact, it was not necessary to analyze since everyone agreed that slaughterhouses were unhealthy. The solution proposed by the President was the closure of slaughterhouses, which greatly surprised the experts. Yet it was the unique solution that can be offered to the services of the Prime Minister to ensure meat consumers' safety. It was obvious that this report will generate strong high-level discussions and lead to a reaction of the authorities which may be either a closure, which seems technically impossible, or cleaning and control of facilities, or even an improvement of the buildings or reconstruction to an adequate standard level. In fact, we read recently (December 2014) in the newspapers that the government chose to clean and improve the slaughterhouses of Abidjan showing the power of the expert group.

Here we could see very well the role of a collective expertise and its effect on the political world. An individual expertise will never impact the decision makers as a collective expertise. The collective expertise is necessarily exact because it takes into account the knowledge of many people who are experts in a particular field while individual expertise is based on the knowledge of one person.

The collective expertise should consider a controlled input mode. Individual people cannot enter a committee of experts. He may only be entered by legal persons, that is to say, associations, ministries or expert themselves.

Conclusion

The use of an integrated, comprehensive and holistic approach is essential for tackling food safety challenges and ensuring the implementation of adequate and operating measures. There is an urgent need for an innovative combination of methods such as (i) the application of HACCP method and its pre-requisites (Good Agricultural Practices, Good Manufacturing Practices, Good Storage Practices) with all key players (producers, processors, enterprises, markets, laboratories, regulatory bodies), (ii) scientific collective expertise on contaminants, (iii) specific campaigns for sensitizing the official Health and Agriculture authorities and regulatory bodies on food safety challenges, either for public health or economics. This would allow making food safety a reality, especially in the specific context of less developed countries.

References

Bonne, R. et al. 2005. Lignes directrices sur le HACCP, les bonnes pratiques de fabrication et les bonnes pratiques d'hygiène pour les PME. Programme CE-ASEAN de coopération économique sur les normes, la qualité et l'évaluation de conformité (Asia/2003/069-236), Comité Européen de Normalisation, Bruxelles.

Brabet, C., E. Salay, O. Freitas-Silva, A.F. Alves, M. Machinski, Jr., E.A. Vargas and N. Zakhia-Rozis. 2005. Maîtrise des mycotoxines dans la filière maïs au Brésil. Cahiers Agricultures, 14 (1): 164–168.

Cea, J. 2006. Update on worldwide regulations for mycotoxins. The MERCOSUR harmonization of limits on mycotoxins with the international regulations. *In*: Advances in Research on Toxigenic Fungi and Mycotoxins in South America Ensuring Food and Feed Safety in a Myco-Globe Context, Conference of the EC MYCO-GLOBE Project, 15–17 March 2006, Villa Carlos Paz, Córdoba, Argentina.

Henry, G., A. Engler, D. Iglesias and G. Gutierrez. 2006. Socio-economic constraints and opportunities affecting the implementation of mycotoxin control measures in Southern Cone grain supply chains. *In*: Advances in Research on Toxigenic Fungi and Mycotoxins in South America Ensuring Food and Feed Safety in a Myco-Globe Context, Conference of the EC MYCO-GLOBE Project, 15–17 March 2006, Villa Carlos Paz, Córdoba, Argentina.

Montet, D. et al. 2010. Identification of topics of common interest for EU and SEA partners in the thematic "Food Quality, Safety and Traceability" towards a strengthening of Science and Technology collaboration between SE Asia and EU in the FP7 programmes. EU-Southeast Asia Expert meeting on Food Quality, Safety & Traceability. Bangkok, Thailand—24–27 February 2009. Quality Assurance and Safety of Crops & Foods, 2(4): 158–164.

Montet, D. et al. (33 authors). 2015. Future topics of common interest for EU and Egypt in Food Quality, Safety and Traceability. Published on line in Quality Assurance and Safety of Crops & Foods, pp. 401–408, DOI: http://dx.doi.org/10.3920/QAS2014.0428.

Montet, D. et al. to be published. Projet 3C Ivoire: Création du Comité National de Coordination des actions pour la Sûreté des Aliments en Côte d'Ivoire, EuropeAid,\DCI/NSAPVD/2010/64, Acteurs non étatiques et autorités locales dans le\Développement, Actions dans les pays partenaires Côte d'Ivoire. Partenaires: Cirad\Montpellier France, Supagro Montpellier France, INP Yamoussoukro Côte d'Ivoire (2011–2014).

13

Strategies to Destroy or Control Foodborne Pathogens

Dam Sao Mai* and Nguyen Thi Kim Anh

Microbiological Criteria and Hazard Analysis and Critical Control Point (HACCP)

Introduction to HACCP

HACCP is a management system in which food safety is addressed through the analysis and control of biological, chemical, and physical hazards from raw material production, procurement and handling, to manufacturing, distribution and consumption of the finished product (US Food and Drug Administration). HACCP was found in 1960s within NASA space program to prevent food poisoning and ensure food safety for astronauts in Columbia spacecraft. In 1971, HACCP began to be applied in the food industry in US and quickly become a food quality control system that developed and applied widely in the world. Nowadays, HACCP is an essential condition in almost all developed countries including US and Europe. The certification from independent institution proves the commitment of food safety.

HACCP is based on preventing pathogens from entering the food supply chain and controlling this contamination after it occurs. Principles of HACCP are agreed in over the world and can be applied for all food production and beverages in distribution and retails. HACCP plan contains seven principles including: (1) Conducting a hazard analysis; (2) Identifying critical control points (CCPs); (3) Setting critical limits for each CCP; (4) Developing CCP monitoring procedures; (5) Performing corrective

Industrial University of HCMC, Institute of Biotechnology and Food Technology, 12 Nguyen Van Bao Str. Go Vap District, HCM City, Vietnam.
* Corresponding author: damsaomai@foodtech.edu.vn

actions; (6) Establishing record-keeping and documentation procedures; and (7) Verifying the system is working as planned.

The risk of failure can be minimized by hazard analysis. Well trained staff can manage the implication of critical control points. However, HACCP application is not simply hazard analysis and critical control points but need to include prerequisite condition and program such as Good Manufacturing Practice (GMP), Sanitation Standard Operating Procedures (SSOP) and other supportive programs as basis for food safety guaranty.

HACCP could be applied for food and beverage companies such as seafood, food, animal feed companies, hotels, restaurants, etc.

HACCP is an effective way to help food and beverage companies identify and prevent risks of food contamination and their products will be qualified as regulation requirement. Products that are proven and labeled with HACCP system will put customers trust in its quality.

The HACCP certification brings to your business a number of advantages:

- Proof of safety process for your food products.
- Your food products are monitored by all reasonable prevention methods in order to ensure about food safety.
- Save time for management and other fees.
- Minimize spoiled products and returned products.
- Enhance the effectiveness.

Benefit of HACCP when put in practice:

- *Enterprise benefit*: Improves the prestige of their product quality, thence raise the competition, control and extend the market, especially activity to export foods. Products that are labeled by HACCP system make firm belief of customers and fellow traders. HACCP certification is a basis of possible success when conducting a negotiation, trading contract signing within country or export to other nations. Also, this is a basis of prior investment and education policies of the state as well as of the foreign partners.
- *Industry benefit*: the competition and marketing will be enhanced by HACCP system. In addition, it reduces cost due to the control of spoiled product; production process and environment condition will be improved; management capacity to ensure the food safety will be improved; business opportunity and food import-export will be increased.
- *State benefit*: public health will be improved and health fee will be decreased, food control will be more effectively; trade development will be more facile; put people trust in food supply.
- *Customer benefit*: reduce risk of foodborne diseases; enhance the public awareness of food safety, raise the belief of food supply; improve living standard concerning health, economy and society.

Condition to apply HACCP

In order to apply HACCP successfully, a company needs to have certain essential criteria:

- *Commitment of leaders*: Enterprise leader board commits to deploy the HACCP application and maintenance through supplying indispensable resources timely according to food safety requirement.
- *Human resources*: Key management personnel must be trained to know about products and produce processing, to experience and understand well HACCP regulations. They must be aware of the role of HACCP in food safety.
- *Factory and equipment*: technology, equipment and factory condition play crucial role in response to law demand and the institution of food safety.

Steps to build up HACCP system:

- Set up HACCP team.
- Describe product.
- Determine the purpose of product using.
- Establish technology processing diagram.
- Check up technology processing diagram.
- Conduct the hazard analysis.
- Define the critical control points.
- Build up the limit of control points.
- Build up the monitor system.
- Define the repairing activities.
- Set up the procedure of file storing.
- Build up the examine and verify procedure.

A good HACCP program needs to be clear, understandable and contain all guide procedures of implementation methods. All forms should be easy to use, to record all examine, monitoring and repairing activities. However, only HACCP is not good enough but other systems like pest control, recall protocol, hygiene and sanitation need to be built up and implemented to make an effective food safety program.

HACCP plans to control microbiological hazards

Foodborne diseases caused by pathogens are on the rise as reported from many countries. Through analysis of hazards and where they can occur, systems and procedures can be implemented to minimize the risk of failure. Critical control points can be managed by trained in-house staff, providing for a truly hands-on quality management system at each and every operation. Most manufactures involved in the food industry are aware of pathogen control program and train their staffs in order to be able to prevent the potential fatal problem from the products. The employees need to be educated in the proper methods and understand what they should do to meet the standards that have been established by the authorized office. Every manufacturer might have their own program to control pathogens, that is, Good Manufacturing Processes

(GMP) and to ensure the facilities are free of contamination risk. All employees should be trained to ensure full compliance and safety in food chain.

Concepts of pathogen control program like different pathogens overview and effects of human ingestion should be covered in the course content. Most common pathogens need to be mentioned, for example *E. coli*, that is the most often bacteria associate with ground beef.

The successful pathogen control program in the control of foodborne pathogens should also be included. From the good example, employees understand how to carry out the program, what to expect from it and they will study what is the role of HACCP in the elimination of these pathogens. The final purpose of these courses is to train staffs how to produce high quality foods that are free of foodborne pathogens. Furthermore, training of managers, owners and others responsible for health and safety is implemented in specialization up to the establishment and individual needs.

Microbiological testing plays an important role in implementing HACCP programs. It is a crucial mechanism for collecting data used in developing and conducting a HACCP plan, because they can help establish standard operation procedures (SOPs) for sanitation and critical limits, and assess the possible occurrence of hazards as well as the validity of the HACCP plan. In addition, microbial testing is useful in implementing an HACCP plan by helping to oversee the effectiveness of sanitation SOPs, the compliance of incoming ingredients with safety criteria, the safety of products being held for corrective action, and the safety of the finished product (Kvenberg and Schwalm, 2000). Foodborne diseases caused by some pathogens such as *Campylobacter* spp., *Salmonella* (nontyphoidal), and *Listeria monocytogenes* occur annually and result in a number of hospitalization and death cases. To reduce or eliminate foodborne disease agents of the food chain from farm to table, it is essential to develop new methods to improve the food supply. An HACCP-based program was designed to reduce the potential of microbial contamination through the production and distribution process was introduced in 1998 for the Greenhouse Vegetable Growers in Ontario, Canada (Powell et al., 2002). The program implementation including individual on-site grower visits, and the microbiological testing of produce and water was successful. It helps improve knowledge, understanding and awareness of growers to the microbial risks associated with fresh produce. The program shows that producers are aware of consumer's concern and try to act to manage risks.

Sanitation Standard Operating Procedures

Pre-harvest control of foodborne pathogens

Outbreaks of a number of foodborne illnesses have been linked to contamination occurring in the pre-harvest stage of food processing that warn us to have concern about the complexity of the farm-to-fork continuum and the relevance of the pre-harvest role. For many food products, it is not easy to prevent transmission of pathogens to consumers after the food is taken away from the farm. Many food products which are eaten raw like fruits, vegetables, nuts and some seafood, it is not usual to eliminate pathogens by cooking. In another hand, some raw food like meat can also be a pathogen source to spread on to other products during preparation before cooking. Food was processed from farm to table following a number of steps that face contamination risk and foodborne diseases. Therefore, pre-harvest food safety is the most critical stage of food production.

Pre-harvest food safety concerns both animals and plants. For plants, it relates to plants in the field while for animals, it involves safety control in the animal production stage. Pre-harvest food safety applies to fruits and vegetables when growing, packaging, and marketing of the raw products. It puts in practice to animal production on farms where animals and their products like eggs and milk are produced. In the food-processing industry, pre-harvest food safety is involved in raw material. However, the food service industry and consumers consume many unprocessed foods. It is unexpected that these groups will develop HACCP plans, but need to be informed about CCP that is associated with the processing more than the raw material treatments.

Pre-harvest food safety control, together with basic hygiene practice, are strategies to prevent foodborne pathogens. In order to implement food safety control in food production environment, there are multiple needs for data like systematic surveillance to provide baseline data on the prevalence of pathogens, and epidemiologic research will help identify effective controls. To eliminate all pathogens from the pre-harvest environment is not possible. Many practice programs for pre-harvest food safety just aim to reduce pathogen numbers to levels that decrease the degree of hazard to public health. Some foods are supplied direct from farm to retail (produce) when other products have intermediate processing procedures (meats) and the difference of their impacts of on-farm measures make the practice for pre-harvest food safety program become very complicated. Also, the pathogens continue to reproduce and disseminate during other steps in the farm-to-table continuum. Furthermore, the cultivation practices are different between developed and developing countries, the organisms of concern on farms may be not the same. Therefore, the understanding of the epidemiology of foodborne pathogens on the farm is the best way to control the risk of contamination. A number of specific pre-harvest food safety research needs were identified, including validation and development of interventions, development of better methods for pathogen detection and enumeration, and investigation of the effects of interventions on microbial community dynamics.

Unhealthy animals may be condemned at inspection before slaughtering, so farmers make efforts to keep animals healthy. Human pathogens that infect animal to cause zoonotic disease are only detected in the laboratory. The strategy to monitor animal production food safety should be considered the same as animal disease control and eradication that include detection, slaughter, quarantine, and biosecurity and farm hygiene. Vaccination is also included but just plays a minor role.

Quality assurance plans have been established for different animal production systems but there is still no available effective program.

Antibiotics and grow-promoting antimicrobials' use as feed additives that cause antimicrobial drug resistance is an issue that needs to be considered. On the one hand, using those drugs makes economic benefits thank to the improvement of growth rate and feed conversion, reduce mortality, improve health and increase the resistance ability to disease. It is reported that dietary antimicrobial growth promotions increased growth performance 72% when 12,153 feeding studies were reviewed (Rosen, 1996). On the other hand, the use of antibiotics for growth promotion may increase the prevalence of antibiotic-resistant bacteria in the animal and it results in an increase in antibiotic-resistance bacteria in the food supply and may cause illness in humans. If the antibiotic-resistance bacteria are pathogenic, or if they can transfer the resistance to other pathogens, they will be a human health risk. Many countries now ban the use antibiotics in animal feeds to eliminate antibiotic-resistant bacterial populations in livestock.

Food-producing husbandry animals

Meat is a product that has a lot of nutrients, and the pH from 6 to 6.5 is suitable for the growth of microorganisms. Animals can be infected with microorganisms by endogenously or exogenously. When meat is infected, the quality of meat will reduce and sometimes can not be used. The number of microorganisms in meat shows the quality of the meat. Spoilage meat is from 10^7 or more bacteria per 1 cm^2 surface of meat or 1 g meat. Although not much microbes were infected in the good health husbandry via the interior way, but some bacteria were still found in the blood of red meat, such as: *Staphylococci, Streptococci, Clostridium*, and *Salmonella*. Healthy cattle are less infected microorganisms endogenously. Nevertheless, we still found some microorganisms in the blood of some red meat such as: *Staphylococci, Streptococci, Clostridium*, and *Salmonella*. The sources of microbial contamination for animals are: the process killing animals, shaving, and slaughter. Microorganisms from components such as feathers, internal organs, dung or instruments infect the meat in the slaughter house. In modern technologies slaughter—slaughter hook, microbial infection reduces significantly. Therefore, reducing microbial contamination to the meat we need to pay attention to the sanitation during the slaughter process.

Ruminants

In cattle, *Escherichia coli* O157:H7 is one member of the normal flora and is not regularly associated with clinical diseases. However, *E. coli* O157:H7 causes many foodborne disease outbreaks in humans with beef and beef products that lead the public health concern. Because cattle is a source of *E. coli* O157:H7, it would be very advantageous to bring cattle to the slaughterhouse in an *E. coli* O157:H7 free status. Contaminated products cause human outbreaks, and also effect the industry due to the product return and also decrease the belief of consumers. Therefore, eliminating *E. coli* O157:H7 from the food supply is essential for not only human health but also for economic perspective.

To control *E. coli* O157:H7 effectively, the intervention strategies at all points of the food chain, from farm to fork, should be implemented. Sanitation attempt after slaughter have been reported to reduce contamination of beef carcasses with *E. coli* O157:H7. However, pre-harvest interventions also decrease the bacteria number before entering in the food processing chain. In fact, at farm, the common intervention measures that have been investigated are general sanitation strategies, dietary component modification, and pre-slaughter feeding strategies. Bovine viral diarrhea virus (BVDV) is an immunosuppressive pathogen on faecal *E. coli* O157:H7 shedding. BVDV is widely distributed among cattle, and identifying a potential on-farm risk factor for *E. coli* O157:H7 will assist the beef and dairy industry by identifying critical control strategies to minimize the threat of human foodborne illnesses associated with *E. coli* O157:H7, as well as ensuring consumer confidence through the provision of wholesome food.

Antibacterial drugs have been used in the dairy industry for treatment and prevention of diseases for several decades. The use of antibiotics results in healthier, more productive animals, lower disease incidence and reduced mortality and morbidity in animals, and production of nutritious, high quality and low cost production for human consumption. It is obvious that the use of antibiotics in dairy cows and other food-producing animal contribute to antibiotic resistance. The development of antimicrobial resistance impacts the disease treatment that affects the human health. It is required to have antibiotics intervention when food safety issue is concerned (Matias et al., 2010).

Swine

It appears that parasites are the most harmful foodborne hazard to human public health rather than bacterial foodborne pathogens. In developed countries, thanks to the changes in pork production system, the incidence of foodborne parasites has been reduced dramatically in recent years. However, it still remains a problem in developing countries. Major parasite foodborne incidences are *Taenia solium, Trichinella spiralis*, and *Toxoplasma gondii*. *T. solium* is a tapeworm that lives only in the human intestines. Pigs are the intermediate host for this parasite after ingesting tapeworm eggs shed in human feces and develop cysts in the muscle and tissues. If people eat undercooked pork that contains cysts, they will be infected. According to FAO report, about 50 million people over the world host the adult tapeworm and among those 50,000 deaths every year showed cysticercoids (Davies, 2011; Eddi et al., 2003; Aubry et al., 1995). *Trichinella spiralis* is another parasite that forms cysts in pig tissue and is responsible of the infections in a broader host range. *Toxoplasma gondii*, together with *Salmonella* and *Listeria*, causes three-quarter of fatal foodborne infections in US (Mead et al., 1999). Pre-harvest intervention has been shown to be effective to reduce presence of *T. solium*, *T. spiralis*, and *T. gondii* in pig farms.

In contrast, intestinal bacteria, such as *Salmonella, Campylobacter, Listeria*, and *Yersinia enterolitica* that are raise the public health concern of consumers, are not controlled effectively by pre-harvest interventions. These pathogens may occur at any stage during harvest and processing or until the product is served at the table (Davies et al., 2004; Reij and Den Aantrekker, 2004).

In pigs, epidemiological researches on *Salmonella* have been carried out much more than any other bacteria such as *Campylobacter, Listeria*, and *Yersinia*. Table 1 that summarized by Davies (2011) shows the Salmonella prevalence in pigs in recent studies in some African and Asian countries.

Because the pre-harvest interventions are not effective, post-harvest control should be considered to reduce risk of pathogen contamination in pigs.

Table 1. Reported *Salmonella* prevalence in pigs in recent studies in countries of Africa and Asia (Davies, 2011).

Location	Sample	n[a]	Prevalence (%)	Reference
Kenya	Fecal (abattoir sampling)	58	8	(Kikuvi et al., 2010)
	Carcass (abattoir sampling)	58	19	
Cameroon	Carcass (abattoir sampling)	30 samples	40% of samples	(Akoachere et al., 2009)
Laos	Gastrointestinal tract (abattoir sampling)	49	76	(Boonmar et al., 2008)
Ethiopia	Various (abattoir sampling)	278	43% of pigs	(Aragaw et al., 2007)
Vietnam	Cecum (abattoir sampling	117	52.1	(Le Bas et al., 2006)
	Carcass	46	95.1	

a = numer of samples

Poultry

Due to the large number of chickens slaughtered everyday, it is impossible to conduct ante mortem one by one, but only inspection on a flock or lot basis is possible. By observing poultry in coops or grouped for slaughter before or after they are removed from trucks, the abnormal chickens can be identified and removed. Abnormalities detected can be clinical signs of some disease such as listeriosis, salmonellosis, or heavy metal toxicosis. Many of the bacteria that are involved in foodborne diseases live in the intestinal tract of poultry, e.g., *Salmonella, Campylobacter*, and *E. coli* O157:H7, shed to the environment and no pathological evidence is expressed. For this reason, removal of visible faecal contamination is important. Contamination may occur when bacteria invade the slaughter instrument contaminating carcass during the de-feathering process or when faeces and ingesta were spilled from the intestinal tract. Contaminated carcass must be condemned or reworked to remove contamination in a proper manner as far as possible. It is important to remove visible faecal contamination.

Traditionally, pathogen control methods in foods are drying, curing, salting, sugaring, heating, and cooling. A step like pasteurizing can affect food quality, such as changes in flavor and texture (Ralston et al., 2002). Multiple hurdles that either kill pathogens or minimize pathogen growth are typically used in the food industry. Multiple hurdles are implemented to control pathogens in production processes for raw products in some meat and poultry firms. Pathogen control strategy depends on how strictly it needs to control pathogens in specific raw meat products.

Chicken meat is known as an important vehicle of foodborne bacteria such as *Salmonella* spp. and there is required an effective control of its contamination during industrial processing. The levels of contamination of chicken carcasses over the processing are various, depending on the different control procedures adopted by different slaughterhouses (Oliver et al., 2011).

Wegener et al. (2004) reported a combination of government and private farm *Salmonella* control program of broiler chickens that had been successfully implemented in Denmark. The programs were carried out in 5 years involving extensive pathogen testing of feed supplies, of birds in quarantine, in the hatchery, on the farm, and in the slaughterhouse. *Samonella* were identified in the production chain and birds and feed that are contaminated were condemned. A strategy to control pathogens depends on specific raw meat products that a firm should consider to design proper interventions. The levels of pathogen control can be different depending on different products within a meat company. For instance, cattle products can be sold in three markets with high-risk raw ground beef, medium-risk roast market, and low-risk processed products. Each of these markets request their own pathogen control program.

Food producing aquatic animals

Fish also is a product which is rich in nutrion. Fish is a cold-blooded animal. There is a viscous layer of protein on the skin of fish. The gills of fish also have many nutrion which is suitable for microorganism growth.

Both fresh water and sea fish contain a lot of protein and nitrogen. They also possess lipid, whose amount depends on the species. Most nitrogen compounds of fish are protein. The non-protein nitrogen is amino acid, baz nitrogen, such as: ammonia and trimethylamine; creatine, taurine, betaine, uric acid, anserine, cRNAosine and histamine. The nutrions of fish and husbandry animals are similar. But the structure of fish meat is untight unlike husbandry animal meat, so the microorganism is easier to be infected inside. Following are the ways of microorganisms infecting fish:

- *Penetrating through the gut*: In dead fish, proteolytic enzymes decompose the protein of the intestine wall to facilitate entry of microorganisms from the fish, digestive organ.
- *Infiltration from epidermal mucus*: fish skin gland secretion is a good environment for microbial growth and penetration into meat.
- *Infiltration from gills*: In dead fish, the blood present is a suitable environment for microbial growth and penetration into the meat.
- Penetrate through wounds caused during fishing.

In addition, during processing, contaminated water also cause microbial contamination into meat and fish.

Thus fish and seafood are foods which are easily infected by microorganisms. *Clostridium botulinum* type E and *Vibrio parahaemolyticus* (especially sea fish) are the strains causing bacterial disease related to fish. Other bacterial diseases in fish and seafood include *C. perfringens, Staphyloccous* spp., *Salmonella.* spp., *Shigella* spp., *V. cholera* and other type of *Vibrio*. Fish skin and intestinal usually have *Pseudomonas, Moraxella, Achromobacter, Flavobacterrium, Vibrio. Micrococcusalso, Bacillus, Clostridium* are small quantities. The growth of these microorganisms causes the spoilage of the product. *Coliforms* have little or no presence. There is no *Salmonella, Shigella* and other enteric bacteria. The presence of these bacteria indicates the environmental contamination from catching and processing. During processing, the product can *be infected* by *Clostridium botulinum* (canned fish). The reason why fish products are infected by microorganisms is because fish is not heat-treated enough or the infection occurs after or during processing. The transport and storage frozen also contribute to microorganism contamination in fish products.

Temperature and pH are the factors limiting the development of bacteria in fish products. These are applied in the sterilization and heat treatment especially intestinal organ (Whipple and Rohovec, 1994).

For frozen freshwater fish the damage is caused by bacteria, while salted fish and dried fish mostly is caused by fungal. The microorganism that cause spoilage in fish are mostly in the group of bacilli, negative gram bacteria, non-spore forming, such as *Pseudomonas* and *Acinetobacter-Moraxella*. Several microbial spoilage in fish can grow well at temperatures 0°C–1°C. Shaw and Shewan showed a large number of *Pseudomonas* spp. which may cause damage to fish at temperature 3°C with a slow speed.

The spoilage of freshwater fish and saltwater fish are happened in the same way; the main differences are the salt concentration requirement of microorganisms in sea water and the chemistry composition of fish namely non-protein nitrogen components. The fish gut is easily vulnerable. If fish is not gutted, the intestinal microorganisms will attack the intestinal wall and get into the flesh through the holes. This process is caused by proteolytic enzymes (available in the gut or bacteria).

The damage of the fish occurs with different speed, this may be due to differences in surface properties of fish. Fish skin is thin in such fish as cod, so the spoilage occurs faster than the fish such as halibut whose epidermis and dermis are thicker. Besides, halibut have a thick viscous layer which contains mucopolysaccharid, free amino acids, trimethylamineoxide, piperidine derivatives and some other antimicrobial components such as antibodies and enzymes.

Pseudomonas spp. and *Salmonella* are mainly microorganisms causing the spoilage of fish. Because the final product is H_2S and reducing trimethylamine-N-oxide (TMAO), the bacteria are considered to be most significant in spoiling fish.

In addition, fish are infected by parasites such as worms including the following:

- *Diphyllobothrium latum.*
- *Clonorchis sinesis.*
- *Opithorchis felineus.*
- *Heterophyes heterophyes.*
- *Metagonimus yokogawai.*

With tropical fish, almost after the fish dies, the bacteria start growing exponentially. If preserved in ice, the amount of microorganisms will increase doubly after a day and after 2–3 weeks the number of microorganism reach 10^5–10^9 cfu/g meat, or 1 cm^2 on the skin. When fish is stored at room temperature, after 24 hours, the number of microorganisms was nearly 10^7–10^8 cfu/g. If tropical fish is preserved in ice, bacteria will be in the latent period from 1–2 weeks, then start growing exponentially. At the time of damage, the amount of bacteria in tropical fish and temperate zone fish are the same. If fish are preserved in ice in anaerobic conditions, the number of cold-tolerant bacteria (such as *Pseudomonas* and *S. putrefaciens*) is often lower than in aerobic conditions. However, with damaged fish the amount of bacteria (*P. phosphoreum*) still reached 10^{7-8} cfu/g.

When the redox potential on the surface of fish is reduced (aerobic bacteria use to metabolize carbohydrates and lactate to CO_2 and H_2), the anaerobic bacteria

(*Alteromonas putrefacien*) will use TMAO-reductase to eliminate TMAO ($(CH_3)_3$ NO) into TMA (trimethylamine), TMA is bad for the flavor of the fish. TMA, NH_3, amines are total vaporable basic nitrogen (TVB), commonly used as chemical criteria for assessing the quality of fish (mainly TMA). Maximum limit of TVB in cold storage fish is 30–35 mg/100 g. The amount of TMA in fresh fish is low. Quality of cold storage fish is good when the content of TMA lower than 1.5 mg in 100 g fish. Maximum limit is 10–15 mg TMA/100 g.

In addition, the bacteria decompose amino acids (cysteine, methionine contain sulfur) that form H_2S, CH_3-SH and $(CH_3)_2S$. These are volatile to cause harmful odor even at very low doses (ppb) that reduces the value sensory of the product.

Fish should be washed immediately after catching. Tanks and storage area should also be cleaned. Depending on the requirements of the market, fish will be gutted to eliminate microorganisms in the digestive tract. During gutting, the air will contact with the cut in the fish belly that can cause oxidation. So it is considered to carry out this section or not. It is difficult to chill the whole big fish because of the thickness of the flesh. During chilling, fish can be damaged before the temperature of the center of fish is 0–2°C. Therefore, this fish needs to be cut into 4–6 cm thickness.

Fruits and vegetables

Fresh fruits and vegetables are increasingly being recognized as vectors for foodborne illnesses. The number of gastrointestinal disease outbreaks has been related to the fresh fruit and vegetables consumption. In US, there were 148 outbreaks caused by contaminated fresh produce recorded during the period from 1990 to 2001 (Smith-DeWaal et al., 2002). Fresh fruit and vegetables are considered high risk foods because they are susceptible to contamination by manure and soil. Also, they are almost underprocessed. *Samonella* and pathogenic *Escherichia coli* are two major pathogens that cause outbreaks linked to fresh fruit and vegetables. The increasing proportion of foodborne disease outbreaks may be due to the raw vegetables consumption, the suboptimal safety produce condition, packing methods that extend the shelf-life but reduce the safety, using contaminated manure and water for plants.

Pre-harvest food safety concerns prevention of microbial contamination of plants in the field. To prevent microbial contamination, a food safety program is of prime importance. Good agriculture practices (GAPs) during growing, harvesting, sorting, packaging and storage of fresh fruit and vegetables are programs to prevent microbial contamination. There are a number of preventive measures, such as microbial quality of water, personal hygiene, packing in the field and packing house, sanitation, manure management, and pesticide residues (Barinas et al., 2010).

Pre-harvest food safety for fresh fruits and vegetables include practices in growing, harvesting, packaging and marketing of raw fruits and vegetables. The harvesters should be cleaned, otherwise they may lead to pathogen contamination of the produce, especially when they harvest manually. The harvest containers should be cleaned and sanitized using chlorine solution.

Post-harvest control of foodborne pathogens

Meat and fish

Live cattle are a source of *E. coli* O157:H7 and it is difficult to eliminate pathogens in the pre-harvest stage. Alternatively, beef production is a highly fragmented industry and it is expensive to develop new interventions for each segment of the industry to control bacteria from cattle. Therefore, an intervention for full integrated system of beef production industry is needed. Intervention costs should be shared among the different segments of the industry. Due to the fact that pre-harvest prevention of *E. coli* O157:H7 is not effective, post-harvest intervention such as using heat to kill bacteria is the most effective method.

For red meat, the heads, viscera, and carcass were examined at one or more postmortem inspection stations. The inspectors observe, palpated and incised the tissues to find if there were any abnormalities. For poultry, they were usually examined at a single station and inspectors perform observations and palpations. Only red meat and poultry are free of visible contaminants like feces or ingesta, damage and other abnormalities are able to be consumed by humans. Feces and ingesta are potential cause of contamination of meat and poultry products with harmful pathogens, such as *E. coli* O157:H7, *Salmonella*, and *Campylobacter*. Therefore, prevention of these contaminants in slaughterhouse is critical for the inspection system.

Fish is consumed as fresh fish or frozen fish or dried fish or in its salted and processed form for the purpose of preservation (canned, smoked...). During the process of preserving fish and meat, to ensure the safety of the product, attention should be paid to the microbial contamination in products. One of the causes of microbial contamination is water. Therefore, the water needs to be treated well and fish should be carefully selected. In technology seafood processing by ripening method, the significant factor are the duration of cooking; temperature of the steam, water and other factors such as slice thickness, the precision of the thermometer, monitoring equipment, other time measurement devices.

Fresh fruits and vegetables

Vegetables are 88% water, 8.6% carbohydrate, 1.9% protein, 0.3% fat and 0.84% ash. Total percentage of vitamins, nucleic acids and other components are less than 1%. Yeast, mold, and bacteria usually grow in vegetables and they cause the damage. *Botrytis* fungus appears on the flowers of strawberries and causes gray rot, *Colletotrichum* is present in the epidermis of bananas and causes ulcers, and *Gloeosporium* causes spotting holes in apples. The biggest damage in commercial vegetables is postharvest, although fungi often invade and cause damage to the product, but the other attacks on specific areas. For example, *Thielaviopsis* penetrate into pineapple stem and cause rot of black, broken black spots in potatoes is caused by *Ceratocystis*. Bacteria have rarely been the first factor causing the spoilage of vegetables and fruits. Most cases are caused by a combination of mold and bacteria. Vegetables have higher protein than fruits and often do not have acid reaction, so vegetables are spoiled easier than fruits by bacteria. The water content is high in vegetables that help the growth of the spoilage bacteria. The pH in most vegetables is consistent with the pH that the majority of the bacteria can be developed. So bacteria is the most popular factor damaging vegetables. The spoilage bacteria in vegetables mainly are *Erwinia* and *Pectobacterium.*

The safety solutions for fruits and vegetables are still less clear. It can be taken as applying a critical control point such as pasteurization of juices, or good agricultural practices (GAPs) to prevent contamination from various sources. However, the effective measures remain unknown. A series of guidelines of GAPs was introduced in 2003 by FDA in collaboration with the USDA to reduce the risk of foodborne diseases from fresh fruits and vegetables, but contamination potential of fresh fruit and vegetables is still exist. Therefore, decontamination methods must further be applied to reduce or prevent the potential for finished produce contamination. Decontamination treatments including thermal processing, chemical processing and irradiation, however, affect the quality of most of fresh fruits and vegetables. Therefore, there is a need to develop alternative methods for fruits and vegetables decontamination.

Some post-harvest interventions that can be considered for fresh produce including fruits and vegetables are in the area of pre-cooling and storage (packing house equipment, management of animal pets in packing and storage facilities, facility sanitation, temperature control, and shipping/vehicles) (Barinas et al., 2010).

Processed foods

Undercooked and/or improperly chilled meat and eggs have been identified as one of the most frequent causes of foodborne diseases in some countries such as the USA. To control the growth of foodborne pathogens, heating and chilling of meat, eggs and meat products are critical processes. It is a complex process to use air chilling method, for carcasses, fresh eggs, and freshly cooked, ready-to-eat meat products. Specification and evaluation of temperature decline rate are of importance to control the microbiological hazards such as *Salmonella* spp., *E. coli* O157:H7, *Clostridium perfringens* contamination to keep food safety.

Preservation/maintenance

Food preservation is vital to safety, extending the products shelf life and maintaining the quality of products, the customers' main concern. Conventional methods of preservation are canning, pickling, drying, using sugar and salt, and smoking. Common methods of shelf life extension are chilling, freezing, modified atmosphere packing, controlled atmosphere packing, vacuum packing and shrink wrapping. Preservation may reduce the microbial contamination or limit the growth of microorganisms. To achieve an effective preservation result, clean and high quality ingredients should be used. The preservation method that is varied among different products impact the product safety and the process facility including equipment, space, and hygiene.

Quality System

Quality management

Quality is a set of inherent characteristics of products meet the published requirements, mandatory is implied. Quality is not self born, quality is not a random result, it is the result of the impact of a series of factors that are closely related to each other. Achieve the desired quality that needs to properly manage factors. Management activities in the field of quality are called quality management. Quality management is the combined operation to control an organization in setting policy, quality objectives and define the business processes, the necessary resources ensuring and improving the quality.

Quality management has been applied in every industry, not only in manufacturing but also in all areas, all types of companies, large and small, whether to participate in the international market or not. Quality management ensures the company doing the right thing to do. If companies want to compete in the international market, they must learn and apply the concepts of quality management effectively. Quality management is the coordinated activities to direct and control the quality. The direction and control of quality often includes policy, objectives, planning, controlling, ensuring and improving quality.

Some specific activities related to quality management:

1. *Quality checking*: It is the measurement, review, testing of one or more characteristics of the product and comparing the results with the requirements for the conclusion of the match or mismatch. When there is production and exchange of goods, this method exists. Quality checking is a passive activity.
2. *Quality control*: It is the operations and technique being used to meet the quality requirements. Quality control is identification and control, adjustment factors affecting directly the quality that maintain or only change these factors in the level of determination. So the quality of the product will be guaranteed.
   The relevant factors are: human resources; equipment; methods; processes; materials and fuels; and environment.
3. *Quality management system*: It is the organizational structure, procedure, processes and necessary resources to implement quality management.
   There are 8 main principles in quality management system: (1) Orientation by customer and market; (2) Leadership; (3) The participation of all members; (4) Process approach; (5) Solution to manage the system; (6) Continuous improvement; (7) Decision based on data; (8) Creating partnerships and achieving benefits with providers.
4. *Quality assurance system*: It is the management system in which all activities are carried out as planned. This system is able to prove the ability of the company that it matches the requirements for quality if necessary.
5. *Total quality control*: It is an effective system. This system embodies different groups into an organization so that the marketing, engineering, manufacturing and services can be conducted most economically to satisfy customer requirement.
6. *Total quality management (TQM)*: It is a way of managing the organization. TQM is based on the principle of participation of all the members, focusing on quality

management in order to achieve long-term success through customer satisfaction and provide benefits for members of the organization and society as well.

ISO 22000

The importance of ISO 22000

ISO 22000 in combination with HACCP criteria and other prevention is a guarantee for food safety.

In recent years, numerous evidences have proven that monitoring food safety is crucial. News from media indicates clearly that the weak points of the supply chain menace on the safety and health of the consumers. Continuing problems and demand for safety for clients make request to have more tools in order to reduce dramatically or eliminate the hazards.

Criteria for food safety have been established after a long history. In ancient times, the purity of beer and wine was examined to protect drinkers. In US, food safety has been considered since the latter half of the 18th century and it becomes law in 1906 (Pharmaceutical and Pure Food Act). Time and again, the prevention approaches become appropriate with food safety control; and from HACCP that is pinnacle in 1995, ISO 22000 was developed. Nowadays, ISO 22000 is a global criterion that applies for food chain agencies. It contains prevention methods to ensure quality in addition to food safety. The aim of ISO 22000 is to provide the practical approach to guarantee that food hazards are reduced and to provide protection measures for clients.

ISO 22000 was established by the contribution of 187 nations in the worlds and was made public for the first time on September 1, 2005. In Vietnam, it was admitted as a national standard in 2008 and called as TCVN ISO 22000:2008. A key difference between ISO 22000 and HACCP is the importance of using PRPs of ISO. PRPs are the general monitor applied for any business activities to maintain hygiene condition when implementing environment treatment. PRPs assign the necessary prerequisite condition in producing safe foods. Base on the related activities, the following requirements should be considered:

- Good Agriculture Practice (GAP).
- Good Hygiene Practice (GHP).
- Good Produce Practice (GPP).
- Good Distribution Practice (GDP).

There are other factors that belong to the prerequisite program such as sanitation and hygiene; monitoring pathogenic organisms; personal hygiene; air, water, energy sources; eliminating sewage and rubbish; monitoring supplier, educating staff, and others.

Prevention in large scale

ISO 22000 is based on the traditional prevention activities, identified and adjusted by experts on quality and food safety. Three concepts were adopted from ISO 9001.

These include: (1) To plan (work well with plan and need to follow and obey the plan); (2) Procedures (consistent during the long period, especially when there are many participants); and (3) Employee's capacity (it is essential to employ competent staff to obtain required outputs). Other concepts are document control, file control, repairing, measurement, etc.

Requirements for food safety

Due to the reason that ISO 22000 includes most of HACCP, understanding HACCP is of crucial importance. HACCP implementation involves 12 steps:
- Establish an HACCP team to ensure HACCP plan is adequate and will be performed properly.
- Describe product in combination with specific HACCP plan.
- Determine the intended use of the product.
- Make a flow chart of processing steps.
- Confirm the chart by direct observation, list the potential hazards that are associated with the chart, perform hazard analysis to identify the hazards that need to be controlled.
- Identify and examine the control manners that are possible to prevent hazards.
- Determine the essential control point at the examination place to ensure food safety.
- Establish the critical limits for each essential control point when satisfactory levels are clearly determined.
- Set up a control system for each essential control point.
- Set up repairing activities that can be done when the critical limit is offended.
- Establish investigation procedures to monitor results.
- Establish a document and file saving system to provide monitoring proof when needed.

One point of difference between HACCP and ISO 22000 is the importance of prerequisite programs (PRPs). These are general controls that are implemented by food business to maintain hygiene in the environment. PRPs assign prerequisite essential conditions to produce safe foods.

ISO 22000 is effective for all organizations regardless of scale. Food safety implementation is hazardous and costly. It is easier and safer to prevent the problems from occurrence. ISO 22000 is harmonious with national standards and other industries that are involved in food safety. If your organization is part of the food industry, you should learn more about ISO 22000.

The following is a brief list of organizations that may consider ISO 22000:

- Food manufacturers.
- Food ingredients manufacturers.
- Cultivate producer.
- Raw material or finished product transporters.
- Packagers.
- Retailers.

- Food businessmen.
- Animal feed producers.
- Food processing equipment producers.
- Packaging material producers.

In order to apply ISO 22000 successfully, you need to set up a clear plan that includes the identification of function, responsibilities and time needed. Benefits of using ISO 22000 should be transmitted to the staff. Educate them to understand the technical aspect of ISO 22000.

ISO 22000 can be applied in all food-producing, food-distribution types in food supply chain, regardless of the scale of business, including:

- Producing and processing animal feeds.
- Functional foods for the old, children and the sick.
- Processing business of vegetable, fruits, meat, eggs, milk, and sea foods.
- Producing and processing business of beverage: soft drink, pure water, beer, alcohol, coffee, tea, etc.
- Producing and processing business of spices.
- Food transporters.
- Producing and processing business of ready-to-eat foods, restaurants.
- Supermarkets, wholesales, retails.
- Packaging material producing business.
- Farms of plants and animal husbandry.

A manufacture that applies food safety management standard ISO 22000 will be considered to have a food safety management system with international standard, resulting in high competition advantages. Consequently, the opportunities of exporting products to strict markets in the world will be more opened. Other possible benefits when applying ISO 22000 are listed as below:

- Standardization of all management activities on produce and business of manufacture.
- This is a replacement of many other criteria, such as GMP, HACCP, EUROGAP, IFS.
- Reduce the selling cost.
- Minimize the outbreaks, lawsuits, and complain from customers.
- Strengthen prestige, belief, and convenience for distributors and customers.
- Improve overall activity of manufacture.
- Integrate conveniently with other management systems, such as ISO 9001 and ISO 14000.

Requirements for those who perform ISO 22000—HACCP

Human resource to perform ISO 22000 and HACCP is a decisive factor for success of ISO 22000—HACCP application. Also, humans are one of highest risks that need to be monitored.

Health status

All workers in food processing firms need to be examined about health status carefully, especially those who work directly with foods. It is essential to perform regular health check ups for employees to ensure only pass employees can work in the food processing places.

Isolation of contamination sources

Employees who are suspected to have infectious diseases such as tuberculosis, viral hepatitis, etc. may contaminate food and must be left out of the food processing chain. Employees who have easy contagious disease, such as flu, viral fever, cholera, typhoid fever, dermatological diseases, wound, spot, boil that can contaminate foods should be off work and be treated until well.

Hygiene regime

Regulations for workers who have direct contact with foods (including raw materials and other ingredients), with surface that foods contact with and with packaging materials, have to follow seriously the following hygiene regulations:

- Clean body before working.
- Use a clean and bright coat, cap, mask, water resistance and uncorrosive gloves, boots.
- Do not wear jewelry (ring, watch, etc.) that may fall into the food or running equipment and cause contamination for foods.
- Wash and/or disinfect hands before working.
- Do not eat, drink, chew gum, smoke, etc. that may contaminate foods.
- Do not throw waste in the food processing area.

Visitors should also follow the requested hygiene regime, such as using coat, hat, mask and boots when needed.

Education, training and investment

- Educate the workers regularly about the awareness of personal hygiene, factory hygiene, and public hygiene.
- Train the management staff about the technique and management duties.
- Invest hygiene equipment for factory and personal. Maintain, repair and check equipment to insure that they are in good condition.

Inspect and monitor

Assigning responsible staff to inspect and monitor hygiene implementation at every step.

References

Akoachere, J.-F.T.K., N.F. Tanih, L.M. Ndip and R.N. Ndip. 2009. Phenotyping characterization of *Salmonella typhimurium* isolates from food-animals and abattoir Drains in Buea, Cameroon. Journal of Health, Population, and Nutrition, 7(5): 612–618.

Aragaw, K., B. Molla, A. Muckle, L. Cole and E. Wilkie et al. 2007. The characterization of *Salmonella* serovars isolated from apparently healthy slaughtered pigs at Addis Ababa abattoir, Ethiopia. Prev. Vet. Med., 82: 252–261.

Aubry, P., D. Bequet and P. Queguiner. 1995. Cysticercosis: a frequent and redoutable parasitic disease, Med. Trop. (Mars), 55: 79–87.

Barinas, M., D. Doohan, R. Downer, A. Kleinschmidt, H. Kneen and T. Kline. 2010. Food safety for fruits and vegetables. The Ohio State University.

Boonmar, S., K. Markvichitr, S. Chaunchom, C. Chanda, A. Bangtrakulnonth, S. Pornrunangwong, S. Yamamoto, D. Suzuki, K. Kozawa, H. Kimura and Y. Morita. 2008. *Salmonella* prevalence in slaughtered buffaloes and pigs and antimicrobial susceptibility of isolates in Vientiane, Lao People's Democratic Republic. J. Vet. Med. Sci., 70(12): 1345–1348.

Davies, P.R., H.S. Hurd, J.A. Funk, P.J. Fedorka-Cray and F.T. Jones. 2004. The role of contaminated feed in the epidemiology and control of *Salmonella enterica* in pork production, Foodborne Pathog. Dis., 1: 202–215.

Davies, P.R. 2011. Intensive swine production and pork safety, Foodborne Pathogens and Disease, 8(2): 189–201.

Eddi, C., A. Nari and W. Amanfu. 2003. *Taenia solium* cysticercosis/taeniosis: potential linkages with FAO activities; FAO support possibilities, Acta Trop., 87: 145–148.

Kikuvi, G.M., J.N. Ombui and E.S. Mitema. 2010. Seotypes and antimicrobial resistance profiles of Salmonella isolates from pigs at slaughter in Kenya. J. Infect. Dev. Count., 4: 243–248.

Kvenberg, J.E. and D.J. Schwalm. 2000. Use of microbial data for hazard analysis and critical control point verification—Food and Drug Administration perspective. J. Food Prot., 63(6): 810–4.

Le Bas, C., T.H. Tran, T.T. Nguyen, D.T. Dang and C.T. Ngo. 2006. Prevalence and epidemiology of *Salmonella* spp. in small pig abattoirs of Hanoi, Vietnam. Ann. N Y Acad. Sci., 1081: 269–272.

Matias, B.G., P.S.A. Pinto, M.V.C. Cossi and L.A. Nero. 2010. *Salmonella* spp. and hygiene indicator microorganism in chicken carcasses obtained at different processing stages in two slaughterhouses. Foodborne Pathogens and Disease, 7(3): 313–318.

Mead, P.S., L. Slutsker, V. Dietz, L.F. McCaig, J.S. Bresee, C. Shapiro, P.M. Griffin and R.V. Tauxe. 1999. Food-related illness and death in the United States. Emerg. Infect. Di., 5: 607–625.

Oliver, S.P., S.E. Murinda and B.M. Jayarao. 2011. Impact of Antibiotic Use in Adult Cows on Antimicrobial Resistance of Veterinary and Human Pathogens: A Comprehensive Review, 8(3): 337–355.

Powell, D., A. Luedtke and K. Blaine. 2002. Pre-harvest issues related to food safety in fruits and vegetables, ISHS Acta Horticulturae 642: XXVI International Horticultural Congress: Horticulture, Art and Science for Life—The Colloquia Presentations.

Ralston, K., C. Brent, Y. Starke, T. Riggins and C.T. Lin. 2002. Consumer food safety behavior: A case study in hamburger cooking and ordering (Agricultural Economic Report No. AER804), Washington, DC: United States Department of Agriculture Economic Research Service. Available on the World Wide Web: http://www.ers.usda.gov/Publications/aer804/.

Reij, M.W. and E.D. Den Aantrekker. 2004. Recontamination as a source of pathogens in processed foods, Int. J. Food Microbiol., 91: 1–11.

Smith DeWaal, C., K. Barlow, L. Alderton and M.F. Jacobson. 2002. Outbreak Alert! Center for Science in the Public Interest.

Rosen, G.D. 1996. The nutritional effects of tetracyclines in broiler feeds. Proceedings of the World Poultry Science Society, 2: 141.

Wegener, H.C., T. Hald, D.L.F. Wong, M. Madsen, H. Korsgaard, F. Bager et al. 2003. *Salmonella* control programs in Denmark, Emerging Infectious Diseases, 9(7): 774–780. Available on the World Wide Web: http://www.cdc.gov/ncidod/EID/vol9no7/03-0024.htm.

Whipple, M.J. and J.S. Rohovec. 1994. The effects of heat and low pH on selected viral and bacterial fish pathogens. Aquaculture, 123: 179–189.

14

Biosecurity in Research Laboratories, Agriculture, and the Food Sector

Md. Asadulghani[1],* and Barbara Johnson[2]

Introduction

United States' President Barack Obama, speaking on his Global Health Initiative said "…we cannot wall ourselves off from the world and hope for the best, nor ignore the public health challenges beyond our borders... we cannot simply confront individual preventable illnesses in isolation. The world is interconnected, and that demands an integrated approach to global health…." (GHI, 2009). "…we must come together to prevent, and detect, and fight every kind of biological danger—whether it's a pandemic like H1N1, or a terrorist threat, or a treatable disease (GHSA, 2011)."

Knowledge about mechanisms of infection, interactions of microbial communities, and evolution and co-evolution processes in many different habitats (e.g., the human microbiome) have significantly advanced within the last few decades. New microorganisms have been identified; the number of newly described microbial species is quickly growing along with enormous progress in fundamental and applied microbiology. Many valuable and unique research collections, some of which contain human, animal and plant pathogens are maintained by institutions, collaborating working groups and international repositories. Culture collections and institutions that house and work with pathogens (opportunistic to highly dangerous) have over the years increasingly developed and advanced and in some countries

[1] Head, Biosafety and BSL3 Laboratory, Biosafety Office, icddr,b, 68 Shaheed Tajuddin Ahmed Sarani, Mohakhali, Dhaka-1212, Bangladesh.

[2] Owner, Biosafety Biosecurity International, 1165 Reston Avenue Herndon, VA 20170, USA.

* Corresponding author: asadulghani@icddrb.org

standardized processes for enhancing biosafety and biosecurity have been developed. In many countries biosecurity is inextricably linked to biosafety; where biosafety encompasses biocontainment to prevent accidental release and safeguard workers' health, biosecurity aims toward the prevention of possible malicious theft, misuse and release of organisms (WHO, 2006). This definition of biosafety and biosecurity is consistent with current World Health Organization (WHO) and American Biological Safety Association (ABSA) usage of these terms. In some countries the concept of biosafety and biosecurity are described using a single word such as bioversicherung in German; bioseguridad in Spanish and in Russia the transliteration for security is used to describe both concepts.

Biosecurity has been practiced in the laboratory setting for many decades and is not a new concept. Addressing biosafety and biosecurity together has been newly coined as biorisk management. Developing a comprehensive biosafety risk assessment or biorisk management matrix to measure a weighted value of each risk has become the norm. The use of proper safety and security procedures and equipment, appropriate facility design and construction, and of paramount importance, the proper training of employees ensures not only their own safety but that of the community and environment (Nordman, 2010). Countries are increasingly developing biosecurity regulatory requirements based on the risk group of pathogens and toxins, creating 'schedules' that list specific pathogens and toxins that should be subject to varying degrees of control and protection.

Biosecurity is implemented not only at the local level but on a global level in terms of national import/export permitting and export control frameworks (BMBL, 2009; CBSG, 2013; The Australia Group, 2014). At the highest level, biosecurity is ultimately governed by the non-proliferation approach of the Biological Weapons Convention (BWC). A total number of 171 countries signed this convention (ratified by 155) and as a result States' parties have enacted specific legislation or other measures to assure domestic compliance with the convention (Rohde et al., 2013). Microorganisms inherently bear a 'dual-use' potential and consequently most microbiologists are more or less affected by dual-use issues (UNODA, 1925; Millett, 2010). According to the Article I of the convention—

Each State Party to this Convention undertakes never in any circumstances to develop, produce, stockpile or otherwise acquire or retain:

Microbial or other biological agents, or toxins whatever their origin or method of production, of types and in quantities that have no justification for prophylactic, protective or other peaceful purposes (GHI, 2009);

Weapons, equipment or means of delivery designed to use such agents or toxins for hostile purposes or in armed conflict (GHSA, 2011).

As it became increasingly clear that biosecurity required special attention, in 2007 an expert group published the Organization for Economic Co-operation and Development (OECD, 2007) Best practice guidelines on biosecurity for Biological Resource Centre (BRC) (OECD). The goal of the OECD guidelines is to balance the methods for reducing the probability that dangerous pathogens can be misused without unduly hindering beneficial research or being financially unsustainable.

On a global scale, biosecurity has an increasing profile due to a range of factors to include the increasing diversity and volume of international trade in animals, plants and their products; the introduction of genetically modified crops and animals (i.e., fish) into the environment; the increase in human interactions with wildlife resulting in newly emergent zoonotic diseases; the significant increase in the numbers of laboratories worldwide conducting research with pathogens; the potential for theft and misuse of materials; and agricultural intensification and environmental change (Jones et al., 2013). While approaches to defining, developing and implementing a biosecurity program may differ in the context of the laboratory setting, agricultural enterprise and food industry; in each case the goal is to protect individual and community health (human and animal).

Laboratory Biosecurity: Preventing the Intentional Misuse of Pathogens

In 2001 letters containing *Bacillus anthracis* spores were mailed to several news media offices and two US Senators, killing five people and infecting 17 others. While the 2001 anthrax letter attacks brought biosecurity back to the forefront of concern in the US, it is not an isolated incident regarding the theft and misuse of pathogens against humans. Other incidents of biocrimes (targeting of a particular individual(s)) and bioterrorism (targeting large numbers of people) have been perpetrated in relatively recent years. The first highly documented event of bioterrorism in the US occurred in 1984 when members of the Rajneeshee cult contaminated salad bars at 8 local restaurants and a supermarket in Dalles, Oregon. Their plan was to make local voters so sick the cult could seize control of the county government. At least 751 people became very ill, and no deaths were reported. An investigation showed the *Salmonella* in the salad bars was the same strain as a vial found in the cults medical lab that was ordered from a culture collection. Ma Anand Sheela and an accomplice were found guilty and sentenced to three to ten years in prison (Flynn, 2009). Numerous other cases where pathogens have been obtained from work or culture collections and used against a person(s) are recounted in a working paper by Dr. Seth Carus to include *Y. pestis*, *Shigella dysenteriae* Type 2 and *Ascaris suum* among others. There are many other cases cited where agents were illegally procured but not deployed. The misuse of biological agents is not only a problem in the US as cases have also been documented in Britain, France, Germany, Poland, Russia, Africa and Japan (Carus, 2001). Incidents in the US led to the incremental strengthening of biosecurity laws and requirements in the US for laboratories and diagnostic facilities working with pathogens and the transfer of materials from labs and culture collections.

Biosecurity in laboratory and culture collection settings combines administrative practices with facility infrastructure to safeguard assets. In this case practices include conducting personnel background checks to ensure they are reliable and responsible, controlling and accounting for inventory, reporting suspicious activities and potential security breaches to include the theft or loss of materials, and the use of technologies and practices to address security threats and vulnerabilities. This integrated approach was first described in the publication, 'Security Considerations for Microbiological and Biomedical Facilities' (Royse and Johnson 2002). Many countries have enacted formal legislative requirements to strengthen laboratory biosecurity and biosafety over the past decades. In countries where biosafety and biosecurity are used as separate terms, great care is taken to ensure legislation is complementary and harmonized. Below are examples describing how and why legislation was enacted in two countries.

The following example describes the US drivers and approach to enacting biosecurity legislation. While many entities in the US practiced biosecurity there was no clear law or guidance with regards to what should be secured, how much or what form of security was required, or what 'biosecurity' meant. Following an incident in 1995 where an individual unlawfully acquired *Y. pestis* by mail order, the US Congress passed the Antiterrorism and Effective Death Penalty Act of 1996 (Public Law 104–132, April 24, 1996) directing Health and Human Services to regulate the transfer of what would be called 'select agents'. The mechanism for upholding the regulatory requirements went into effect in 1997. With guidance from the Centers for Disease Control (CDC), many entities began implementing procedures to ensure transfers of material were going to approved collaborators and were in fact received in 1996. It is important to note that this legislation regulated shipment, but not necessarily possession of select agents.

While a bill was drafted in the Congress to address possession of select agent the US suffered its first anthrax mail attack on a US citizen in September 2001. In October of 2001, Congress passed legislation that substantially expanded the scope of the Select Agent Program. The USA PATRIOT Act of 2001 (Public Law 107–56, October 26, 2001, Section 817), expanded and amended Chapter 10 Title 18 of criminal law prohibiting the knowing possession of a biological agent, toxin or delivery system of a type or in a quantity that under the circumstances is not reasonably justified by a prophylactic, protective, bona fide research or other peaceful purpose. The language exempted naturally occurring agents (ASM, 2014). In June of 2002, Congress passed the Public Health Security and Bioterrorism Preparedness and Response Act of 2002 (EPA, 2002) requiring entities that possessed biological select agents and toxins (BSAT) to register with CDC or US Department of Agriculture (USDA) Animal and Plant Health Inspection Service (APHIS) and implement considerably more stringent security requirements to control access to BSAT to include clearing the names of persons with access by the Department of Justice. Many entities conducting work with BSAT opted to stop work with the materials (some destroying uniquely valuable collections, while others transferred materials to CDC or APHIS) due to the high costs associated with complying with the new requirements. Costs cited included infrastructure modifications, hiring of staff to implement the program, and the increased administrative burden. The CDC and APHIS also incurred the responsibility for auditing all entities possessing BSAT. This required the development of inspection standards, the hiring and training of inspectors, and communication of expectations with the entities. With the announcement that new CFR would be drafted in the future, the American Biological Safety Association formed a Biosecurity Task Force and submitted a white paper for consideration by CDC, USDA and US lawmakers (Johnson, 2003). The paper stated the focus of biosecurity should be on hiring trustworthy, reliable personnel, and implementing countermeasures that addressed vulnerabilities. The white paper and the Royse publication aimed to inform future national biosecurity legislation with intent that biosecurity not hinder biomedical and scientific research and that the solutions address actual risks as opposed to perceived risks.

The Bioterrorism Act was implemented through a series of regulations starting in 2003, with the final Codes of Federal Regulations (CFR) 42, CFR 73 (human pathogens), 9 CFR 121 (animal pathogens), and 7 CFR 331 (plant pathogens) becoming effective in the spring of 2005. In 2012, the CDC and APHIS published final changes to the CFRs creating a subset of BSAT called 'Tier 1' agents which would require even greater safeguarding. These were believed to represent BSAT with the greatest risk of deliberate misuse and the greatest potential for mass casualties or devastating effects to the economy, critical infrastructure and erosion of public confidence.

On a parallel track the Biosafety in Microbiological and Biomedical Laboratories (BMBL) (BMBL, 2009) underwent a series of revisions to harmonize with the new laws and CFRs. In December 2002, Appendix F of the BMBL 4th edition was updated to reflect security and emergency response guidance for laboratories working with Select Agents. In 2009 Appendix F was again revised (BMBL, 2009) to focus exclusively on Select Agent Laboratories (BMBL) and includes current references for the USDA and CDC Select Agent Programs (CDC, 2011). A new section was added on the principles of biosecurity, as security is important in all laboratories and not just for those conducting work with select agents. Section VI describes how a biosecurity program is developed, a risk assessment and management approach to biosecurity, and the elements of a biosecurity program. In addition to developing Section VI, the USDA and CDC commissioned a team of experts to develop a national biosafety and biosecurity training program for labs conducting work at Biosafety Level (BSL) 2 as those pathogens can cause disease in people and economically important agricultural species (Personal Communication with Joseph Kozlovak, USDA).

The nation of Singapore provides a second example of how a country may enact biosafety and biosecurity legislation and the drivers for enacting legislation. While the US approach was largely prompted by concerns over intentional misuse, Singapore's approach was more influenced as a result of emerging infectious diseases. Singapore initially enacted the Animals and Plants Act, 1965 and the Infectious Diseases Act, 1976, to safeguard animal, plant and human health. Several events were key to leading to the development of expanded biosafety and biosecurity legislation to include Nipah virus infections in Malaysia and Singapore, 1998, anthrax letters in the US, 2001, and the Singapore Economic Development Board's drive to develop biomedical industry, 2001. The Strategic Goods (Control) Act, 2002 was enacted to control the transfer and brokering of strategic goods, strategic goods technology, goods and technology capable of being used to develop, produce, operate, stockpile or acquire weapons capable of causing mass destruction, to include biological agents. The Act requires entities to register on-line through Tradenet and obtain a permit from the Agro-Veterinary Authority (AVA) or Ministry of Health (MOH) before it may import certain biological agents or toxins. Following the SARS laboratory acquired infection of 2003 a WHO and MOH inspection team recommended the development of a national standard for biosafety and biosecurity, creation of a structure to oversee lab certification and development of a tracking system for importation, exportation to and from Singapore and movement or transfer between local labs. The Biological Agents and Toxins Act (BATA) was passed by the Parliament in October 2005 and enacted in January 2006.

It is administered by the Biosafety Legislative Branch of the MOH, Singapore. In brief, the BATA provides for the regulation of biological agents and toxins, identifies duties and obligations of facility operators and carriers and provides for enforcement authority by MOH (Ling, 2011).

In 2006 the Genetic Modification Advisory Committee (GMAC) released the Singapore Biosafety Guidelines for Research on Genetically Modified Organisms (GMOs) to ensure the safe containment, handling and transport of GMOs used in research and to provide a common framework for assessment and notification of research on GMOs. The scope covers experiments that involve the construction and/ or propagation of all biological entities (cells, organisms, prions, viroid's or viruses) which have been made by genetic manipulation and are of a novel genotype and which are unlikely to occur naturally or which could cause public health or environmental hazards. Taken together, the Acts and Guidelines provide comprehensive biosafety and biosecurity coverage for industries in Singapore working with biological materials, and safeguard the community and environment. As the number of containment labs in Singapore has increased since 2003 and the country enjoys notoriety as an international biotech/biomedical research hub, one could conclude while the Acts and guidelines improved safety and security they have not hindered or burdened the research enterprise.

These two examples demonstrate the different approaches of enacted legislations and the impact on entities conducting work with biological materials. It also demonstrates there can be unintended consequences of legislation and deciding what type of legislation is appropriate to enact should not be done using a cookie-cutter approach. Rather it is a deliberative process based on specific conditions and needs and optimally informed by industry leaders and professional organizations with representation from scientists and biosafety professionals. Any legislation and means of enforcement should be country specific with careful consideration given to the types of biological and biomedical enterprises in operation, elements that are needed to address gaps based on a sound risk-benefit assessment and hazard analysis, sustainability, cultural compatibility and it's actual as opposed to perceived effect on science and safety. Approaches to implementing a biosecurity program have been described (OECD, 2007; GHSA, 2011; BMBL, 2009; CDC, 2011; Ling, 2011; Royse and Johnson et al., 2002; FAO, 2007) and OEDC guidelines provide several case studies on costs associated with implementing programs (OECD, 2007), Below is a synopsis of considerations in developing a laboratory biosecurity program.

OEDC defined biosecurity as, "Institutional and personal security measures and procedures designed to prevent the loss, theft, misuse, diversion, or intentional release of pathogens, or parts of them, and toxin-producing organism, as well as such toxins that are held, transferred and/or supplied by BRCs". Biorisk assessment identifies assets, threats, vulnerabilities and countermeasures and considers the likelihood and severity should theft, loss or release occurs. The process of weighing policy alternatives, considering risk assessment and other factors relevant for biosecurity, and selecting appropriate prevention and control actions is termed as "risk management". At the institutional level, discussions of biorisk and management would be interactive between the director, key staff (representatives of scientists, engineering, security, safety, etc.) and local first responders (fire, ambulance, law enforcement, etc.). Figure 1 depicts the interactive and dynamic process undertaken by institutions in the biosecurity risk assessment and mitigation process. At the national level, in situations where national guidelines are in development, or where gaps may exist, institutions may move toward harmonizing best practices by developing a network approach where information, as appropriate, is shared as are lessons learned and optimal practices among the members of the network. This approach not only assists in creating a baseline for biosecurity but also serves as a means of communicating risk within and among institutions, emergency responders and the broader public.

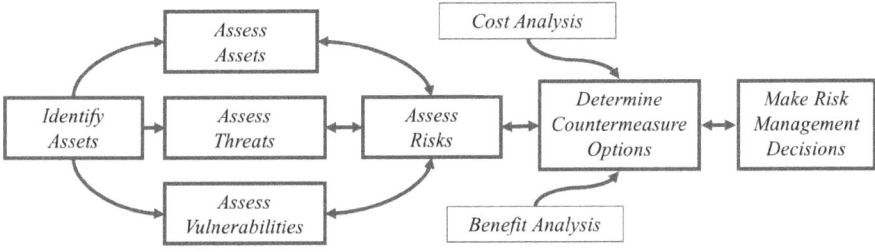

Figure 1. Biosecurity Risk Assessment Considerations.
Source: US Health and Human Services and US Department of Agriculture. Biosafety for BSL-2 Laboratories Training.

Figure 2 provides a broad overview of topics to be covered in a biosecurity risk assessment. Each component depicted in the boxes would be redefined to provide more detail. For example, in further defining risks associated with the agents (assets) themselves one would consider following items for inclusion, though the list may

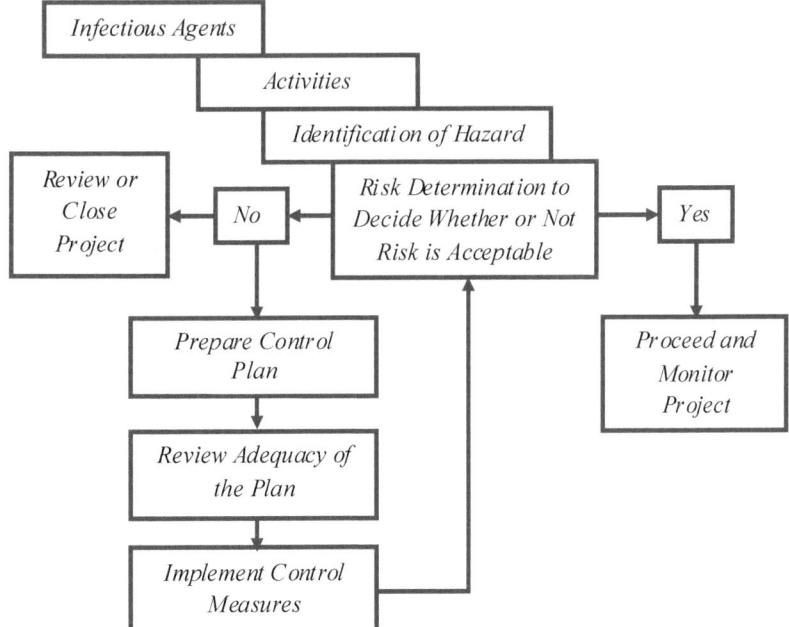

Figure 2. Decision matrix for project approval.

not be exhaustive; hazards associated with the material (infectious/intoxicating dose, severity of illness, communicability, availability of effective treatment/vaccine). In assessing the risks, one would consider risks to personnel in the context of risks associated with lab procedure (known and not previously described routes of infection and hazards inherent in procedures); the risk of agent escape (environmental stability, community spread, risk to flora/fauna, ability to establish through locals vectors or reservoirs and impact on the public/animal health system). The result of considering each subcomponent would lead to a generally accepted decision matrix as to whether the work should proceed depicted in Fig. 2 (Abad, 2014).

Laboratory biosecurity is more than the safeguarding of dangerous pathogens and toxins from individuals or organizations, which would use them for malicious intent. In addition, materials with historical, medical, epidemiological, commercial and scientific value across biomedical enterprises are worth protecting. Examples of biological materials that require this management are not limited to pathogens and toxins, but include cell lines, hospital pathology collections, vaccines and biologically derived therapeutics, non-pathogenic agents (i.e., used in food and antibiotic production), genetically modified organisms (GMOs), nucleic acid and products of synthetic biology. Measures to protect biological materials, toxins, and critical information, from theft or misuse can be achieved using a combination of the following methods after taking into account a site-specific analysis of assets, threats, vulnerabilities and sustainable/available countermeasures.

Technical security measures are often referred to as physical security or engineered security and include items or equipment designed to decrease vulnerabilities to a particular threat and may include but not be limited to architectural barriers, stand-off distance of the facility from roads or parking areas, fences, lights, closed circuit television (CCTV), motion detectors with lights or CCTV, self-closing locking doors, alarms, and combinations of biometric, code and card key access. Access controls are used to limit access to restricted areas to authorized individuals and can also their log entry and exit to areas. A graded approach to physical security is typically implemented where additional controls are added to areas as the value of the asset or requirement for restricted access increases. Areas deemed low risk (high value assets not present) may include public access areas, grounds and general warehouse and would employ routine property protection measures; moderate risk areas begin to limit access and include laboratories, sensitive or administration offices or corridors surrounding exclusion areas and may employ locking doors; while exclusion zones, are most greatly protected and include high containment laboratories, areas housing pathogen collections or proprietary materials and computer network hubs, LAN systems and other critical infrastructure.

Operational security measures include SOPs, personnel reliability programs, Employee Assistance Programs (EAP), enlightened leadership, development of a code-of-conduct for the life sciences, on-site or local security forces and emergency assistance (fire, police, medical). In its simplest form a personnel reliability program is designed to ensure only trustworthy individuals have access to assets. Components of the program can include education level and training requirements, demonstration of laboratory proficiency, conducting background checks (references, criminal offenses, financial rating, etc.), entry and random drug screening, monitoring compliance with safety and security SOPs, and requirements to report observed rule-breaking, or compromises in safety and security. More intricate programs have been implemented successfully, but can be labor intensive and expensive to operate (Higgins et al., 2013). The personnel reliability program is developed with the assistance of the institutes management, human resources and legal personnel, and occupational health staff. There are many personnel sensitive issues in implementing the plan, applying it evenly, ensuring it is a graded plan that identifies 'who' should be in the program (based upon access to assets), and safeguarding personnel records.

An EAP is a valuable tool in preventing problems by proactively providing employees with a means to confidentially work through personal problems that might adversely impact their job performance, health, or overall well-being. Problems may include family health issues, loss of a loved one, substance abuse or any other condition that could negatively impact the worker. EAPs generally have provisions for short-term counseling to help resolve problems and are provided at no cost to the employee.

Enlightened leadership is essential as it develops and fosters a culture of biosafety and biosecurity, clearly communicates its expectations, applies rules fairly and demonstrates equal commitment to all staff. This approach to management empowers the worker to 'do the right thing', whether it be report an incident, request time out of the lab due to fatigue or illness, or seek EAP support without fear of reprisal. Keeping in mind qualified, trained staffs are organization's biggest asset, like the EAP, enlightened humane leadership prevents or reduces workplace problems and improves employee retention.

Many organizations and individuals have called for the development of a code-of-conduct for the life sciences (Rohde et al., 2013; Johnson, 2006; NSABB, 2010; ICGEB, 2005; IAP, 2005). While an agreed upon international code would be optimal, codes started at national and even institutional levels would be of great benefit in informing life scientists and the community of collective responsibilities for safe science. The essence of a code would provide guiding principles and governance mechanisms; promote awareness, safety, security, education, accountability, and research oversight; and champion research merit and integrity, the concept of 'do no harm', ethics, and instill the obligation to report wrong doing. A code should be short, easily understood, culturally acceptable, endorsed by scientific professional organizations, relevant, and taught to all aspiring to work in the life sciences.

A thorough description of components of a biosecurity program has been developed by the OECD (2007); WHO (2006) and US CDC and USDA (BMBL, 2009; CDC, 2011). In brief, the components of biosecurity programs include developing and implementing a program plan and SOPs that at a minimum address:

- Program management,
- Physical security-access control and monitoring,
- Personnel management,
- Inventory and accountability,
- Information security,
- Transport of biological materials,
- Accident, injury and incident response plans,
- Reporting and communication,

- Training and practice drills, and
- Security updates and re-evaluations.

Agricultural Biosecurity: Protecting Animals and Plants

Historically, the term the "biosecurity" described efforts to prevent the spread of infectious disease in economically important agricultural commodities such as crops, livestock and poultry. Animal health biosecurity is concerned with import and export health controls and domestic animal health programs. Biosecurity measures are those practices and procedures that should be taken to keep pathogens from entering the farm and to prevent the transmission of disease within infected farms to neighboring farms. In the context of the farm, biosecurity largely consists of management practices to control movement of personnel, visitors, animals, supplies and equipment onto the farm as well as close observation of animals health (culling sick animal) and providing appropriate veterinary care at various stages in their lives. Vaccines have been successfully used as a component of control and eradication of endemic animal diseases. They are also used to provide ring protection to high value animal herds, and as an emergency measure to slow virus transmission in areas where a disease outbreak is occurring (CFSPH, 2011). Infrastructure and processes may be enhanced to include the provision of germicidal foot baths for staff upon entry, dedicated clothing is worn in place of street cloths, the environment is controlled to optimize growing conditions, and more sophisticated security systems prevent unauthorized access. Import controls are designed to ensure imported animals, animal products and genetic material, feedstuff and other commodities are free of pathogens, vectors and reservoirs of disease that could have an adverse effect on the agricultural industry and consumer. This is especially true with regards to preventing the introduction of exotic animal diseases.

Increasing laboratory capacity and the availability of accurate and rapid diagnostic tools for epidemiological surveillance has been one of the cornerstones of the global biosecurity response to Highly Pathogenic Avian Influenza (HPAI) detection, surveillance and eradication. The use of globally accessible on-line platforms like Program for Monitoring Emerging Diseases (ProMED-mail), publishes and transmits via the internet on an average of seven daily reports of infectious disease outbreaks with commentary from a staff of expert moderators, has made disease surveillance reportable in real-time commensurate with the ability to reliably detect and diagnose disease. A number of countries sharing boarders coordinate their veterinary efforts to combat the spread of animal disease. The Food and Agricultural Organization (FAO) notes, "A specific response to the inevitability of new and emerging diseases is the establishment of "disease-free" geographical areas within countries or regions ("regionalization") to allow the trade of disease free animals and their products while preventing introduction of infected animals and products into the 'safe' production zone (FAO, 2007).

With growing prosperity and technological advances there has been a rapid expansion of consumption of animal products in many countries. The increase in livestock production, such as, privately owned small poultry flocks increases risk of disease transmission to humans as human-animal interaction is close and small privately owned flocks may not receive the same level of veterinary care or be maintained in good sanitary conditions as compared with large producers. Biosecurity goals subtly transition from animal biosecurity to food biosecurity in the move from farm to fork. One example involves balancing the treatment or prevention of animal disease with the need to prevent introducing potential chemical hazards such as residues of veterinary drugs and pesticides into the food chain (Rohde et al., 2013). Many countries have strict laws that clearly define the period of time that must elapse between treatment (i.e., with an antibiotic or drug), and time an animal or product can be sold to market. In the US once animals, poultry or their products (i.e., milk and eggs) enter the food industry the Department of Agriculture and the Food and Drug Administration are the competent authorities responsible for ensuring animals are healthy prior to slaughter, milk and eggs are safe for consumption, and sanitary conditions and processes in the abattoir are appropriate and adhered to. The division of labor between the USDA and FDA is very complicated. But briefly, in overarching simplified terms the USDA is responsible for the safety of meat, poultry and processed egg products, while the FDA inspects shelled eggs, milk and milking parlors, and all other foods.

Another example involves the control of zoonotic diseases (diseases spread from animals to humans) such as Nipah virus which spread from swine to humans in Malaysia and Singapore (Chua, 2003), HPAI transmission from poultry to humans in several Asian countries (WHO, 2005), the emergence and spread of SARS Co-V originating in infected animals in live-markets in China and subsequently spreading through Asia and a number of other countries (Li et al., 2005), and MERS Co-V transmission from camels to humans first reported in Saudi Arabia with cases now reported in several countries in the Arabian Peninsula (Azhar et al., 2014). The SARS Co-V example also demonstrates how wildlife introduced for sale the food chain can result in human disease (Lau et al., 2005). The shipment or importation of wildlife in the marketplace has added to biosecurity and zoonotic risks to the agricultural industry and population (Karesh et al., 2005; NSF, 2009).

The other part of agricultural biosecurity, involves protecting the plants grown for animal (livestock) and human food. This involves the implementation of sanitary and phytosanitary measures and best agricultural practices. Sanitary and phytosanitary practices is defined by the World Trade Organization (WTO, 2015) and include measures that straddle food biosecurity (food is safe to eat) and agricultural practices (safety and security during growth, harvest and processing) as applied to crops and animals. On the agricultural side of the equation, in addition to protecting against pests (insects and infectious agents), noxious weeds and animals, climate change can also impact survival and yield of plants. As farms have become large and equipment is designed specifically for a crop the use of monoculture practices has become prevalent in some geographical areas with land devoted to single crops and year-to-year production of the same crop species on the same land. This practice can lead to increased susceptibility to pests, difficultly in recycling nutrients to the land, and increases in the use of pesticides and fertilizers.

As is the case for animals, import controls help to reduce the introduction of alien pests, pathogens and weeds, however it is not one hundred percent successful. A current example is citrus greening in Florida which was first documented in 2005 and started with the arrival of the Asian citrus psyllid in the Port of Miami. As the tiny insect bites the fruit it spreads the bacteria responsible for the disease from tree to tree causing the fruit to turn bitter and drop before ripening. The disease is decimating the states citrus production and University of Florida agricultural analysts concluded that between 2006 and 2012, citrus greening cost Florida's economy $4.5 billion and 8,000 jobs. There is no cure, it ultimately kills the trees it infects and citrus greening has now spread to groves in Texas and California.

To prevent crop damage, noxious weeds and pests are typically controlled with periodic applications of pesticides during production, storage or transport. The U.S. EPA determines the maximum amount of pesticide residue that can remain in or on a particular food by evaluating the potential health and environmental risks of the pesticide. In the US, most residue levels are well below legal limits as residues break down over time and are mechanically removed by washing and processing prior to sale in consumer markets. None-the-less there is a strong shift toward organic produce (grown in the absence of pesticides or chemical fertilizers) and consumers employ practices to further reduce residues (washing, scrubbing or peeling produce and discarding the outermost leafs).

In the US industry has responded by producing genetically modified (GM) plants (most notably corn, soybean and cotton) that have improved yield and quality, are either resistant to pests so do not require pesticides or are herbicide-tolerant enabling the substitution of glyphosate for more toxic and persistent herbicides. US farmers have adopted the use of GM cops as they have increased yields, decreased pesticide use and cost, and have seen savings in management time and other practices. Some strategies are not without risk however, and the USDA has noted that "overreliance on glyphosate and a reduction in the diversity of weed management practices have contributed to the evolution of glyphosate resistance in some weed species" (Fernandez-Cornejo et al., 2014). There are also fears that GE crops may reduce biodiversity by passing their traits to native crops (i.e., GM corn cross pollinating native strains of maize).

In terms of climate change, farmlands in many countries over the past decade have experienced unusual weather patterns in the form of long-term droughts, extreme and unseasonal rainfall patterns, and unusually hot or cold changes in growing season. Taken together the changes result in crop loss due to mold, pathogens, lack of water, destruction by sleet/rain and increased susceptibility to pests (Qualset and Shands, 2005). While farmers may be able to deploy fans, smudge pot heaters and use other methods to keep cops from freezing or limit damage, there is often no recourse when fields are underwater shortly before harvest due to torrential rains, or a season of corn is lost to prolonged summer drought. Climate related crop loss is a growing problem.

Malicious acts to disrupt agricultural operations is also of concern. While there are reporting chains if an malicious act occurs (local law enforcement or FBI), the USDA has proactively published information to assist farmers in making their operations less attractive as targets by reducing vulnerability to prior to harvest (Qualset and Shands, 2005). The voluntary guidelines and checklist is designed to assist the farmer in assessing and mitigating security risks to their assets and covers topics in a farm-relevant manner to include awareness, training, technical security, barriers, inventory control, reporting incidents, visitor procedures and other topics.

Food Biosecurity: Keeping People Safe

For the purpose of this chapter, we will adopt the FAO definition for food safety as, "biosecurity systems for food safety must control hazards of biological, chemical and physical origin in imported food, food produced domestically and food that is exported (FAO, 2007)". As mentioned earlier, there is overlap in biosecurity from farm to fork and food security begins on the farm with those practices that minimize entrance of pesticide and drug residues (i.e., veterinary), pathogens and toxins in materials destined to be part of the food chain through accidental, neglectful or malicious means. In the today's world of globalized trade and increased demand for animal proteins, it also means ensuring products are of high quality, unadulterated and safe for consumption for even the most vulnerable populations. To meet this need many countries have made significant improvements to import controls, country assessments, use of certified labs for testing imported foods, hiring and training more food inspectors, development of more sensitive tests for pathogens and toxins, and requiring stricter labelling practices (consumer transparency). One recent example of improving food safety for importing countries is to identify exporters that may pose a higher risk of shipping animal products containing drug residues. The US FDA conducts foreign country assessments that examine a country's industry and regulatory infrastructure to control aquaculture and animal drugs. Not only is this of benefit to the consumers of the importing nation, but it can assist the exporter as well. "Evaluating a country's laws for, and implementation of, control of animal drug residues helps FDA direct foreign inspection and border surveillance resources more effectively and efficiently and allows FDA to work directly with countries to resolve drug residue problems (FDA, 2013)". The results of this program have varied from the issuance of specific country-wide import alerts for specific aquaculture products; increased sampling and testing under the food safety compliance program; or identification of countries with exceptional practices and safe exports. This system and similar systems are in use by many countries today as food and food products grown and processed on any given country make their way to the marketplace in other countries.

Global food tracking capability is improving, but not fast enough for consumers unfortunate enough to consume and be stricken with illnesses linked to unintentional contamination with *Salmonella Enteritidis* (CDC, 2015) or *E. coli* O104:H4 (BBC, 2011) after eating bean sprouts; Shiga toxin-producing *Escherichia coli* O157:H7 in ground beef (CDC, 2014); intentional adulteration of infant formula with melamine (WHO, 2008); Hepatitis A linked to berries, tomatoes, seafood and other products (EFSA, 2014) or salmonella in cantaloupes, strawberries and other fresh produce. In some instances contamination occurs in the field and may be linked to water or fertilizer sources, in others it occurs during processing. Improvements are being made in tagging produce lots from the farm of origin, reducing investigation time in what was months to pinpoint the source and alert consumers, to a fraction of that time.

Unfortunately, in many countries it is difficult if not impossible to have routine access to safe clean water to drink, much less process foods. In these environments the potential presence of pathogens is commonplace and the water used in preparation, hands and utensils for handling the food, and the foodstuffs themselves may be contaminated (IFT, 2002). Of the billion meals served globally, WHO estimates that

1.8 million people died in 2005 from the effects of foodborne illness. This is not a problem unique to the developing world as foodborne diseases have been estimated to cause approximately 76 million illnesses, 325,000 hospitalisations and 5000 deaths in the United States each year (Mead et al., 1999).

Among the developing nations, Bangladesh has not yet developed any Administration System. A food biosafety and biosecurity policy for the country is yet to be formulated, although the country has a National Food and Nutrition Policy that concentrates on food safety. Significant activities in food safety and quality control are ongoing in the country. Ministries, Departments and Agencies are involved in these activities with a major responsibility of the Ministry of Health and Family Welfare (MOHFW) that has a unique infrastructure to deliver its services throughout the country. A Management Information System on food safety and food borne illnesses is to some extent integrated with the Primary Health Care Programme. The country has also signed the World Trade Organization (WTO) Agreement (WHO, 2009).

Recently the Consumers in Bangladesh have become victim of serious adulteration in food and food safety has become an important topic. Different Ministries, Departments, Agencies are directly or indirectly responsible for enforcement of food laws, rules and regulations. There are several laws in Bangladesh for maintaining health and safety standards. Among them the oldest is "The Bangladesh Pure Food Ordinance, 1959" and the most recent are Bangladesh Safe Food Act, 2013 and Safe Food Regulations-2014 (WTO, 2014). Under the Bangladesh Pure Food Ordinance, 1959 and the Bangladesh Pure Food Rules, 1967, there are a total of 107 different generic, mandatory food standards. Bangladesh Standard Testing Institute (BSTI) is the Standardization body in the country. There are 50 mandatory generic food standards of BSTI. In addition, there are some 250 optional standards for different foodstuff. A total of 16 laboratories are identified to be responsible for qualitative and quantitative assessment of food items. Food safety Programme in Bangladesh is a collaborative programme of the government and WHO is being implemented in Bangladesh since 1994, under the Food Safety Programme (WHO, 2009).

Bangladesh is enforcing tougher food safety laws effective starting in February 2015. Carrying a five-year jail term and a fine of Tk 20 million (230,000 EUR; 260,000 USD) for food adulteration, the Safe Food Act comes into force from Feb 1. The Safe Food Act will replace the Pure Food Ordinance-1959. According to the new ordinance, one can directly file a case against an unsafe food producer, processor or seller. Safe Food Regulations-2014 was formulated in accordance with the Act (WTO, 2014).

Contamination that occurs during processing has resulted in millions of pounds of ground meat and other food materials being recalled in past years over the potential for bacterial contamination making them unfit for human consumption. The prevalence of *salmonella* sp. on the surface of poultry is so common that consumers are trained to regard all raw poultry products as potentially contaminated and educated on safe preparation and cooking practices. While the deployment of rapid diagnostic tests have successfully prevented products from being brought to market place, the economic cost of disposing of tainted product has be staggering.

We have discussed the development of GM crops as a means of reducing pesticide residue and conferring traits for robustness and growth. In Consumer acceptance of GM foods is hotly debated, accepted more prevalently in some countries than others and is impacted by a consumer's socio-economic status. For example, consumers in the US are less likely to shun products containing GM foods as compared to consumers in the European Union (EU). Reviews of studies have shown more affluent consumers are willing to pay higher prices for non-GM foods and reason that organic 'natural' foods are safer, more nutritious and grown using environmental stewardship practices. However, there is no empirical evidence to say that either product is better. Claims made are dependent a number of variables such as a farmers willingness to embrace eco-friendly practices that foster sustainability, and the awareness that the degree to which consumers are willing to pay more for natural foods is highly variable by country, not just income bracket (Dangour et al., 2009; Fernandez-Cornejo et al., 2014). Chassy points out, "From a food safety standpoint, organic foods have been observed to have higher bacterial counts than their conventional counterparts and have been associated with foodborne disease outbreaks. The exclusive use of organic fertilizers such as composted manure in their cultivation may be responsible for the higher number of outbreaks and pathogens associated with organic products (IFT, 2002)". Taken at face value this would be an egregious statement as there are many producers that use best practices in sanitation and post-harvest processing that minimize if not mitigate the risks stated above. As is the case in any industry, ensuring biosecurity requires a thorough assessment of risk, vulnerability, threat/hazard and means to address the risk. As producers create 'second attribute' GM crops the public perception changes, especially in the developing world. Second attribute crops are those that are created to provide additional nutritive value to correct deficiencies in diets of people in developing nations. Perhaps the most well-known is Golden rice (Potrykus, 2010), a strain that produces pro-vitamin A addressing a serious public health problem affecting millions of children and pregnant women around the world. Other GM foods containing additional nutrients, Genetic Enhancement (GE) of the Omega-3 content, or the addition of cholesterol-fighting compounds made the GM product more appealing to consumers in developed countries as well (Fernandez-Cornejo et al., 2014).

Conclusion

The world is interconnected, although there are nations' demarcations borders, civilizations are not separated by an absolute barrier thus allowing a hazardous biological or chemical danger to cross between countries. Much advancement has occurred in the areas of biosafety compared with that of biosecurity, though each has been practiced in the laboratory setting in a number of countries for many decades. Today we recognize biosecurity as being a vital part of the strategy, efforts, and plans to protect human, animal, and environmental health against biological threats. We have described various aspects of biosecurity, which include legislation and guidelines, application of principles and practices, technical and administrative controls, and diagnostic methods and communication in the overarching context of risk assessment and management.

The inherent risk that while working with pathogens and toxins to benefit humankind, some would misuse the materials to cause harm is referred to as the 'dual use dilemma'. In the context of the laboratory the primary consideration of biosecurity is to prevent intentional misuse or release of biological agents which can cause harm to individuals, the community or the environment.

Agricultural biosecurity largely consists of management practices providing appropriate care, import controls, enhanced infrastructure and access controls, and the implementation of sanitary and phytosanitary measures and best agricultural practices to stop the introduction and transmission of disease, pests and other material hazards to livestock and crops. Highly specific, sensitive, rapid diagnostic tests, laboratory infrastructure and reporting networks are key improvements to safeguarding animal and crop security and health.

Biosecurity systems for food safety must control hazards of biological, chemical and physical origin through accidental, neglectful or malicious means in imported food, food produced domestically and food that is exported. The current trend of globalized trade and increased demand for animal proteins, has resulted in countries making significant improvements to import controls, country-specific food security assessments, use of certified laboratories for testing imported foods, hiring and training more food inspectors, development of more sensitive tests for pathogens and toxins, and requiring stricter labelling practices (consumer transparency). The cultivation of GM crops may provide relief from hunger and specific nutritional deficiencies, and assist with reduction of pesticide use and increasing yield for growers. These benefits must be considered in the backdrop of the potential for decreased biodiversity and ensuring the product is safe for human and livestock consumption. Finally and vitally important is the need to globally facilitate good sanitary practices and access to a reliable, safe water sources as it is a cornerstone in the foundation of providing safe food to the population.

References

Abad, X. 2014. CWA 15793: When the Biorisk Management is the Core of a Facility. Biosafety, Volume 3, Issue 2, http://dx.doi.org/10.4172/2167-0331.1000119.

Azhar, EI, S.A. El-Kafrawy, S.A. Farraj, A.M. Hassan, M.S. Al-Saeed, A.M. Hashem and T.A. Madani. 2014. Evidence for camel-to-human transmission of MERS coronavirus. N. Engl. J. Med., 370: 2499–2505.

BBC News. German tests link bean sprouts to deadly *E. coli*. 10 June 2011. http://www.bbc.co.uk/news/world-europe-13725953.

[BMBL] Biosafety in Microbiological and Biomedical Laboratories. 2009. 5th Edition. http://www.cdc.gov/biosafety/publications/bmbl5/BMBL.pdf.

Carus, W.S. WORKING PAPER: Bioterrorism and Biocrimes, The Illicit Use of Biological Agents Since 1900. February 2001 Revision. www.dtic.mil/cgi-bin/GetTRDoc?AD=ADA402108.

[CBSG] Canadian Biosafety Standards and Guidelines. 2013. First Edition. http://canadianbiosafetystandards.collaboration.gc.ca/cbsg-nldcb/index-eng.php?page=23.

[CDC] Centers for Disease Control. Multistate Outbreak of *Salmonella enteritidis* Infections Linked to Bean Sprouts. January 23, 2015. http://www.cdc.gov/salmonella/enteritidis-11-14/.

[CDC] Center for Disease Control. Multistate Outbreak of Shiga toxin-producing *Escherichia coli* O157:H7 Infections Linked to Ground Beef (Final Update). Posted June 20, 2014 http://www.cdc.gov/ecoli/2014/O157H7-05-14/.

[CDC] Centers for Disease Control and Prevention. 2011. Select Agent Programme. Office of the Public health Preparedness and Response. http://www.cdc.gov/phpr/documents/DSAT_brochure_July2011.pdf.

[CFSPH] Center for Food Security and Public Health, Iowa State University. The Foreign Animal Disease Preparedness and Response Plan (FAD PReP)/National Animal Health Emergency Management

System (NAHEMS) Guidelines. 2011. http://www.cfsph.iastate.edu/pdf/fad-prep-nahemsappendix-a-vaccination-for-foot-and-mouth-disease.

Chua, K.B. 2003. Nipah virus outbreak in Malaysia. J. Clin. Virol., Apr., 26(3): 265–75.

Dangour, A.D., S.K. Dodhia, A. Hayter, E. Allen, K. Lock and R. Uauy. 2009. Nutritional quality of organic foods: a systematic review. Am. J. Clin. Nutr., 90: 680–685.

[EFSA] European Food Safety Authority. Outbreak of hepatitis A in EU/EEA countries. Second update 11 April 2014. http://ecdc.europa.eu/en/publications/Publications/ROA-Hepatitis%20A%20virusItaly%20Ireland%20Netherlands%20Norway%20France%20Germany%20Sweden%20United%2Kingdom%20-%20final.pdf.

[EPA] Environmental Protection Agency. 2002. Title IV—drinking water security and safety. http://www.epa.gov/safewater/watersecurity/pubs/security_act.pdf.

[FAO] Food and Agriculture Organization. 2007. FAO biosecurity toolkit, Part 1, Biosecurity Principles and Components. ftp://ftp.fao.org/docrep/fao/010/a1140e/a1140e01.pdf.

[FDA] US Food and Drug Administration. 2013. The Imported Seafood Safety Program. http://www.fda.gov/Food/GuidanceRegulation/ImportsExports/Importing/ucm248706.htm.

Fernandez-Cornejo, J., S. Wechsler, M. Livingston and L. Mitchell. Genetically Engineered Crops in the United States. USDA Economic Research Report Number 162, February 2014.

Flynn, D. Salmonella Bioterrorism: 25 Years Later. October 7, 2009. Food Safety News. http://www.foodsafetynews.com/2009/10/for-the-first-12/#.VMgIA3Z0zIU.

[GHI] Global Health Initiative. 2009. http://www.whitehouse.gov/the_press_office/Statement-by-thePresident-on-Global-Health-Initiative.

[GHSA] Global Health Security Agenda. 2011. http://www.whitehouse.gov/the-pressoffice/2011/09/21/remarks-president-obama-address-united-nations-general-assembly.

Higgins, J.J, P. Weaver, J.P. Fitch, B. Johnson and R.M. Pearl. Implementation of a Personnel ReliabilityProgram as a Facilitator of Biosafety and Biosecurity Culture in BSL-3 and BSL-4 Laboratories. Biosecurity and Bioterrorism: Biodefense Strategy, Practice, and Science. Volume 11, Number 2, 2013.

[IAP] Inter Academy Panel. 2005. Statement on Biosecurity. http://www.interacademies.net/10878/13912.aspx.

[ICGEB] International centre for genetic engineering and biotechnology. 2005. Building blocks for a codeof conduct for scientists, in relation to the safe and ethical use of biological sciences. http://www.unog.ch/80256EDD006B8954/(httpAssets)/B5383C068566DB44C12571950049D57/$file/050613-PM-ICBEG.pdf.

[IFT] Institute of Food Technologists. IFT Expert Report on Emerging Microbiological Food Safety Issues (2002) IFT, available at http://members.ift.org/IFT/Research/IFTExpertReports/microsfs_report.htm.

Johnson, B. 2006. A Code of Conduct for Biological Scientists: An Important Topic for Action. Applied Biosafety, 11(1): 45–47.

Johnson, B. ABSA Biosecurity Task Force White Paper: Understanding Biosecurity (January 2003), http://www.absa.org/0301bstf.html.

Jones, B.A., D. Grace, R. Kock, S. Alonso, J. Rushton, M.Y. Said, D. McKeever, F. Mutua, J. Young, J. McDermott and D.U. Pfeiffer. 2013. Zoonosis emergence linked to agricultural intensification and environmental change. Proceedings of the National Academy of Sciences of the United States of America (PNAS), 110(21): 8399–8404.

Karesh, W.B., R.A. Cook, E.L. Bennett and J. Newcomb. Wildlife Trade and Global Disease Emergence. Emerging Infectious Diseases • Vol. 11, No. 7, July 2005. http://wwwnc.cdc.gov/eid/article/11/7/pdfs/05-0194.pdf.

Lau, S.K.P., P.C.Y. Woo, K.S.M. Li, Y. Huang, H.W. Tsoi, B.H. Wong, S.S. Wong, S.Y. Leung, K.H. Chan and K.Y. Yuen. 2005. Severe acute respiratory syndrome coronavirus-like virus in Chinesehorseshoe bats. PNAS, September 27, 2005, 102(39): 14041.

Li, W., Z. Shi, M. Yu, W. Ren, C. Smith, J.H. Epstein, H. Wang, G. Crameri, Z. Hu, H. Zhang, J. Zhang, J. McEachern, H. Field, P. Daszak, B.T. Eaton, S. Zhang and L.-F. Wang. 2005. Bats Are Natural Reservoirs of SARS-Like Coronaviruses, Science 28 October 2005, 310(5748): 676–679.

Ling, A.E. 2011. Presentation: Singapore's response to biorisk events at home and abroad. National Academies of Sciences Conference: Anticipating Biosecurity Challenges of the Global Expansion of High Containment Biological Laboratories. Istanbul, Turkey.

Mead, P.S., L. Slutsker, V. Dietz, L.F. McCaig, J.S. Bresee, C. Shapiro, P.M. Griffin and R.V. Tauxe. 1999. Food-related illness and death in the United States. Emerg. Infect. Dis., 5: 607–625.

Millett, P.D. 2010. The Biological Weapons Convention: From International Obligations to Effective National Action. Applied Biosafety, 15(3).

Nordmann, B.D. Issues in biosecurity and biosafety. Int. J. Antimicrob. Agents. 2010 Nov, 36 Supp l1: S66–9. doi: 10.1016/j.ijantimicag.2010.06.025.

[NSABB] National Science Advisory Board for Biosecurity. Enhancing Responsible Science Considerationsfor the Development and Dissemination of Codes of Conduct for Dual Use Research. http://osp.od.nih.gov/sites/default/files/resources/COMBINED_Codes_PDFs.pdf. 2010.

[NSF] National Science Foundation. Press Release 09-087. Wildlife Trade Threatens Public Health and Ecosystems. April 29, 2009. http://www.nsf.gov/news/news_summ.jsp?cntn_id=114706.

[OECD] Organization for Economic Co-operation and Development. 2007. Best Practice Guidelines for Biological Resource Centres. http://www.oecd.org/sti/biotech/38777417.pdf.

Potrykus, I. 2010. Lessons from the 'Humanitarian Golden Rice' project: regulation prevents development of public good genetically engineered crop products. New Biotechnology, 27(5): 466–472.

Qualset, C.O. and H.L. Shands. 2005. Safeguarding the Future of U.S. Agriculture. http://ucce.ucdavis.edu/files/repositoryfiles/SafeguardingFutureUSAg-54956.pdf.

Rohde, C., D. Smith, D. Martin, D. Fritze, J. Stalpers. 2013. Code of conduct on biosecurity for biological resource centres: procedural implementation. Int. J. Syst. Evol. Microbiol. Jul, 63(Pt 7): 2374–82.

Royse, C. and B. Johnson. 2002. Security Considerations for Microbiological and Biomedical Facilities, in Anthology of Biosafety V - BSL4 Laboratories. Chapter 6, http://www.absa.org/0200royse.html.

The Australia Group. 2014. List of Human and Animal Pathogens And Toxins For Export Control. http://www.australiagroup.net/en/human_animal_pathogens.html.

UNODA (United Nations office for Disarmament Affairs). 1925. Convention on the Prohibition of the Development, Production and Stockpiling of Bacteriological (Biological) and Toxin Weapons and on Their Destruction. http://www.un.org/disarmament/WMD/Bio/pdf/Text_of_the_Convention.pdf

[WHO] World Health Organization. Questions and Answers on melamine. 2008. http://www.who.int/csr/media/faq/QAmelamine/en/.

[WHO] The World Health Organization. Laboratory Biosecurity Guidance of 2006 (WHO/CDS/EPR/2006.6) (WHO, 2006).

[WHO] World Health Organization. WHO INFOSAN Information Note No. 7/2005 (Rev. 1. 5 Dec.)—Avian Influenza. 4 November 2005.

[WHO] World Health Organization. 2009. Sixty-second world health assembly Geneva, 18–22 may 2009, resolutions and decisions annexes. http://apps.who.int/gb/ebwha/pdf_files/WHA62-REC1/WHA62_REC1-en.pdf.

[WTO] World Trade Organization. 2015. Sanitary and phytosanitary measures: text of the agreement. http://foodsafetyasiapacific.net/wp-content/uploads/.

[WTO] World Trade Organization. 2014. Emerging Food Safety Issues in WTO Global Scenario in Asia. http://foodsafetyasiapacific.net/wp-content/uploads/2014/02/12-status-in-ASEAN1.pdf.

Index